THEORY OF RELATIVITY

by

W. PAULI

Translated from the German by
G. FIELD

Dover Publications, Inc., New York

Published in Canada by General Publishing Company, Ltd., 30 Lesmill Road, Don Mills, Toronto, Ontario.
Published in the United Kingdom by Constable and Company, Ltd.

This Dover edition, first published in 1981, is an unabridged and unaltered republication of the English translation as originally published by Pergamon Press, Ltd. in 1958. The work originally appeared in German: "Relativitätstheorie," *Encyklopädie der matematischen Wissenschaften,* Vol. V19, B. G. Teubner, Leipzig, 1921.
This edition is published by special arrangement with Pergamon Press Ltd., Headington Hill Hall, Oxford OX3 OBW England.

International Standard Book Number: 0-486-64152-X
Library of Congress Catalog Card Number: 81-66089

Manufactured in the United States of America
Dover Publications, Inc.
180 Varick Street
New York, N.Y. 10014

PREFACE

THIRTY-FIVE years ago this article on the theory of relativity, written by me at the rather young age of 21 years for the *Mathematical Encyclopedia*, was first published as a separate monograph together with a preface by Sommerfeld, who as the editor of this volume of the Encyclopedia was responsible for my authorship. It was the aim of the article to give a complete review of the whole literature on relativity theory existing at that time (1921). Meanwhile, the production of textbooks, reports and papers on the theory of relativity has grown into a flood, which rose anew at the 50th Anniversary of the first papers of Einstein on relativity, in the same year 1955 in which all physicists were mourning his death.

In this situation any idea of completeness regarding the now existing literature in a revised new edition of the book had to be given up from the beginning. I decided therefore, in order to preserve the character of the book as an historical document, to reprint the old text in its original form, but to add a number of notes at the end of the book, which refer to certain passages of the text. These notes give to the reader selected information about the later developments connected with relativity theory and also my personal views upon some controversial questions.

Especially in the last of these notes on unified field theories, I do not conceal to the reader my scepticism concerning all attempts of this kind which have been made until now, and also about the future chances of success of theories with such aims. These questions are closely connected with the problem of the range of validity of the classical field concept in its application to the atomic features of Nature. The critical view, which I uttered in the last section of the original text with respect to any solution on these classical lines, has since been very much deepened by the epistemological analysis of quantum mechanics, or wave mechanics, which was formulated in 1927. On the other hand Einstein maintained the hope for a total solution on the lines of a classical field theory until the end of his life. These differences of opinion are merging into the great open problem of the relation of relativity theory to quantum theory, which will presumably occupy physicists for a long while to come. In particular, a clear connection between the general theory of relativity and quantum mechanics is not yet in sight.

Just because I emphasize in the last of the notes a certain contrast between the views on problems beyond the original frame of special and general relativity held by Einstein himself on the one hand, and by most of the physicists, including myself, on the other, I wish to conclude this preface with some conciliatory remarks on the position of relativity theory in the development of physics.

There is a point of view according to which relativity theory is the end-point of "classical physics", which means physics in the style of Newton–Faraday–Maxwell, governed by the "deterministic" form of causality in space and time, while afterwards the new quantum-mechanical style of the laws of Nature came into play. This point of view seems to me only partly true, and does not sufficiently do justice to the great influence of Einstein, the creator of the theory of relativity, on the general way of thinking of the physicists of today. By its epistemological analysis of the consequences of the finiteness of the velocity of light (and with it, of all signal-velocities), the theory of special relativity was the first step away from naive visualization. The concept of the state of motion of the "luminiferous aether", as the hypothetical medium was called earlier, had to be given up, not only because it turned out to be unobservable, but because it became superfluous as an element of a mathematical formalism, the group-theoretical properties of which would only be disturbed by it.

By the widening of the transformation group in general relativity the idea of distinguished inertial coordinate systems could also be eliminated by Einstein as inconsistent with the group-theoretical properties of the theory. Without this general critical attitude, which abandoned naive visualizations in favour of a conceptual analysis of the correspondence between observational data and the mathematical quantities in a theoretical formalism, the establishment of the modern form of quantum theory would not have been possible. In the "complementary" quantum theory, the epistemological analysis of the finiteness of the quantum of action led to further steps away from naive visualizations. In this case it was both the classical field concept, and the concept of orbits of particles (electrons) in space and time, which had to be given up in favour of rational generalizations. Again, these concepts were rejected, not only because the orbits are unobservable, but also because they became superfluous and would disturb the symmetry inherent in the general transformation group underlying the mathematical formalism of the theory.

I consider the theory of relativity to be an example showing how a fundamental scientific discovery, sometimes even against the resistance of its creator, gives birth to further fruitful developments, following its own autonomous course.

I am grateful to the Institute for Advanced Study in Princeton for affording me the opportunity of writing the Supplementary Notes, pp. 207–232, during my stay there early in 1956. And I should like to thank my colleagues at Princeton with whom I discussed many of the problems in these notes.

I gratefully acknowledge the excellent help of the translator, Dr. Gerard Field, of the Department of Mathematical Physics, University of Birmingham.

Zurich, 18 *November* 1956   W.P.

ACKNOWLEDGMENTS IN THE ORIGINAL ARTICLE

I wish to express my warm gratitude to Geheimrat Klein, for the great interest he has shown in this article, for his active help in proof-reading, and for his valuable advice on many occasions. My thanks are also due to Herr Bessel-Hagen, for his careful proof-reading of part of this article.

CONTENTS

vii

Part IV. General theory of relativity

Part V. Theories on the nature of charged elementary particles

PREFACE
by A. Sommerfeld to the German special edition in book form

In view of the apparently insatiable demand, especially in Germany, for accounts of the Theory of Relativity, both of a popular and of a highly specialized kind, I felt I ought to advise the publishers to arrange for a separate edition of the excellent article by Herr W. Pauli, jr., which appeared in the *Encyklopädie der mathematischen Wissenschaften*, Vol. V. Although Herr Pauli was still a student at the time, he was not only familiar with the most subtle arguments in the Theory of Relativity through his own research work, but was also fully conversant with the literature of the subject.

In its whole lay-out, the article is made to fit into the framework of the *Mathematical Encyclopedia*. Certain references to earlier articles have had to be retained, of course, in this special edition and are hardly likely to trouble the reader. One of these, in particular, is H. A. Lorentz's article on Electron Theory, which presages in its final section the Theory of the Deformable Electron and thus itself represents a milestone in the development of the Theory of Relativity. In keeping with the general character of the Encyclopedia, the mathematical relationships are presented in a completely general and abstract way; especially Part II deals with the mathematical tools of the theory of invariants and multi-dimensional spaces. At the same time, in keeping with the aims of this particular volume of the Encyclopedia, which is devoted to physics, physical applications are always kept to the fore and the possibility of confirmation by experiment is never lost sight of. In Part I, for instance, Ritz's well-known counter-proposal to the Theory of Relativity is presented and is criticized in the light of experimental evidence with a thoroughness which is commensurate with the stature of its originator.

In its comprehensive discussion of experimental data the present article differs from Weyl's great systematic treatment of Space-Time Theory. The latter naturally aims only at expressing Weyl's special point of view, which is in part opposed to that of Einstein; Weyl's theory and Mie's ideas, which it elaborates, are presented in a critical manner in the last Part. On the other hand, Pauli's article differs from Laue's textbook in that the proofs are not generally written out in full, but are only indicated in their main essentials. Whereas Laue's textbook has to restrict itself in many ways in its choice of material, the article aims at including all of the more valuable contributions to the theory which have appeared up to the end of 1920. Beyond this, the author's own opinions are to be found in many places throughout the article.

It is to be hoped that this special edition will be welcomed as a useful addition to the existing literature on Relativity and that it will help physicists and mathematicians alike to gain a deeper understanding of the subject.

Munich, 30 *July* 1921   A. SOMMERFELD.

BIBLIOGRAPHY†

1. Fundamental papers

E. Mach, *Die Mechanik in ihrer Entwicklung historisch-kritisch dargestellt* (Leipzig 1883).

B. Riemann, *Über die Hypothesen, die der Geometrie zugrunde liegen.* Newly edited and annotated by H. Weyl (Berlin 1920) [reprinted from *Nachr. Ges. Wiss. Göttingen*, **13** (1868) 133].

Lorentz–Einstein–Minkowski, *Das Relativitätsprinzip.* A collection of papers (Leipzig 1913, 3rd revised edn., 1920).

H. Minkowski, *Zwei Abhandlungen über die Grundgleichungen der Elektrodynamik*, (Leipzig 1910) [the first paper reprinted from *Nachr. Ges. Wiss. Göttingen* (1908) 53; the second from *Math. Ann.*, **68**, (1910) 526].

A. Einstein and M. Grossmann, *Entwurf einer verallgemeinerten Relativitätstheorie und einer Theorie der Gravitation* (Leipzig 1913) [reprinted from *Z. Math. Phys.*, **63** (1914) 215].

A. Einstein, *Die Grundlagen der allgemeinen Relativitätstheorie* (Leipzig 1916) [reprinted from *Ann. Phys.*, *Lpz.*, **49** (1916) 769].

2. Textbooks

M. v. Laue, *Das Relativitätsprinzip*, (Leipzig 1911; 3rd edn., 1919, Vol. 1, *Das Relativitätsprinzip der Lorentz-Transformation*; 4th edn., 1921).

H. Weyl, *Raum-Zeit-Materie.* Lectures on the general theory of relativity (Berlin 1918; 3rd edn., 1920; 4th edn., 1921), (quoted from the 1st and 3rd edns.).

A. S. Eddington, *Space, Time and Gravitation*, (Cambridge 1920).

A. Kopff, *Grundzüge der Einsteinschen Relativitätstheorie* (Leipzig 1921).

E. Freundlich, *Die Grundlagen der Einsteinschen Gravitationstheorie*, (Berlin 1916).

A. Einstein, *Über die spezielle und die allgemeine Relativitätstheorie* (Braunschweig 1917) (for the general reader).

M. Born, *Die Relativitätstheorie Einsteins und ihre physikalischen Grundlagen* (Berlin 1920) (for the general reader).

3. Papers on specific topics.

H. Poincaré, six lectures given at Göttingen 22–28 April, 1909; sixth lecture, *La mécanique nouvelle*, (Leipzig 1910).

P. Ehrenfest, *Zur Krise der Lichtäther-Hypothese.* Inaugural lecture delivered at Leyden (Berlin 1913).

H. A. Lorentz, *Das Relativitätsprinzip.* Three lectures given at the Teyler Foundation, Haarlem (Leipzig 1914).

† See suppl. note 1.

A. Einstein, *Äther und Relativätstheorie*. Lecture given at Leyden on 5 May 1920 (Berlin 1920).

F. Klein, *Gesammelte mathematische Abhandlungen*, Vol. 1 edited by R. Fricke and A. Ostrowski (Berlin 1921) (in particular the chapter 'Zum Erlanger Programm [1872]').

A. Brill, *Das Relativitätsprinzip*, (Leipzig 1912; 4th edn., 1920).

E. Cohn, *Physikalisches über Raum und Zeit*, (Leipzig 1913).

H. Witte, *Raum und Zeit im Lichte der neueren Physik*, (Braunschweig 1914; 3rd edn., 1920).

4. Papers of philosophical content

M. Schlick, *Raum und Zeit in der gegenwärtigen Physik, zur Einführung in das Verständnis der allgemeinen Relativitätstheorie* (Berlin 1917; 3rd edn., 1920).

H. Holst, *Vort fysiske Verdensbillede og Einsteins Relativitetstheori* (Copenhagen 1920).

H. Reichenbach, *Relativitätstheorie und Erkenntnis a priori* (Berlin 1920).

E. Cassirer, *Zur Einsteinschen Relativitätstheorie* (Berlin 1921).

J. Petzold, *Die Stellung der Relativitätstheorie in der geistigen Entwicklung der Menschheit* (Dresden 1921).

The following articles in the *Enzykl. math. Wiss.* form supplements to the present work: On the astronomical side, F. Kottler, 'Gravitation und Relativitätstheorie' (contribution to article **VI** 2, 22 by S. Oppenheim). On the mathematical side, R. Weitzenböck, 'Neuere Arbeiten über algebraische Invariantentheorie, Differentialinvarianten', III 3, 10; and L. Berwald, 'Differentialinvarianten der Geometrie. Riemannsche Mannigfaltigkeiten und ihre Verallgemeinerungen', **III** 3, 11.

PART I. THE FOUNDATIONS OF THE SPECIAL THEORY OF RELATIVITY

1. Historical background (Lorentz, Poincaré, Einstein)

The transformation in physical concepts which was brought about by the theory of relativity, had been in preparation for a long time. As long ago as 1887, in a paper still written from the point of view of the elastic-solid theory of light, Voigt[1] mentioned that it was mathematically convenient to introduce a local time t' into a moving reference system. The origin of t' was taken to be a linear function of the space coordinates, while the time scale was assumed to be unchanged. In this way the wave equation

$$\Delta\phi - \frac{1}{c^2}\frac{\partial^2\phi}{\partial t^2} = 0$$

could be made to remain valid in the moving reference system, too. These remarks, however, remained completely unnoticed, and a similar transformation was not again suggested until 1892 and 1895, when H. A. Lorentz[2] published his fundamental papers on the subject. Essentially physical results were now obtained, in addition to the purely formal recognition that it was mathematically convenient to introduce a local time t' in a moving coordinate system. It was shown that all experimentally observed effects of first order in v/c (ratio of the translational velocity of the medium to the velocity of light) could be explained quantitatively by the theory when the motion of the electrons embedded in the aether was taken into account. In particular, the theory gave an explanation for the fact that a *common* velocity of medium and observer relative to the aether has no influence on the phenomena, as far as quantities of first order are concerned.[3]

However, the negative result of Michelson's interferometer experiment[4],

---

[1] W. Voigt, 'Über das Dopplersche Prinzip', *Nachr. Ges. Wiss. Göttingen* (1887) 41. Voigt's formulae are obtained by substituting $\kappa = \sqrt{(1-\beta^2)}$ in Eqs. (1), below.

[2] H. A. Lorentz, 'La théorie électromagnétique de Maxwell et son application aux corps mouvants', *Arch. néerl. Sci.*, **25** (1892) 363; *Versuch einer Theorie der elektrischen und magnetischen Erscheinungen in bewegten Körpern* (Leyden 1895).

[3] Fizeau's result purported to show the effect of the earth's motion on the change in the azimuth of polarization when polarized light is obliquely incident on a glass plate. This contradicted both the relativity principle and Lorentz's theory and was later shown to be wrong by D. B. Brace (*Phil. Mag.*, **10** (1908) 591) and B. Strasser (*Ann. Phys., Lpz.*, **24** (1907) 137). It should further be mentioned that in Lorentz's theory it might be possible to obtain first-order effects of the "aether wind" by considering gravitation. Thus, as mentioned by Maxwell, the motion of the solar system relative to the aether would produce first-order differences in the times of the eclipses of the Jupiter satellites; but it was found by C. V. Burton (*Phil. Mag.*, **19** (1910) 417; cf. also H. A. Lorentz, 'Das Relativitätsprinzip', *3 Haarlemer Vorträge* (Leipzig 1914), p. 21) that the inherent observational errors would be as large as the expected magnitude of the effect. Observations on the satellites would therefore not help in deciding for or against the old aether theory.

[4] A description of this experiment is given by Lorentz in article V14 of the *Encykl. math. Wiss.* (Leipzig 1904).

concerned as it was with an effect of second order in v/c, created great difficulties for the theory. To remove these, Lorentz[5] and, independently, FitzGerald put forward the hypothesis that all bodies change their dimensions when moving with a translational velocity v. This change of dimension would be governed by a factor $\kappa \sqrt{[1 - (v^2/c^2]}$ in the direction of motion, with κ as the corresponding factor for the transverse direction; κ itself remains undetermined. Lorentz justified this hypothesis by pointing out that the molecular forces might well be changed by the translational motion. He added to this the assumption that the molecules rest in a position of equilibrium and that their interaction is purely electrostatic. It would then follow from the theory that a state of equilibrium exists in the moving system, provided all dimensions in the direction of motion are shortened by a factor $\sqrt{[1 - (v^2/c^2)]}$, with the transverse dimensions unaltered. It now remained to incorporate this "Lorentz contraction" in the theory, as well as to interpret the other experiments[6] which had not succeeded in showing the influence of the earth's motion on the phenomena in question. There was first of all Larmor who, as early as 1900, set up the formulae now generally known as the Lorentz transformation, and who thus considered a change also in the time scale[7]. Lorentz's review article[8], completed towards the end of 1903, contained several brief allusions which later proved very fruitful. He conjectured that if the idea of a variable electromagnetic mass was extended to all ponderable matter, the theory could account for the fact that the translational motion would produce only the above-mentioned contraction and no other effects, even in the presence of molecular motion. This would also explain the Trouton and Noble experiment. In addition, he raised the important question whether the size of the electrons might be changed by the motion.[9] However, in the introduction to his article, Lorentz still maintained the principle that the phenomena depended not only on the relative motion of the bodies, but also on the motion of the aether.[9a]

We now come to the discussion of the three contributions, by Lorentz[10], Poincaré[11] and Einstein[12], which contain the line of reasoning and the developments that form the basis of the theory of relativity. Chronologically, Lorentz's paper came first. He proved, above all, that Maxwell's equations are invariant under the coordinate transformation[13]

[5] H. A. Lorentz, 'De relative beweging van de aarde en dem aether', *Versl. gewone Vergad. Akad. Amst.*, **1** (1892) 74.

[6] F. T. Trouton and H. R. Noble, *Philos. Trans.*, A **202** (1903) 165; Lord Rayleigh, *Phil. Mag.*, **4** (1902) 678.

[7] J. J. Larmor, *Aether and Matter* (Cambridge 1900) pp. 167–177.

[8] Article V14 of the *Encykl. math. Wiss.* (Leipzig 1904), final §§ 64 and 65.

[9] *Ibid.*, p. 278.

[9a] *Ibid.*, p. 154.

[10] H. A. Lorentz, 'Electromagnetic phenomena in a system moving with any velocity smaller than that of light', *Proc. Acad. Sci., Amst.*, **6** (1904) 809 (*Versl. gewone Vergad. Akad., Amst.*, **12** (1904) 986).

[11] H. Poincaré, 'Sur la dynamique de l'électron', *C.R. Acad. Sci., Paris*, **140** (1905) 1504; 'Sur la dynamique de l'électron', *R.C. Circ. mat. Palermo*, **21** (1906) 129.

[12] A. Einstein, 'Zur Elektrodynamik bewegter Körper', *Ann. Phys., Lpz.*, **17** (1905) 891.

[13] To obtain (1) from Larmor's and Lorentz's formulae, one has to replace their x by $x—vt$, since they first make the usual transition to the moving system.

$$x' = \kappa\frac{x - vt}{\sqrt{(1 - \beta^2)}}, \qquad y' = \kappa y, \qquad z' = \kappa z, \qquad t' = \kappa\frac{t - (v/c^2)x}{\sqrt{(1 - \beta^2)}}, \qquad (1)$$

$$\left(\beta = v/c\right)$$

provided the field intensities in the primed system are suitably chosen. This, however, he proved rigorously only for Maxwell's equations in charge-free space. The terms which contain the charge density and current are, in Lorentz's treatment, not the same in the primed and the moving systems, because he did not transform these quantities quite correctly. He therefore regarded the two systems as not completely, but only very approximately, equivalent. By assuming that the electrons, too, could be deformed by the translational motion and that all masses and forces have the same dependence on the velocity as purely electromagnetic masses and forces, Lorentz was able to derive the existence of a contraction affecting all bodies (in the presence of molecular motion as well). He could also explain why all experiments hitherto known had failed to show any influence of the earth's motion on optical phenomena. A less immediate consequence of his theory is that one has to put $\kappa = 1$. This means that the transverse dimensions remain unchanged during the motion, if indeed this explanation is at all possible. We would like to stress that even in this paper the relativity principle was not at all apparent to Lorentz. Characteristically, and in contrast to Einstein, he tried to understand the contraction in a causal way.

The formal gaps left by Lorentz's work were filled by Poincaré. He stated the relativity principle to be generally and rigorously valid. Since he, in common with the previously discussed authors, assumed Maxwell's equations to hold for the vacuum, this amounted to the requirement that all laws of nature must be covariant with respect to the "Lorentz transformation"[14]. The invariance of the transverse dimensions during the motion is derived in a natural way from the postulate that the transformations which effect the transition from a stationary to a uniformly moving system must form a group which contains as a subgroup the ordinary displacements of the coordinate system. Poincaré further corrected Lorentz's formulae for the transformations of charge density and current and so derived the complete covariance of the field equations of electron theory. We shall discuss his treatment of the gravitational problem, and his use of the imaginary coordinate ict, at a later stage (see §§ 50 and 7).

It was Einstein, finally, who in a way completed the basic formulation of this new discipline. His paper of 1905 was submitted at almost the same time as Poincaré's article and had been written without previous knowledge of Lorentz's paper of 1904. It includes not only all the essential results contained in the other two papers, but shows an entirely novel, and much more profound, understanding of the whole problem. This will now be demonstrated in detail.

[14] The terms 'Lorentz transformation' and 'Lorentz group' occurred for the first time in this paper by Poincaré.

2. The postulate of relativity

The failure of the many attempts[15]† to measure terrestrially any effects of the earth's motion on physical phenomena allows us to come to the highly probable, if not certain, conclusion that the phenomena in a given reference system are, in principle, independent of the translational motion of the system as a whole. To put it more precisely: there exists a triply infinite set[16] of reference systems moving rectilinearly and uniformly relative to one another, in which the phenomena occur in an identical manner. We shall follow Einstein in calling them "Galilean reference systems"—so named because the Galilean law of inertia holds in them. It is unsatisfactory that one cannot regard *all* systems as completely equivalent or at least give a logical reason for selecting a particular set of them. This defect is overcome by the *general* theory of relativity (see Part IV). For the moment we shall have to restrict ourselves to Galilean reference systems, i.e. to the relativity of uniform translational motions.

Once the postulate of relativity is stated, the concept of the aether as a *substance* is thereby removed from the physical theories. For there is no point in discussing a state of rest or of motion relative to the aether when these quantities cannot, in principle, be observed experimentally. Nowadays this is all the less surprising as attempts to derive the elastic properties of matter from electrical forces are beginning to show success. It would therefore be quite inconsistent to try, in turn, to explain electromagnetic phenomena in terms of the elastic properties of some hypothetical medium.[17] Actually, the mechanistic concept of an aether had already come to be superfluous and something of a hindrance when the elastic-solid theory of light was superseded by the electromagnetic theory of light. In this latter the aether substance had always remained a foreign element. Einstein[18] has recently suggested an extension of the notion of an aether. It should no longer be regarded as a substance but simply as *the totality of those physical quantities which are to be associated with matter-free space*. In this wider sense there does, of course, exist an aether; only one has to bear in mind that it does not possess any mechanical properties. In other words, the physical quantities of matter-free space have no space coordinates or velocities associated with them.

It might seem that the postulate of relativity is immediately obvious, once the concept of an aether has been abandoned. Closer reflection shows

---

[15] In addition to the references in footnote 6, the following should be mentioned: the repetition of Michelson's experiment by E. W. Morley and D. C. Miller, *Phil. Mag.*, **8** (1904) 753 and **9** (1905) 680. (See also its discussion by J. Lüroth, *S.B. bayer. Akad. Wiss.*, **7** (1909); E. Kohl, *Ann. Phys., Lpz.*, **28** (1909) 259 and 662; M. v. Laue, *Ann. Phys., Lpz.*, **33** (1910) 156); further attempts to find double refraction due to the earth's motion: D. B. Brace, *Phil. Mag.*, **7** (1904) 317, **10** (1905) 71 and *Boltzmann-Festschrift* (1907) 576; an experiment by F. T. Trouton and A. O. Rankine to determine the change in the electric resistance of a wire corresponding to its orientation with respect to the earth's motion, *Proc. Roy. Soc.*, **8** (1908) 420; see also a review article by J. Laub on the experimental basis of the relativity principle, *Jb. Radioakt.*, **7** (1910) 405.

† See suppl. note 2.

[16] We shall not consider trivial displacements of the origin and coordinate axes.

[17] This point was made by M. Born, *Naturwissenschaften*, **7** (1919) 136.

[18] A. Einstein, 'Äther und Relativitätstheorie', lecture delivered at Leyden (Berlin 1920).

however that this is not so.[19] Naturally we cannot subject the whole universe to a translational motion and then investigate whether the phenomena are thereby altered. Our statement will therefore only be of heuristic value and physically meaningful when we regard it as valid for any and every closed system. But when is a system a closed system? Would it be sufficient to stipulate that all masses should be far enough removed?[20] Experience tells us that this is sufficient for uniform motion, but not for a more general motion. An explanation for the preferred rôle played by uniform motion is to be given at a later stage (see Part IV, § 62). Summarizing, we can say the following: The postulate of relativity implies that a uniform motion of the centre of mass of the universe relative to a closed system will be without influence on the phenomena in such a system.

3. The postulate of the constancy of the velocity of light. Ritz's and related theories

The postulation of relativity is still not sufficient for inferring the covariance of all laws of nature under the Lorentz transformation. Thus, for instance, classical mechanics is perfectly in accord with the principle of relativity, although the Lorentz transformation cannot be applied to its equations. As we saw above, Lorentz and Poincaré had taken Maxwell's equations as the basis of their considerations. On the other hand, it is absolutely essential to insist that such a fundamental theorem as the covariance law should be derivable from the simplest possible basic assumptions. The credit for having succeeded in doing just this goes to Einstein. He showed that only the following single axiom in electrodynamics need be assumed: *The velocity of light is independent of the motion of the light source.* If this is a point source, then the wave fronts are in all cases spheres with their centres at rest. For conciseness we shall denote this by "constancy of the velocity of light", although such a designation might give rise to misunderstandings. There is no question of a *universal* constancy of the velocity of light in vacuo, if only because it has the constant value c only in Galilean systems of reference. On the other hand its independence of the state of motion of the light source obtains equally in the general theory of relativity. It proves to be the true essence of the old aether point of view. (See § 5 concerning the equality of the *numerical values* of the velocity of light in all Galilean systems of reference.)

As will be shown in the next section, the constancy of the velocity of light, in combination with the relativity principle, leads to a new concept of time. For this reason W. Ritz[21] and, independently, Tolman[22],

---

[19] Cf. A. Einstein, *Ann. Phys., Lpz.*, **38** (1912) 1059.

[20] In a different context, H. Holst has pointed out that it is necessary to take distant masses into account, even in the special theory of relativity (cf. footnote 43, p. 15).

[21] W. Ritz, 'Recherches critiques sur l'électrodynamique générale', *Ann. Chim. Phys.*, **13** (1908) 145 [*Coll. Works*, p. 317]; 'Sur les théories électromagnétiques de Maxwell-Lorentz', *Arch. Sci. Phys. Nat.*, **16** (1908) 209 [*Coll. Works*, p. 427]; 'Du rôle de l'éther en physique', *Riv. Sci., Bologna*, **3** (1908) 260 [*Coll. Works*, p. 447]; see also P. Ehrenfest, 'Zur Frage nach der Entbehrlichkeit des Lichtäthers', *Phys. Z.*, **13** (1912) 317; 'Zur Krise der Lichtäther-hypothese', lecture delivered in Leyden, 1912 (Berlin, 1913).

[22] R. C. Tolman, *Phys. Rev.*, **30** (1910) 291 and **31** (1910) 26.

Kunz[23] and Comstock[24] have raised the following question: Could one not avoid such radical deductions and yet retain agreement with experiment, by rejecting the constancy of the velocity of light and retaining only the first postulate? It is clear that one would then have to abandon not only the idea of the existence of an aether but also Maxwell's equations for the vacuum, so that the whole of electrodynamics would have to be constructed anew. Only Ritz has succeeded in doing this in a systematic theory. He retains the equations

$$\operatorname{curl} \mathbf{E} + \frac{1}{c}\dot{\mathbf{H}} = 0, \qquad \operatorname{div} \mathbf{H} = 0,$$

so that the field intensities can be derived, just as in ordinary electrodynamics, from a scalar and vector potential

$$\mathbf{E} = -\operatorname{grad} \phi - \frac{1}{c}\dot{\mathbf{A}}, \qquad \mathbf{H} = \operatorname{curl} \mathbf{A}.$$

The equations

$$\phi(P,t) = \int \frac{\rho \, dV_{P'}}{[r_{PP'}]_{t'=t-(r/c)}} \quad , \quad \mathbf{A}(P,t) = \int \frac{(1/c)\rho \mathbf{v} \, dV_{P'}}{[r_{PP'}]_{t=\,'-(r/c)}}$$

of ordinary electrodynamics are now, however, replaced by

$$\phi(P,t) = \int \frac{\rho \, dV_{P'}}{[r_{PP'}]_{t'=t-[r/(c+v_r)]}} \quad , \quad \mathbf{A}(P,t) = \int \frac{(1/c)\rho \mathbf{v} \, dV_{P'}}{[r_{PP'}]_{t'=t-[r/(c+v_r)]}} .$$

This corresponds to the principle that it is only the velocity of a light wave *relative to its source*, and analogously the velocity of an electromagnetic disturbance *relative to the electron*, which is equal to c. We shall call all theories which are based on this assumption "emission theories". Since the relativity principle is automatically satisfied by all such theories, they can all explain Michelson's interferometer experiment. It remains now to investigate whether they are also compatible with the results of other optical experiments.

It has first to be noted that the emission theories are not consistent with the electron-theoretical explanation of reflection and refraction, for which it is essential that the spherical waves emitted by the dipoles in the body should interfere with the incident wave. If we now think of the body as at rest, and the light source moving relative to it, then according to Ritz the waves emitted by the dipoles will have a velocity (i.e. c) different from that of the incident wave. Interference is therefore not possible. A further important point is that additional, artificial, hypotheses are needed to enable emission theories to explain Fizeau's experiment (see § 6), which is so fundamental to the optics of moving media. Let us investigate more closely what the emission theories have to say on the

[23] J. Kunz, *Amer. J. Sci.*, **30** (1910) 1313.
[24] D. F. Comstock, *Phys. Rev.*, **30** (1910) 267.

Doppler effect. A simple argument shows that the frequency changes in just the way it is required to by the aether theory, whereas the wavelength remains the same as for a source at rest, because of the change in velocity[22a]. The question therefore arises whether, in the usual astronomical observations of the Doppler effect, it is a change in wave length or in frequency which is established? It may be assumed in favour of the emission theories that, for experiments carried out with prisms, it is a change in frequency. Much more difficult to decide is the case of observations made with diffraction gratings. Tolman takes the view that it is here a question of wave length, which would disprove the emission theories. Stewart[25], on the other hand, is of the opposite opinion. It is not possible to reach a straightforward decision in this matter, since the concept of diffraction in the emission theories is, in any case, not a very clear one. The various emission theories diverge in their predictions regarding the Doppler effect for a moving mirror. According to J. J. Thomson[26] and Stewart[25], the moving mirror is equivalent to the mirror image of the light source, as far as the velocity of the reflected light ray is concerned. According to Tolman, the mirror acts like a new light source present on its surface. Finally, according to Ritz[21a] the velocity of the reflected ray is equal to that of a parallel ray, emitted by the original light source. For a source at rest and the mirror in motion one would thus, according to Thomson and Stewart, not expect a wave-length Doppler effect, whereas with Tolman the effect would be half that of ordinary optics, and according to Ritz it would be equal to it. Now the Doppler effect of the wave length of light reflected from a moving mirror has recently been determined interferometrically in a number of experiments[27], with the unquestionable result that it agrees with the value required by classical optics. The assumptions of Thomson and Stewart and of Tolman have thus been proved wrong. Furthermore, Q. Majorana[28] determined the Doppler effect for a moving light source, again interferometrically, and found it to be identical with the classical value. As pointed out by Michaud[29], in particular, Majorana's experiment does not disprove Ritz's theory, and this for the following reason: Let L be a light source moving with velocity v away from a mirror S which is at rest, and let A be a fixed point in front of the mirror (see Fig. 1). The essence of Majorana's experiment lies in the change in optical path length $AS = l$ of the light before and after reflection, as the velocity of the light source increases from zero to v. Before reflection we have the velocity equal to $c-v$, the frequency $\nu_1 = \nu[1-(v/c)]$, so that $\lambda_1 = (c-v)/\nu_1 = \lambda$. On reflection at the mirror S, which is at rest, the

[22a] This was first pointed out by Tolman (Cf. footnote 22, p. 5, *ibid.*).

[25] O. M. Stewart, *Phys. Rev.*, **32** (1911) 418.

[26] J. J. Thomson, *Phil. Mag.*, **19** (1910) 301.

[21a] Cf. footnote 21, p. 5, W. Ritz and P. Ehrenfest, *ibid.*; see also R. C. Tolman, *Phys. Rev.*, **35** (1912) 136. Whenever the "Ritz theory" is mentioned in what follows, the term is taken to include the above-mentioned prescription which is not free from arbitrariness.

[27] A. A. Michelson, *Astroph. J.*, **37** (1913) 190; Ch. Fabry and H. Buisson, *C.R. Acad. Sci., Paris*, **158** (1914) 1498; Q. Majorana, *C.R. Acad. Sci., Paris*, **165** (1917) 424, *Phil. Mag.*, **35** (1918) 163 and *Phys. Rev.*, **11** (1918) 411.

[28] Q. Majorana, *Phil. Mag.*, **37** (1919) 190.

[29] P. Michaud, *C.R. Acad. Sci., Paris*, **168** (1919) 507.

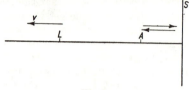

Fig. 1

frequency remains unaltered, but the velocity becomes $c+v$, so that, to first order in small quantities, the wave length becomes $\lambda_2 = (c+v)/\nu_1 = \lambda\,[1+(2v/c)]$. The required change in the total path length is therefore

$$\Delta = \frac{2v}{c}\,l = \frac{v}{c}\,2l$$

just as in the classical theory. It can be shown quite in general that for quantities of first order there is no difference between Ritz's and ordinary or relativistic optics, provided one deals with closed light paths. In other words, terrestrial experiments can only be made to decide in favour of one or other of the two viewpoints, if they include second-order effects.[30] Michelson's interferometer experiment could, according to La Rosa[31] and Tolman[32] serve as "experimentum crucis", provided it is not carried out with a terrestrial light source but makes use of light coming from the sun. In contrast to the theory of relativity, Ritz's theory would then demand a shift of the interference fringes on rotation of the apparatus.[†]

There are, however, first-order effects which could disprove Ritz, provided one deals not with closed but with open light paths. While it is not possible to perform such measurements terrestrially, astronomical observations are certainly feasible. Comstock[24a] already pointed to possible effects with twin stars. De Sitter[33] later discussed the problem quantitatively and came to the following conclusions: If the velocity of light were not assumed constant, then for circular orbits of spectroscopic twin stars the time dependence of the Doppler effect would correspond to that of an eccentric orbit. Since the actual orbits have very small eccentricity, this leads one to conclude that the velocity of light is, to a large degree, independent of the velocity v of the twin star. If one assumes the expression for the light velocity to be of the form $c+kv$, then k must be < 0.002. One can now consider this result in conjunction with the above-mentioned difficulties encountered by the emission theories to explain Fizeau's experiment and to give an atomistic interpretation of refraction. It can then be safely said that the postulate of the constancy of the

[30] This has been remarked by Ehrenfest, cf. footnote 21, p. 5, *Phys. Z., loc. cit.*

[31] M. La Rosa, *Nuovo Cim.*, (6) **3** (1912) 345 and *Phys. Z.*, **13** (1912) 1129.

[32] R. C. Tolman, *Phys. Rev.*, **35** (1912) 136.

[†] See suppl. note 3.

[24a] Cf. footnote 24, p. 6, *ibid.*

[33] W. de Sitter, *Proc. Acad. Sci., Amst.*, **15** (1913) 1297 and **16** (1913) 395; *Phys. Z.*, **14** (1913) 429 and 1267; see also the review article by P. Guthnik, *Astr. Nachr.*, **195** (1913) No. 4670, and E. Freundlich's objection [*Phys. Z.*, **14** (1913) 835], refuted by de Sitter's second paper. See also W. Zurhellen, *Astr. Nachr.*, **198** (1914) 1.

velocity of light has been proved to be correct, whereas attempts by Ritz and others to account for the Michelson experiment have been shown untenable.

4. The relativity of simultaneity. Derivation of the Lorentz transformation from the two postulates. Axiomatic nature of the Lorentz transformation

At first sight it appears as if the two postulates were incompatible. For, let us take a light source L which moves relative to an observer A with velocity v, and consider a second observer B at rest with respect to L. Both observers must then see as wave fronts spheres whose centres are at rest relative to A and B, respectively. In other words, they see *different* spheres. This contradiction disappears, however, if one admits that space points which are reached by the light simultaneously for A, are not reached simultaneously for B. This brings us directly to the relativity of simultaneity. Here, it will first of all be necessary to say what is meant by the synchronization of two clocks at different places. The following definition was chosen by Einstein. A light ray is emitted from point P at time t_P, is reflected at Q at time t_Q, and returns to P at time t'_P. The clock at Q is then considered synchronized with that at P if $t_Q = (t_P + t'_P)/2$. Einstein uses light for regulating the clocks because the two postulates enable us to make definite statements about the mode of propagation of the light signals. Naturally, one could think of other ways of comparing the clocks, such as transporting them, or using mechanical or elastic couplings, etc. Only it must be stipulated that no such method should lead to a contradiction with the optical regulation method.

We can now derive the transformation formulae which connect the coordinates, x, y, z, t and x', y', z', t', of two reference systems K and K' in uniform relative motion. The x-axis is chosen to lie along the direction of motion in such a way that K' moves relative to K with velocity v in the positive x-direction. All writers start with the requirement that the transformation formulae should be linear. This can be justified by the statement that a uniform rectilinear motion in K must also be uniform and rectilinear in K'. Furthermore it is to be taken for granted that finite coordinates in K remain finite in K'. This also implies the validity of Euclidean geometry and the homogeneous nature of space and time. It follows from the two postulates that the equation

$$x^2 + y^2 + z^2 - c^2 t^2 = 0 \qquad (2)$$

entails the corresponding equation

$$x'^2 + y'^2 + z'^2 - c^2 t'^2 = 0. \qquad (2')$$

Since the transformation is to be a linear one, this is only possible if

$$x'^2 + y'^2 + z'^2 - c^2 t'^2 = \kappa(x^2 + y^2 + z^2 - c^2 t^2)$$

where κ is a constant depending on v. If one also bears in mind that any motion parallel to the x-axis must remain so after the transformation, formulae (1) of § 1 will be seen to follow immediately. It is, however, still necessary to show that κ can be put equal to 1. Einstein's procedure consists in applying transformation (1) once more, with the velocity in the opposite direction.

$$x'' = \kappa(-v)\frac{x' + vt'}{\sqrt{(1 - \beta^2)}}, \qquad y'' = \kappa(-v)y', \qquad z'' = \kappa(-v)z',$$

$$t'' = \kappa(-v)\frac{t' + (v/c^2)x'}{\sqrt{(1 - \beta^2)}}.$$

Then

$$x'' = \kappa(v)\kappa(-v)x, \qquad y'' = \kappa(v)\kappa(-v)y, \qquad z'' = \kappa(v)\kappa(-v)z,$$

$$t'' = \kappa(v)\kappa(-v)t.$$

Since K'' is at rest relative to K, it must be identical with it. Hence

$$\kappa(v)\kappa(-v) = 1.$$

It was already pointed out in § 1 that $\kappa(v)$ corresponds to the change in the transverse dimensions of a rod, and this must, for reasons of symmetry, be independent of the *direction* of the velocity. Hence $\kappa(v) = \kappa(-v)$. The above relation then gives $\kappa(v) = 1$, since κ must be positive. Poincaré arrives at this conclusion in a similar way. He considers the totality of all linear transformations which transform Eq. (2) into itself (this totality naturally forms a group) and demands that it should contain as subgroups

(a) the one-parameter group of translations parallel to the x-axis (the group parameter here is the velocity v),

(b) the ordinary displacements of the coordinate axes.

Once again, $\kappa = 1$ follows, since Einstein's symmetry requirement $\kappa(v) = \kappa(-v)$ is contained in (b). We have thus obtained the definite result that

$$x' = \frac{x - vt}{\sqrt{(1 - \beta^2)}}, \qquad y' = y, \qquad z' = z, \qquad t' = \frac{t - (v/c^2)x}{\sqrt{(1 - \beta^2)}}, \qquad \text{(I)}$$

$$x'^2 + y'^2 + z'^2 - c^2t'^2 = x^2 + y^2 + z^2 - c^2t^2. \qquad \text{(II)}$$

The transformation which is the inverse of (I) can be obtained by replacing[34] v by $-v$:

$$x = \frac{x' + vt'}{\sqrt{(1 - \beta^2)}}, \qquad y = y', \qquad z = z', \qquad t = \frac{t' + (v/c^2)x'}{\sqrt{(1 - \beta^2)}}. \qquad \text{(Ia)}$$

---

[34] For some applications it is useful to know the transformation formulae also for the general case where the x-axis is not in the direction of the velocity v. They can be obtained by splitting $\mathbf{r}$ into components $\mathbf{r}_{\|}$ (along the direction of the velocity $\mathbf{v}$ of K' relative to K) and $\mathbf{r}_{\perp}$ (perpendicular to $\mathbf{v}$). It follows from (I), first of all, that

$$\mathbf{r}'_{\|} = \frac{\mathbf{r}_{\|} - \mathbf{v}t}{\sqrt{(1 - \beta^2)}}, \qquad \mathbf{r}'_{\perp} = \mathbf{r}_{\perp}, \qquad t' = \frac{t - (\mathbf{v}\cdot\mathbf{r}_{\|})/c^2}{\sqrt{(1 - \beta^2)}};$$

The simple structure of formulae (I) makes one wonder whether they could not have been derived from general group-theoretical considerations, without having to assume the invariance of (2). To what extent this is in fact possible has been shown by Ignatowsky, and Frank and Rothe.[35] One need assume no more than the following conditions:

(a) the transformations must form a one-parameter homogeneous linear group;

(b) the velocity of K relative to K' is equal and opposite to that of K' relative to K;

(c) the contraction of lengths at rest in K' and observed in K is equal to that of lengths at rest in K and observed in K'.

This already suffices to show that the transformation formulae must be of the form

$$x' = \frac{x - vt}{\sqrt{(1 - \alpha v^2)}}, \qquad t' = \frac{t - \alpha vx}{\sqrt{(1 - \alpha v^2)}}. \tag{3}$$

Nothing can, naturally, be said about the sign, magnitude and physical meaning of α. From the group-theoretical assumption it is only possible to derive the general form of the transformation formulae, but not their physical content. Incidentally, it is to be noted that (3) contains the transformation formulae of ordinary mechanics

$$x' = x - vt, \qquad t' = t, \tag{4}$$

which can be obtained by putting $\alpha = 0$. Following P. Frank, these latter are now generally given the name "Galilean transformations." It is obvious that, by putting $c = \infty$, they can equally well be derived from (1).

5. Lorentz contraction and time dilatation

The Lorentz contraction is the simplest consequence of the transformation formulae (1), and thus also of the two basic assumptions. Take a rod lying along the x-axis, at rest in reference system K'. The position coordinates of its ends, x_1' and x_2' are thus independent of t' and

$$x_2' - x_1' = l_0 \tag{5}$$

but since

$$\mathbf{r}_{\parallel} = \frac{(\mathbf{r} \cdot \mathbf{v})\mathbf{v}}{v^2}, \qquad \mathbf{r}_{\perp} = \mathbf{r} - \mathbf{r}_{\parallel} = \mathbf{r} - \frac{(\mathbf{r} \cdot \mathbf{v})\mathbf{v}}{v^2}, \qquad \mathbf{r}' = \mathbf{r}'_{\parallel} + \mathbf{r}'_{\perp},$$

this can also be written

$$\mathbf{r}' = \mathbf{r} + \frac{1}{v^2}\left(\frac{1}{\sqrt{(1 - \beta^2)}} - 1\right)(\mathbf{r} \cdot \mathbf{v})\mathbf{v} - \frac{\mathbf{v}t}{\sqrt{(1 - \beta^2)}},$$
$$t' = \frac{t - (1/c^2)(\mathbf{r} \cdot \mathbf{v})}{\sqrt{(1 - \beta^2)}}. \tag{1a}$$

These formulae can be found in a paper by G. Herglotz, *Ann. Phys., Lpz.*, **36** (1911) 497, Eq. 9.

[35] W. v. Ignatowsky, *Arch. Math. Phys., Lpz.*, **17** (1910) 1 and **18** (1911) 17; *Phys. Z.*, **11** (1910) 972 and **12** (1911) 779; P. Frank and H. Rothe, *Ann. Phys., Lpz.*, **34** (1911) 825 and *Phys. Z.*, **13** (1912) 750.

is the rest length of the rod. On the other hand, we might determine the length of the rod in system K in the following way. We find x_1 and x_2 as functions of t. Then the distance between the two points which coincide simultaneously with the end points of the rod in system K will be called the length l of the rod in the moving system:

$$x_2(t) - x_1(t) = l. \tag{6}$$

Since these positions are not taken up simultaneously in system K', it cannot be expected that l equals l_0. In fact, it follows from (I) that

$$x_2' = \frac{x_2(t) - vt}{\sqrt{(1 - \beta^2)}}, \qquad x_1' = \frac{x_1(t) - vt}{\sqrt{(1 - \beta^2)}},$$

so that

$$l_0 = \frac{l}{\sqrt{(1 - \beta^2)}}, \quad l = l_0\sqrt{(1 - \beta^2)}. \tag{7}$$

The rod is therefore contracted in the ratio $\sqrt{(1-\beta^2)} : 1$, as was already assumed by Lorentz. Since the transverse dimensions of a body remain unaltered, the same formula applies to the contraction of its volume:

$$V = V_0\sqrt{(1 - \beta^2)}. \tag{7a}$$

We have seen that this contraction is connected with the relativity of simultaneity, and for this reason the argument has been put forward[36] that it is only an "apparent" contraction, in other words, that it is only simulated by our space-time measurements. If a state is called real only when it can be determined in the same way in all Galilean reference systems, then the Lorentz contraction is indeed only apparent, since an observer at rest in K' will see the rod without contraction. But we do not consider such a point of view as appropriate, and in any case the Lorentz contraction is in principle observable. In this connection a "thought experiment" of Einstein's[37] is instructive. It shows that the determination of the simultaneity of spatially separated events, which is necessary for the observation of the Lorentz contraction, can be carried out entirely with the help of measuring rods, without the use of clocks. For let us think of using two rods A_1B_1 and A_2B_2 of the same rest length l_0, which move relative to K with equal and opposite velocity v. We mark the points A^* and B^* in K at which the points A_1 and A_2 and B_1 and B_2, respectively, overlap. (For reasons of symmetry the overlappings take place simultaneously in K.) The distance A^*B^*, as measured by rods at rest in K, will then have the value

$$l = l_0\sqrt{(1 - \beta^2)}.$$

It therefore follows that the Lorentz contraction is not a property of a single measuring rod taken by itself, but is a reciprocal relation between

[36] V. Varićak, *Phys. Z.*, **12** (1911) 169.
[37] A. Einstein, *Phys. Z.*, **12** (1911) 509.

two such rods moving relatively to each other, and this relation is in principle observable.

Analogously, the time scale is changed by the motion. Let us again consider a clock which is at rest in K'. The time t' which it indicates in K' is its proper time τ and we can put its coordinate x' equal to zero. It then follows from (I a) that

$$t = \frac{\tau}{\sqrt{(1 - \beta^2)}}, \qquad \tau = \sqrt{(1 - \beta^2)}t. \tag{8}$$

Measured in the time scale of K, therefore, a clock moving with velocity v will lag behind one at rest in K in the ratio $\sqrt{(1 - \beta^2)}:1$. While this consequence of the Lorentz transformation was already implicitly contained in Lorentz's and Poincaré's results, it received its first clear statement only by Einstein.

The time dilatation gives rise to an apparent paradox which was already mentioned in Einstein's first paper and later discussed in more detail by Langevin[38], Laue[39] and Lorentz[40]. Consider two synchronized clocks C_1 and C_2 at a point P. If one of them, say C_2, is now set in motion at time $t = 0$ and made to move with constant speed v along an arbitrary curve, reaching point P' after time t, then it will no longer be synchronous with C_1 afterwards. On arrival at P' it will show the time $t\sqrt{(1 - \beta^2)}$ instead of t. The same result will hold, in particular, when P and P' coincide, i.e. when C_2 returns to its initial position. We can neglect the effect of acceleration on the clock, so long as we are dealing with a Galilean reference system. If we take the special case where C_2 is moved along the x-axis to a point Q and then back again to P, with discontinuous velocity changes at P and Q, then the effect of the acceleration will certainly be independent of t and can easily be eliminated. The paradox now lies in the following statement: Let us describe the process in terms of a reference system K^*, always at rest with respect to C_2. Clock C_1 will then move relative to K^* in the same way as C_2 moves relative to K. Yet, at the end of the motion, clock C_2 will have lost compared with C_1, i.e. C_1 will have gained compared with C_2. The paradox is resolved by observing that the coordinate system K^* is not a Galilean reference system and that in such a system the effect of acceleration cannot be neglected, since the acceleration is not produced by an external force, but, in the terminology of Newtonian mechanics, by an inertial force. Of course, a complete explanation of the problem can only be given within the framework of the general theory of relativity (see Part IV, § 53 (β)); for the four-dimensional formulation of the clock paradox, see Part III, § 24). It is further to be noted that the time regulation by means of transporting clocks, as mentioned in the previous section, is not possible without imposing some restriction. It will only furnish the correct result when the time indications of the clocks are extrapolated to zero transport velocity.

[38] P. Langevin, 'L'évolution de l'espace et du temps', *Scientia*, **10** (1911) 31.
[39] M. v. Laue, *Phys. Z.*, **13** (1912) 118.
[40] H. A. Lorentz, 'Das Relativitätsprinzip', *3 Haarlemer Vorlesungen* (Leipzig 1914), pp. 31 and 47.

It is obvious that experiments which are intended to show the effect of the motion of the coordinate system as a whole on phenomena within it must, according to the theory of relativity, show a negative result. It is nevertheless instructive to investigate how such experiments are seen from a system K which is at rest. For this purpose we shall discuss the Michelson interferometer experiment. Let l_1 be the length, measured in K, of the interferometer arm parallel to the direction of motion, and l_2 that of the arm at right angles to it. The times t_1 and t_2 taken by the light to traverse the arms are then given by

$$ct_1 = \frac{2l_1}{1 - \beta^2}, \qquad ct_2 = \frac{2l_2}{\sqrt{(1 - \beta^2)}}.$$

Now, because of the Lorentz contraction, we have

$$l_1 = l_0 \sqrt{(1 - \beta^2)}, \text{ while } l_2 = l_0,$$

so that

$$ct_1 = ct_2 = \frac{2l_0}{\sqrt{(1 - \beta^2)}}.$$

It would therefore seem that an observer travelling with K' measures a velocity of light

$$c' = c \sqrt{(1 - \beta^2)}, \tag{9}$$

different from that measured by an observer in K. This is the point of view put forward by Abraham[41]. According to Einstein, however, the time dilatation

$$t' = t \sqrt{(1 - \beta^2)}$$

has still to be taken into account, so that

$$ct_1' = ct_2' = 2l_0,$$

and thus the velocity of light is the same in K' as in K. According to Abraham there is no time dilatation. Abraham's point of view is consistent with Michelson's experiment, but it contradicts the postulate of relativity, since it would in principle admit of experiments which would allow one to measure the "absolute" motion of a system.[42]

Let us discuss the difference between Einstein's and Lorentz's points of view still further. Einstein showed in particular that the distinction between "local" and "true" times disappears with a more profound formulation of the concept of time. Lorentz's local time is shown to be simply the time in the moving system K'. There are as many times and

[41] M. Abraham, *Theorie der Elektrizität*, Vol. 2 (2nd edn., Leipzig 1908), p. 367.
[42] One might also mention here "thought experiments" by W. Wien [*Würzb. phys. med. Ges.*, (1908) 29 and *Taschenb. Math. Phys.*, **2** (1911) 287] and by G. N. Lewis and R. C. Tolman [*Phil. Mag.*, **18** (1909), footnote on p. 516] which serve as illustrations of the term $(v/c^2)x/\sqrt{(1 - \beta^2)}$ in the transformation formula for t.

spaces as there are Galilean reference systems. It is also of great value that Einstein rendered the theory independent of any special assumptions about the constitution of matter.

Should one, then, in view of the above remarks, completely abandon any attempt to explain the Lorentz contraction atomistically? We think that the answer to this question should be No. The contraction of a measuring rod is not an elementary but a very complicated process. It would not take place except for the covariance with respect to the Lorentz group of the basic equations of electron theory, as well as of those laws, as yet unknown to us, which determine the cohesion of the electron itself. We can only postulate that this is so, knowing that then the theory will be capable of explaining atomistically the behaviour of moving measuring rods and clocks. However, the equivalence of two coordinate systems in relative motion will always have to be kept in mind.

The epistemological basis of the theory of relativity has recently been undergoing a close examination from the side of philosophy.[43] In this connection the opinion has been expressed that the theory of relativity has thrown overboard the concept of causality. We take the view that it is perfectly satisfactory from the standpoint of the theory of knowledge to say that the relative motion is the cause of the contraction, since this latter is not the property of a *single* measuring rod, but a relation between two such rods. Also, it is unnecessary to refer, as Holst does, to all matter present in the universe, in order to satisfy the causality condition.

6. Einstein's addition theorem for velocities and its application to aberration and the drag coefficient. The Doppler effect

It is immediately apparent that the manner in which velocities are combined in the old kinematics would not lead to the correct result in relativistic kinematics. For instance, it is obvious that on adding a velocity v ($< c$) to c one should obtain c, and not $c+v$. The rules which have to be used here are fully contained in the transformation formulae (I). Let an arbitrary motion

$$x' = x'(t'), \qquad y' = y'(t'), \qquad z' = z'(t')$$

be given in K'. There is then a motion

$$x = x(t), \qquad y = y(t), \qquad z = z(t)$$

in K which corresponds to it. We want to know the connection between the velocity components in K',

$$\frac{dx'}{dt'} = u'_x = u' \cos\alpha', \quad \frac{dy'}{dt'} = u'_y, \quad \frac{dz'}{dt'} = u'_z, \quad u' = \sqrt{(u'^2_x + u'^2_y + u'^2_z)},$$

and the corresponding quantities in K,

$$\frac{dx}{dt} = u_x = u \cos\alpha, \quad \frac{dy}{dt} = u_y, \quad \frac{dz}{dt} = u_z, \quad u = \sqrt{(u_x^2 + u_y^2 + u_z^2)}.$$

[43] See in particular J. Petzold, *Z. positivistische Philos.*, **2** (1914) 40; *Verh. dtsch. phys. Ges.*, **20** (1918) 189 and **21** (1918) 495; *Z. Phys.*, **1** (1920) 467; M. Jakob, *Verh. dtsch. phys. Ges.*, **21** (1919) 159 and 501; H. Holst, *Math.-fys. Medd.*, **2** (1919) 11; *Z. Phys.*, **1** (1920) 32 and **3** (1920) 108.

From (Ia) one obtains

$$dx = \frac{dx' + v\,dt'}{\sqrt{(1 - \beta^2)}}, \qquad dy = dy', \qquad dz = dz', \qquad dt = \frac{dt' + (v/c^2)dx'}{\sqrt{(1 - \beta^2)}}$$

and hence, dividing by the last equation,

$$u_x = \frac{u'_x + v}{1 + (vu'_x/c^2)}, \qquad u_y = \frac{\sqrt{[1 - (v^2/c^2)]}u'_y}{1 + (vu'_x/c^2)},$$

$$u_z = \frac{\sqrt{[1 - (v^2/c^2)]}u'_z}{1 + (vu'_x/c^2)}. \tag{10}$$

These relations are also found in Poincaré's paper which was quoted above.
It follows straightaway that

$$u = \frac{\{u'^2 + v^2 + 2u'v\cos\alpha' - [(u'v\sin\alpha')/c]^2\}^{\frac{1}{2}}}{1 + (u'v\cos\alpha')/c^2}, \tag{11}$$

which can also be written

$$\sqrt{\left(1 - \frac{u^2}{c^2}\right)} = \frac{\sqrt{[1 - (v^2/c^2)]}\sqrt{[1 - (u'^2/c^2)]}}{1 + (u'v\cos\alpha')/c^2} \tag{11a}$$

and

$$\tan\alpha = \frac{\sqrt{[1 - (v^2/c^2)]}u'\sin\alpha'}{u'\cos\alpha' + v}. \tag{12}$$

The inverse formulae are obtained by replacing v by $-v$. Thus the commutative law holds for the absolute values, but not for the directions, of
the velocities. For the special cases where the velocities to be added are
parallel or perpendicular to one another the rules can be deduced immediately from formulae (10).

It is further seen from (11 a) that a combination of velocities less than c
will always result in a velocity less than c. Also, material bodies cannot
move with a relative velocity greater than c; this follows from the fact
that transformation (I) would in such a case produce imaginary values for
the coordinates. But one can make a further statement. Suppose it were
possible that an effect in a system K is propagated with velocity greater
than c. There would then exist a system K' (moving relative to K with
velocity less than c) in which an event causing another, later, event in
K would happen *after* it. For we can put $u_y = u_z = 0$, $u > c$, and then,
using the inverse formula to (10), we obtain

$$u' = \frac{u - v}{1 - (uv/c^2)} < 0$$

on choosing $(c/u) < (v/c) < 1$. This would be tantamount to upsetting the
concepts of cause and effect, and it can therefore be concluded that it is
impossible to send out signals with a velocity greater than that of light.[44]

[44] A. Einstein, *Ann. Phys., Lpz.*, **23** (1907) 371.

Thus, in the theory of relativity, the velocity of light plays in many respects the rôle of an infinitely great velocity. To prevent the kind of misunderstandings which have occasionally arisen, we should like to stress here particularly that the theorem of the non-existence of velocities greater than c applies, by its very derivation, only to Galilean reference systems.

Let us now consider in more detail the case in which one of the velocities to be compounded is equal to c, while leaving the direction of the light ray arbitrary; thus we have $u' = c$. It then follows from (11) that $u = c$, i.e. (velocity of light) + (velocity less than c) gives again velocity of light. The relation (12) then leads for this case to

$$\tan \alpha = \frac{\sqrt{(1 - \beta^2)} \sin \alpha'}{\cos \alpha' + \beta}. \tag{13}$$

This is the relativistic formula for the aberration of light, which was already deduced by Einstein in his first paper. A more rigorous proof of it will be given later. The formula agrees with the classical formula for quantities of first order. An inherent simplification is brought about by the theory of relativity in so far as the two cases, moving light source—observer at rest; light source at rest—moving observer, become identical.

A second important application of Einstein's addition theorem for velocities consists in the explanation of the Fresnel drag coefficient. It was first considered by Laue[46] after an incomplete attempt had been made by J. Laub[45]. Just as in the case of aberration, the theory of relativity cannot furnish a new result when compared with Lorentz's[47] electron-theoretical explanation, at least where first-order quantities are concerned, which alone are observable. The relativistic derivation has however the great advantage that it is simpler than the electron-theoretical one and, in particular, that it demonstrates the independence of the final result from any special assumptions about the mechanism of the refraction of light. The general approach is different, too. Formerly, Fizeau's experiment was actually looked upon as proving the existence of an aether at rest. The interpretation then put forward[48] was that the light waves were propagated relative to the moving medium not with velocity c/n, but with velocity $(c/n) - (v/n^2)$. This involves an application of non-relativistic kinematics which is not justified from the relativistic point of view. Rather should one say that, for an observer moving with the medium, light is propagated as usual with velocity c/n in all directions. Just because of this, its velocity of propagation as seen by an observer moving with velocity v relative to the medium is not $(c/n)+v$ but some other velocity V, determinable from (10). Let us restrict ourselves to the case where the direction of the light ray coincides with that of the

[45] J. Laub, *Ann. Phys., Lpz.*, **23** (1907) 738.

[46] M. v. Laue, *Ann. Phys., Lpz.*, **23** (1907) 989.

[47] See the treatment in *Encykl. math. Wiss.*, V14 (Leipzig 1904) §60. A simplified derivation of the drag coefficient from an electron-theoretical point of view is given by H. A. Lorentz in *Naturw. Rdsch.*, **21** (1906) 487.

[48] See e.g. H. A. Lorentz, *Encykl. math. Wiss.*, V13 (Leipzig 1914), §21, p. 103.

motion of the observer relative to the medium (for the general case, which is to be dealt with in Part III, § 36 (γ), the addition theorem has to be used with care). We thus put $u'_x = u' = (c/n)$, $u_x = u = V$, and the first of equations (10) gives

$$V = \frac{(c/n) + v}{1 + (v/cn)} \simeq \frac{c}{n} + v\left(1 - \frac{1}{n^2}\right),\tag{14}$$

when first-order terms only are retained. For dispersive media a correction term becomes necessary on the right-hand side of the equation, as was already pointed out by Lorentz[49]. For it can be seen from the derivation of the formula that n denotes the refractive index corresponding to the wavelength λ' which is observed in the moving system K'. Because of the Doppler effect (the theory of which will be discussed immediately below) λ' is related to the wavelength λ (valid in system K) in the following way:

$$\lambda' = \lambda\left(1 + \frac{v}{u'}\right) = \lambda\left(1 + \frac{nv}{c}\right).$$

(Once again we restrict ourselves to first-order quantities.) Thus we have

$$\frac{c}{n(\lambda')} = \frac{c}{n(\lambda)} - \frac{c}{n^2}\cdot\frac{dn}{d\lambda}\cdot\lambda\frac{nv}{c},$$

and by writing n instead of $n(\lambda)$, we finally obtain

$$V = \frac{c}{n} + v\left(1 - \frac{1}{n^2} - \frac{\lambda}{n}\cdot\frac{dn}{d\lambda}\right).\tag{14a}$$

Zeeman[50] was able to prove experimentally the existence of this additional term.

The experimental arrangement has recently undergone various modifications, such as making the light emerge not from fixed, but from moving, surfaces, and also perpendicularly to the direction of motion of the body in which the drag was determined. Moving solids of glass and quartz were used, instead of a liquid as in Fizeau's experiment. In this case the theory of the Fizeau experiment has to be modified, and different formulae result.[51] In addition, the translational motion was replaced by one of

[49] H. A. Lorentz, *Versuch einer Theorie der elektrischen und optischen Erscheinungen in bewegten Körpern* (Leyden 1895) p. 101.

[50] P. Zeeman, *Versl. gewone Vergad. Akad. Amst.*, **23** (1914) 245 and **24** (1915) 18.

[51] Such experiments were carried out by G. Sagnac, *C. R. Acad. Sci., Paris*, **157** (1913) 708 and 1410; *J. Phys. théor. appl.*, (5) **4** (1914) 177 [theory discussed by M. v. Laue, *S. B. bayer. Akad. Wiss.* (1911) 404 and *Das Relativitätsprinzip*, (3rd edn., 1919)]; F. Harress, *Dissertation* (Jena, 1911), and report by O. Knopf, *Ann. Phys., Lpz.*, **62** (1920) 389 [for the theoretical interpretation see P. Harzer, *Astr. Nachr.*, **198** (1914) 378 and **199** (1914) 10; A. Einstein, *Astr. Nachr.*, **199** (1914) 9 and 47]; finally, P. Zeeman, *Versl. gewone Vergad. Akad. Amst.*, **28** (1919) 1451, and P. Zeeman and A. Snethlage, *Versl. gewone Vergad. Akad. Amst.*, **28** (1919) 1462; *Proc. Acad. Sci., Amst.*, **22** (1920) 462 and 512; the theory of *all* these experiments is developed in detail by M. v. Laue, *Ann. Phys., Lpz.*, **62** (1920) 448.

rotation. Particularly worth mentioning is Sagnac's experiment, in which all parts of the apparatus are rotated together, because it shows that the rotation of a reference system relative to a Galilean system can be determined by means of optical experiments within the system itself. The result of this experiment is in perfect agreement with the theory of relativity. Already before this, Michelson[51a] had proposed a similar experiment to demonstrate optically the rotation of the earth, and Laue[51a] discussed this proposal very thoroughly from the theoretical point of view. It is essentially the optical analogue of the Foucault pendulum.[†]

Of the phenomena which are basic to the optics of moving bodies, the third and last to be discussed here will be the Doppler effect, although it has nothing to do with the addition theorem for velocities. Consider a very distant source of light L, at rest in system K. An observer moves with a second system K' in the positive x-direction with velocity v relative to K. In K, the line joining source and observer makes an angle α with the x-axis, and the z-axis is taken to be normal to the plane determined by these two directions. Then in K the phase of the light is determined by

$$\exp 2\pi i\nu \left[t - \frac{x\cos\alpha + y\sin\alpha}{c} \right],$$

where ν is the normal frequency of the light source. As will be discussed in more detail in Part III, § 32 (δ), the phase must be an invariant. Therefore

$$\exp 2\pi i\nu' \left[t' - \frac{x'\cos\alpha' + y'\sin\alpha'}{c} \right] = \exp 2\pi i\nu \left[t - \frac{x\cos\alpha + y\sin\alpha}{c} \right].$$

It follows from (I) that

$$\nu' = \nu \frac{1 - \beta\cos\alpha}{\sqrt{(1 - \beta^2)}}, \tag{15}$$

$$\cos\alpha' = \frac{\cos\alpha - \beta}{1 - \beta\cos\alpha}, \qquad \sin\alpha' = \frac{\sin\alpha \sqrt{(1 - \beta^2)}}{1 - \beta\cos\alpha}, \tag{16}$$

from which one obtains

$$\tan\alpha' = \frac{\sin\alpha \sqrt{(1 - \beta^2)}}{\cos\alpha - \beta} \tag{16 a}$$

and

$$\tan\frac{\alpha'}{2} = \sqrt{\left(\frac{1 + \beta}{1 - \beta} \right)} \tan\frac{\alpha}{2}. \tag{16 b}$$

We shall give here also the transformation formula for the solid angle $d\Omega$ of a pencil of rays. Since

$$\frac{d\Omega'}{d\Omega} = \frac{d(\cos\alpha')}{d(\cos\alpha)},$$

[51a] A. A. Michelson, *Phil. Mag.*, 8 (1904) 716; M. v. Laue, *S.B. bayer. Akad. Wiss., math.-phys. Kl.*, (1911) 405.
[†] See suppl. note 4.

by differentiation of

$$1 + \beta \cos \alpha' = \frac{1 - \beta^2}{1 - \beta \cos \alpha} \tag{16 c}$$

it follows immediately that

$$d\Omega' = \frac{1 - \beta^2}{(1 - \beta \cos \alpha)^2} d\Omega. \tag{17}$$

Formula (15) expresses the Doppler effect, (16 a) is the inverse of Eq. (13). We have thus obtained a new, as well as a more rigorous, derivation for the relativistic aberration formula. As was to be expected, we also find that the expression for the Doppler effect agrees with the classical expression up to first-order terms, which alone are open to experimental verification. As in the case of aberration, the theory of relativity introduces an intrinsic simplification, since the two cases (light source at rest, moving observer; moving source, observer at rest) which are different in the old theory and for sound, become identical here.

It is characteristic for the theory of relativity that the Doppler effect does not vanish even when the motion of the light source is at right angles to the direction from which it is observed ($\cos \alpha = 0$). Rather we find in this case, using (15), that

$$\nu' = \frac{\nu}{\sqrt{(1 - \beta^2)}}. \tag{17a}$$

This transverse Doppler shift towards the red is perfectly in keeping with the time dilatation postulated for every moving clock (§ 5). Soon after Stark had observed the Doppler effect in light emitted by canal-ray particles it was suggested by Einstein[52] that the transverse Doppler effect might possibly be verified by observations on canal rays. So far it has not proved possible to carry out such an experiment, as it is extremely difficult to make α exactly equal to 90° and to separate the relativistic transverse from the usual longitudinal Doppler effect.†

[52] A. Einstein, *Ann. Phys., Lpz.*, **33** (1907) 197.
    † See suppl. note 5.

Part II. MATHEMATICAL TOOLS

7. The four-dimensional space-time world (Minkowski)

We showed in Part I that the two postulates of relativity and of the constancy of the velocity of light can be combined into the single requirement that all physical laws should be invariant under the Lorentz transformation. From now on we shall take to mean by the Lorentz transformation the totality of all $(\infty)^{10}$ linear transformations which satisfy the identity (II). Each such transformation can be made up of rotations of the coordinate system (to which may possibly also be added reflections) and the special Lorentz transformation of the type (I).[53] Mathematically speaking, therefore, the special theory of relativity is the theory of invariants of the Lorentz group.

The work of Minkowski[54] has been fundamental for the development of the theory. He managed to give the theory an extraordinarily elegant form by making consistent use of two facts:

(*a*) If, instead of the ordinary time t, the imaginary quantity $u = ict$ is introduced, the behaviour of the space and time coordinates is formally completely equivalent in the Lorentz group, and thus also in the physical laws which are invariant with respect to this group. In fact, the invariant which is characteristic for the Lorentz transformation

$$x^2 + y^2 + z^2 - c^2 t^2,$$

goes over into

$$x^2 + y^2 + z^2 + u^2. \tag{18}$$

It is therefore expedient from the beginning not to separate space and time, but to consider the four-dimensional space-time manifold. We shall follow Minkowski by calling it, in short, "world".

(*b*) Since expression (18) is invariant under the Lorentz transformations and is also quadratic in the coordinates, it would seem natural to define it as the *square of the distance* of the world point $P(x, y, z, u)$ from the

[53] As soon as one goes over from a transformation of the coordinates proper to one of their differentials, there is no transformation which corresponds to shifts of the origin (see next section). See § 22 about the restrictions on admissible transformations of the Lorentz group due to reality requirements, and about time reversal.

[54] H. Minkowski: (I) 'Das Relativitätsprinzip', lecture delivered before the Math. Ges. Göttingen, on 5 Nov. 1907, published in *Jber. dtsch. Mat. Ver.*, **24** (1915) 372 and in *Ann. Phys., Lpz.*, **47** (1915) 927. (II) 'Die Grundgleichungen für die elektromagnetischen Vorgänge in bewegten Körpern', *Nachr. Ges. Wiss. Göttingen* (1908) 53, and *Math. Ann.*, **68** (1910) 472, also separately (Leipzig 1911). (III) 'Raum und Zeit', lecture delivered at the Congress of Scientists, Cologne, 21 Sept. 1908, published in *Phys. Z.*, **10** (1909) 104, also reprinted in the collection *Das Relativitätsprinzip* (Leipzig 1913). These will be quoted as Minkowski I, II, and III.

As a precursor of Minkowski one should mention Poincaré (Cf. footnote 11, p. 2, *R.C. Circ. mat. Palermo*, *loc. cit.*). He already introduced on occasion the imaginary coordinate $u = ict$ and combined, and interpreted as point coordinates in R_4, those quantities which we now call vector components. Furthermore, the invariant interval plays a rôle in his considerations.

origin, in analogy to the corresponding square of the distance $x^2+y^2+z^2$ in ordinary space. *With this, a world geometry (metric) is determined which is closely related to Euclidean geometry.* The two geometries are not completely identical because of the imaginary character of one of the coordinates. The latter property implies, for instance, that two world points whose distance from each other is zero do not necessarily coincide; such matters will be discussed in more detail in § 22. Notwithstanding these geometrical differences, we can regard the Lorentz transformations as *orthogonal* linear transformations of the world coordinates and as (imaginary) rotations of the world coordinate axes, in analogy with the rotations of a coordinate system in R_3. Moreover, just as the ordinary vector and tensor calculus can be looked upon as an invariant theory of the orthogonal linear coordinate transformations in R_3, so the invariant theory of the Lorentz group takes the form of a four-dimensional vector and tensor calculus[55]. Summarizing, therefore, we can express the *second* aspect which is essential to Minkowski's representation of the theory, in the following way: *Because the Lorentz group leaves a quadratic form of the four world coordinates invariant, the invariant theory of this group can be represented geometrically and it then appears as a natural generalization of the ordinary vector and tensor calculus for a four-dimensional manifold.*

8. More general transformation groups

In order to be able to develop the mathematical tools which will be necessary for the general theory of relativity, we shall now anticipate some of its formal results.

In the general theory of relativity it is no longer possible to define, in the simple manner of relation (18), the interval between two world points which are a finite distance apart. But here, too, the square of the distance ds between two *infinitesimally near* points can be written as a quadratic form of the coordinate differentials. The coordinates will be denoted by x^1, x^2, x^3, x^4 (instead of x, y, z, u), or in short by x^i. The coefficients of this form will correspondingly be denoted by g_{ik} and, following Einstein, we shall leave out the summation signs by stipulating that every index occurring twice will have to be summed over from 1 to 4. We can then write

$$ds^2 = g_{ik}dx^i dx^k \qquad (g_{ik} = g_{ki}). \qquad (19)$$

The sums on the right-hand side are to be carried out in such a way that i and k each independently take the values 1 to 4. In (19) the combinations ik for which $i \neq k$ will therefore each occur twice, those for which $i = k$ only once. With this convention, for instance, a differentiation of the quadratic form

$$J = g_{ik}u^i u^k$$

with respect to u^i gives

$$\frac{\partial J}{\partial u^i} = 2g_{ik}u^k, \qquad (20)$$

[55] The first formulations of such a tensor calculus can be found in Minkowski's papers, quoted above. A systematic presentation was first given by Sommerfeld: A. Sommerfeld, *Ann. Phys., Lpz.*, **32** (1910) 749 and **33** (1910) 649.

which is also in accordance with Euler's theorem

$$u^i \frac{\partial J}{\partial u^i} = 2J.$$

For the line element (19) the g_{ik} can in general be arbitrary functions of the coordinates. Correspondingly, the general theory of relativity deals with the invariant theory of the group of all point transformations

$$x'^k = x'^k(x^1, x^2, x^3, x^4),$$

once the quantities g_{ik} have been introduced *explicitly*.

We shall now give a summary of those transformation groups which are the most important for physics, amplifying in part what has been said above. In this we shall follow the "Erlanger Programm" of F. Klein[55a]. Each of the groups enumerated contains the previous ones as subgroups, with the exception of (B').

(A) The group of orthogonal linear transformations (Lorentz group) which leave the square of the distance

$$s^2 = x_1^2 + x_2^2 + x_3^2 + x_4^2$$

invariant. The inhomogeneous transformations may or may not be included, as desired. If however the Lorentz group is defined as the group of linear transformations of the *differentials* of the coordinates, which leave the *infinitesimal* quantity

$$ds^2 = dx_1^2 + dx_2^2 + dx_3^2 + dx_4^2$$

invariant, then it only consists of $(\infty)^6$ homogeneous transformations. But for some applications it is just the shifts in origin which are of importance. In addition one has to distinguish between the proper orthogonal transformations, with functional determinant $+1$, and the wider group which also contains the mixed orthogonal transformations with functional determinant -1. The former type of transformations can be made to go over into the identity transformation in a continuous manner, the latter are related to reversals.

(B) The affine group, which contains all linear transformations.

(B') The group of affine transformations which transform the equation of the light cone

$$x_1^2 + x_2^2 + x_3^2 + x_4^2 = 0$$

into itself, so that

$$x_1'^2 + x_2'^2 + x_3'^2 + x_4'^2 = \rho(x_1^2 + x_2^2 + x_3^2 + x_4^2)$$

where ρ is an arbitrary function of the coordinates. See §§ 28 and 65 (δ)

[55a] F. Klein, 'Programm zum Eintritt in die philosophische Fakultät,' (Erlangen 1872); reprinted in *Math. Ann.*, **43** (1893) 63. See also his lecture 'Über die geometrischen Grundlagen der Lorentz-Gruppe', *Jber. dtsch. Mat.Ver.*, **19** (1910) 281, and *Phys. Z.*, **12** (1911) 17. See also remarks in Klein's *Gesammelte mathematische Abhandlungen*, Vol. 1 (Berlin 1921), pp. 565-567.

for its applicability to Maxwell's equations and for its rôle in Nordström's theory of gravitation.

(C) The projective group of linear fractional transformations. This was mainly used by mathematicians in earlier investigations in non-Euclidean geometry. For physics it is of minor importance (see, however, § 18).

(D) The group of all point transformations to which the differential form (19) is adjoint. Its invariant theory is the tensor calculus of the general theory of relativity.

(E) See Part V, § 65, for the still wider group of Weyl.

9. Tensor calculus for affine transformations[56]

To avoid the inconvenience of writing the same formulae in different ways in the special and general theories of relativity, we shall use the affine group as a basis for our considerations from the beginning and not restrict ourselves to orthogonal transformations. Seen geometrically, this means that we admit oblique (but not curvilinear) coordinate systems. The g_{ik} are constants, but do not always have the normalized values $g_{ik} = \delta_i{}^k$ as in the orthogonal systems. The quantities $\delta_i{}^k$ are here defined by

$$\delta_i{}^k = \begin{cases} 0 \text{ for } i \neq k \\ 1 \quad ,, \quad i = k. \end{cases} \tag{21}$$

The tensor calculus can now be set up in a variety of ways. Either the tensor components are interpreted as projections of certain geometrical entities, or they can be characterized, in a purely algebraic way, by their behaviour under the coordinate transformations. Minkowski regarded only the four-vector geometrically, while arriving at the concept of a second-rank skew-symmetric tensor (or, as he puts it, vector of the second kind), first introduced by him, in a purely algebraic way. Through the influence of Sommerfeld's papers[55] the geometrical method became the prevalent one and remained so until the replacement of the Lorentz group by more general transformation groups. Thus, no geometrical considerations are contained in the paper by Ricci and Levi-Civita[56] which was basic for the tensor calculus of general point transformations†, apart from

[56] In addition to the references quoted in § 7, see also: H. Grassmann, *Ausdehnungslehre* (Berlin 1862); M. v. Laue, *Das Relativitätsprinzip* (1st edn. 1911, 3rd edn. 1919); H. Weyl, *Raum—Zeit—Materie*, (1st edn. 1918, 2nd edn. 1919, 3rd edn. 1920) [*Space–Time–Matter* (London, 1922)]; G. Ricci and T. Levi-Civita, 'Méthodes de calcul différentiel absolu et leurs application', *Math. Ann.* **54** (1901) 135; A. Einstein, 'Die formale Grundlage der allgemeinen Relativitätstheorie,' *S.B. preuss. Akad. Wiss.* (1914) 1030, and 'Die Grundlage der allgemeinen Relativitätstheorie', *Ann. Phys., Lpz.*, **49** (1916) 769, also bound separately (Leipzig 1916). A different terminology is used by G. N. Lewis, *Proc. Amer. Acad. Arts Sci.*, **46** (1910) 165 and by E. B. Wilson and G. N. Lewis, *ibid.* **48** (1912) 387, cf. also the report by G. N. Lewis, *Jb. Radioakt.*, **7** (1910) 321. Also H. Kafka, *Ann. Phys., Lpz.*, **58** (1919) 1; H. Lang, *Dissertation* (Munich 1919) and *Ann. Phys., Lpz.*, **61** (1920) 32. See also C. Runge, *Vektoranalysis* (Leipzig 1919), on reciprocal vector systems. He restricts himself however to R_3. For matters dealt with in Part II see also R. Weitzenböck, *Encykl. math. Wiss.*, III E 7, 2nd part, section C. It should be observed that the tensor calculus for affine transformations, as presented here, differs only in its terminology from the invariant theory of forms which is used in algebra.
† This paper formed the starting point for Einstein's work[56].

an attempt to interpret the contra- and co-variant components of a vector geometrically. Only in later papers by Hessenberg, Levi-Civita and Weyl[57] do we find the geometrical aspect stressed again to a greater extent. It is fully brought out also in the dissertation by Lang[56]. The purely algebraic representation has the advantage of simplicity and clarity, the geometrical one that of being "anschaulich". We shall use the former to start off with, but shall later, in special cases, give geometrical interpretations to the concepts and theorems we shall develop.

The quantities $a_{iklm...}{}^{rst...}$ in which the indices can, independently, take on the values 1, 2, 3, 4 are called tensor components. They are, in particular, called covariant components for the indices $iklm$ ... and contravariant for the indices rst ..., if the following conditions are fulfilled: For an affine transformation

$$x'^i = \alpha_k{}^i x^k \tag{22}$$

with its inverse

$$x^k = \bar{\alpha}_i{}^k x'^i, \tag{23}$$

and where the coefficients $\bar{\alpha}_i{}^k$ satisfy

$$\alpha_r{}^i \bar{\alpha}_k{}^r = \alpha_i{}^r \bar{\alpha}_r{}^k = \delta_i{}^k, \tag{24}$$

the tensor components should transform as[57a]

$$a'_{iklm...}{}^{rst...} = a_{\iota\kappa\lambda\mu...}{}^{\rho\sigma\tau...} \bar{\alpha}_i{}^\iota \bar{\alpha}_k{}^\kappa \bar{\alpha}_l{}^\lambda ... \alpha_\rho{}^r \alpha_\sigma{}^s .... \tag{25}$$

The summation convention is to be used here for indices occurring twice (see § 14 for a generalization of this definition for arbitrary coordinate transformations). The number of indices which the components have, is called the *rank* of the tensor. Tensors of first rank are also called vectors. The simplest example of such a vector would be the (contravariant) coordinates x^i of a point. Also, the quantities $\delta_i{}^k$ defined by (21) form, according to (24), the components of a tensor, covariant with respect to index i and contravariant with respect to index k. This tensor $\delta_i{}^k$ has moreover the property that its components take on the same numerical values in all coordinate systems.

By adding two tensors [of the same rank] one obtains a new tensor of equal rank; by multiplying them, a tensor of higher rank. Thus, for instance,

$$a_i + b_i = c_i$$

$$a_i b_k = c_{ik}, \qquad a_i b^k = c_i{}^k.$$

By contraction (summation over corresponding upper and lower indices), one obtains a tensor of lower rank. Thus the tensor of second rank $t_i{}^k$ gives

---

[57] See references in §§ 10 and 14, footnotes 58 a (p. 27) and 65, 66, 67 (p. 37), resp.

[57a] We would consider it more correct if the labels "contravariant" and "covariant" were interchanged, corresponding to the historically older nomenclature "cogredient" and "contragredient". Quantities which transform like the coordinates would then be called covariant. However, we adopt here the terminology which is now in general use and which was originated by Ricci and Levi-Civita and used by Einstein and Weyl (Cf. footnote 56, p. 24, *ibid.*).

rise to the invariant $t = t_i{}^i$ (where the summation sign is left out, according to our convention). Multiplication and contraction can also be combined. One can, for instance, first form the tensor

$$s_i{}^k = a_i b^k$$

by multiplying a_i and b^i, and then obtain the invariant

$$s = s_i{}^i$$

by contraction. This can however be obtained *directly* from the vectors a_i and b^i by means of the operation

$$s = a_i b^i.$$

In the same way, a tensor of second rank a_{ik} and a vector x^k can be combined to give the vector

$$y_i = a_{ik} x^k$$

and the invariant

$$J = a_{ik} x^i x^k.$$

The rule which we employed here can also be inverted: If $a_i x^i$ is an invariant for any *arbitrary* vector x^i, then the a_i are the covariant components of a vector; if $a^{ik} = a^{ki}$ and $a^{ik} x_i x_k$ is an invariant for any *arbitrary* vector x_i, then the a^{ik} are the contravariant components of a tensor of second rank, etc. The generalization of these results to tensors of arbitrary rank is immediately obvious.

A tensor is called symmetrical or skew-symmetrical [(antisymmetrical)] in the indices i and k if, on interchanging i and k, its components are not altered or only change sign, respectively (for instance, $a_{ik} = a_{ki}$ and $a_{ik} = -a_{ki}$, respectively). It can easily be verified that these relations are independent of the particular coordinate system chosen. It is however essential that the two indices should *both* be either upper or lower indices.

The quantities g_{ik} introduced in (19) also form a tensor, as can be seen from the invariance[58] of $g_{ik} x^i x^k$. This tensor is of the greatest importance both in geometry and physics and is called the *fundamental* (or *metric*) *tensor*. One can obtain new tensor components from the g_{ik} as follows. Take the determinant of the g_{ik},

$$g = \det |g_{ik}| \tag{26}$$

and divide the minor of a particular g_{ik} by g. In this way ten quantities g^{ik} ($g^{ik} = g^{ki}$) are obtained, which satisfy the relations

$$g_{i\alpha} g^{k\alpha} = \delta_i{}^k. \tag{27}$$

It should also be mentioned at this point that

$$\det |g^{ik}| = \frac{1}{g}. \tag{26 a}$$

---

[58] In the region of validity of the affine group, the two points (of which the square of the distance is determined by $g_{ik} x^i x^k$) need not be assumed to be infinitesimally close to each other, cf. § 10.

We now assert that the g^{ik} are the contravariant components of a tensor of second rank. For proof, take the contravariant components a^k of a vector and multiply them by g_{ik}. By contraction one then obtains

$$a_i = g_{ik} a^k. \tag{28}$$

The inverse of this system of equations is then

$$a^i = g^{ik} a_k, \tag{28 a}$$

and since the components a_k are quite arbitrary, the tensor character of the g^{ik} follows from the theorem quoted above.

We call the quantities a_i and a^i the covariant and contravariant components of the *same* vector. Correspondingly, one can define the raising and lowering of indices for tensors of higher rank, and can regard the resulting quantities as belonging to the *same* tensor. Thus, for instance,

$$a_{ik} = g_{ir} g_{ks} a^{rs} = g_{ir} a^r{}_k, \qquad a^{ik} = g^{ir} g^{ks} a_{rs} = g^{ir} a_r{}^k. \tag{28 b}$$

The raising and lowering of indices does not affect the correctness of a given relation between tensors, but one has always to sum over corresponding upper and lower indices when contracting, for instance

$$J = a_i b^i = a^i b_i,$$

$$c_i = a_{ik} b^k = a_i{}^k b_k, \qquad c^i = a^{ik} b_k = a^i{}_k b^k. \tag{29}$$

This concludes the rules of tensor algebra. Tensor *analysis*, (i.e. rules for deriving new tensors by differentiating tensors with respect to the coordinates) for the affine group can be seen to follow directly from tensor algebra. One has only to note that the operators $\partial/\partial x^k$ behave formally in all respects like the covariant components of a vector. The ordering and geometrical interpretation of such operators can only be discussed within the framework of the tensor calculus of general transformation groups.

10. Geometrical meaning of the contravariant and covariant components of a vector[58a]

A vector may be represented geometrically by a line of given length (hence it could also be called a "line" tensor). Its contravariant components are then given by the parallel projections of the line on to the co-ordinate axes. If the initial point of the vector is taken to be at the origin, these components are identical with the coordinates of the end point of the vector. On going over to a new coordinate system, these coordinates transform, according to what was said in the previous section, just as required by the definition of the contravariant components of a vector. Analogously to vectors in R_3, the sum of two vectors can be represented by the diagonal of the vector parallelogram.

We now have to introduce distances and angles, and for this purpose we

[58a] Our treatment in this section follows closely that of G. Hessenberg, *Math. Ann.*, **78** (1917) 187.

shall consider a Cartesian (orthogonal) system (X_1, X_2, X_3, X_4). The square of the length of a vector x with components X_i is given by[59]

$$x^2 = \sum_i X_i{}^2, \tag{30}$$

and we say that two vectors are normal to each other when

$$x \cdot y = \sum_i X_i Y_i \tag{31}$$

is equal to zero. In general $x \cdot y$ is called the scalar product of the vectors x, y. The invariance property of this definition with respect to orthogonal transformations follows from the invariance of (30) if we consider the relation

$$(\lambda x + \mu y)^2 = \lambda^2 x^2 + 2\lambda\mu \, x \cdot y + \mu^2 y^2.$$

Since this form in λ and μ is non-negative[59], it also follows that

$$(x \cdot y)^2 - x^2 y^2 \leqslant 0$$

(the equality sign only holds when x and y are parallel, i.e. when $x = ay$). We can therefore define the angle between two directions by

$$\cos(x, y) = \frac{x \cdot y}{\sqrt{(x^2 y^2)}}. \tag{32}$$

The geometrical meaning of the scalar product is the same as in ordinary three-dimensional space: it is equal to the product of the orthogonal projection of vector x on to the direction of y, and the length of y. This can be seen immediately by choosing the orthogonal coordinate system in such a way that one of the coordinate axes has the direction of y, which is always possible.

To obtain the expressions for the length and the scalar product of vectors in an arbitrary oblique coordinate system, we shall start by characterizing such a system by its four base vectors e_k ($k = 1$ to 4), whose contravariant components in this coordinate system are given by

$$\begin{aligned}
e_1 &= (1, \quad 0, \quad 0, \quad 0) \\
e_2 &= (0, \quad 1, \quad 0, \quad 0) \\
e_3 &= (0, \quad 0, \quad 1, \quad 0) \\
e_4 &= (0, \quad 0, \quad 0, \quad 1).
\end{aligned} \tag{33}$$

Their lengths can, in general, be different from unity. Thus the length of a vector is measured in units which are equal for all coordinate systems, while the parallel projections on the axis are measured in units which may, in general, even be different for different axes of the same coordinate system. Each vector x can be written in the form

$$x = x^k e_k. \tag{34}$$

---

[59] We assume here that in (30) all squares are positive and that the coordinates are real. In § 22 we shall discuss different conditions which hold in the actual space–time world.

The expressions for distance and for the scalar product are then immediately given by

$$x^2 = (x^i e_i) \cdot (x^k e_k) = e_i \cdot e_k \, x^i x^k = g_{ik} x^i x^k \tag{35}$$

$$x \cdot y = (x^i e_i) \cdot (y^k e_k) = e_i \cdot e_k \, x^i y^k = g_{ik} x^i y^k \tag{36}$$

with

$$g_{ik} = e_i \cdot e_k. \tag{37}$$

We have thus obtained a geometrical meaning for the quantities g_{ik}, too.

We now introduce the four vectors which are reciprocal to the vectors e_k. They are defined by the relations

$$e_i \cdot e^*_k = \delta_i{}^k, \tag{38}$$

i.e. the vectors e^*_i are perpendicular to the spaces formed by taking three of the vectors e_k at a time, and are also suitably normalized. If we denote by x_i the parallel projections of x on the reciprocal axes, measured in the corresponding units, we have

$$x = x_k e^*_k. \tag{39}$$

To obtain the connection between the x_i and the x^i, we multiply the equation

$$x_k e^*_k = x^k e_k \tag{39a}$$

scalarly by e_i. Using (37) and (38), we then obtain

$$x_i = g_{ik} x^k. \tag{40}$$

In other words, *the parallel projections of a vector x on the reciprocal axes, measured in the reciprocal units, are its covariant components*[60]. If, on the other hand, we multiply (39 a) scalarly by e^*_i, we obtain

$$x^i = e^*_i \cdot e^*_k \, x_k$$

and therefore

$$g^{ik} = e^*_i \cdot e^*_k, \tag{41}$$

since $x^i = g^{ik} x_k$. By either squaring the expressions (34) and (39) for x separately, or multiplying them together, we get

$$x^2 = g_{ik} x^i x^k = g^{ik} x_i x_k = x_i x^i. \tag{35a}$$

---

[60] Ricci and Levi-Civita, as well as Lang (Cf. footnote 56, p. 24, *ibid.*), interpret the covariant vector components as *orthogonal* projections on the original axes. In this case, however, a factor has to be added which destroys the simplicity and symmetry of the formulae. For it follows from (39), by scalar multiplications with e_i, that $x_i = e_i \cdot x$. The orthogonal projection of x on e_i is therefore equal to

$$\frac{x_i}{|e_i|} = \frac{x_i}{\sqrt{(e_i \cdot e_i)}} = \frac{x_i}{\sqrt{g_{ii}}} \qquad \text{(from (37))}.$$

(The index i is *not* to be summed over.) Lastly, it should be noted that in a Cartesian coordinate system, in which the g_{ik} have the values $\delta_i{}^k$, the difference between contravariant and covariant components disappears and the basic vectors e_i of the system become identical with their reciprocal vectors.

In the same way, using

$$x = x^k e_k = x_k e^{*k}, \qquad y = y^k e_k = y_k e^{*k},$$

we obtain for the scalar product

$$x \cdot y = g_{ik} x^i y^k = g^{ik} x_i y_k = x_i y^i = x^i y_i. \tag{36 a}$$

We still have to see how the base vectors e_i behave under a coordinate transformation. Let e'_i be the base vectors of the new (primed) coordinate system, then, for any vector x, we have

$$x = x'^i e'_i = x^k e_k.$$

Using (22) and (23), it follows that

$$e'_i = \bar{\alpha}_i{}^k e_k \tag{42}$$

and

$$e_k = \alpha_k{}^i e'_i. \tag{43}$$

Thus the $\bar{\alpha}_i{}^k$ are the components of the new base vectors, referred to the old coordinate system, and the $\alpha_k{}^i$ are the components of the old base vectors in the new system. Also, the transformation formulae for the fundamental tensor g_{ik}, derived from (25), are confirmed here by (37) and (42).

11. "Surface" and "volume" tensors. Four-dimensional volumes

After the line, the next-highest geometrical entity is the surface. Analogously to the connection between vectors ("line" tensors) and lines, there exist tensors of second rank which we shall call "surface" tensors. Such tensors can be obtained by considering two vectors x, y which together span a two-dimensional parallelepiped. Its projections parallel to the axes on to the six two-dimensional coordinate planes, measured in units of the six parallelepipeds of the base vectors e_i, are given by

$$\xi^{ik} = x^i y^k - x^k y^i. \tag{44}$$

They form the contravariant components of a skew-symmetric tensor of second rank and obey the relations

$$\xi^{ik} = - \xi^{ki}. \tag{45}$$

Taking, instead, the reciprocal axes and the unit surfaces formed by the e_i^*, one obtains the covariant components

$$\xi_{ik} = x_i y_k - x_k y_i. \tag{44 a}$$

Any skew-symmetric tensor of second rank whose components satisfy (45) is called a "surface" tensor. Not every such tensor can be written in the form (44)—for the ξ^{ik} of this special form satisfy the relation

$$\xi^{12} \xi^{34} + \xi^{13} \xi^{42} + \xi^{14} \xi^{23} = 0. \tag{46}$$

But it can be written as the sum of two tensors of the form (44). The invariant quantity

$$J = \tfrac{1}{2} \xi_{ik} \xi^{ik} \tag{47}$$

represents the *area* of the parallelogram. More generally, if ξ_{ik} and η_{ik} are two "surface" tensors of the special type (44), then the invariant

$$J = \tfrac{1}{2}\xi_{ik}\,\eta^{ik} \tag{48}$$

is equal to the orthogonal projection of the parallelepiped ξ_{ik} on η_{ik} multiplied by the magnitude of η_{ik}. The corresponding invariants for general "surface" tensors are the sums of the products of such surface quantities[60a]. See § 12 on the significance of the left-hand side of (46) in the theory of invariants for the case of the general "surface" tensor.

A "volume" tensor is represented by a three-dimensional space, spanned by three vectors x, y, z. Its components are given by the determinants

$$\xi^{ikl} = \begin{vmatrix} \xi^i & \eta^i & \zeta^i \\ \xi^k & \eta^k & \zeta^k \\ \xi^l & \eta^l & \zeta^l \end{vmatrix} \qquad \xi_{ikl} = \begin{vmatrix} \xi_i & \eta_i & \zeta_i \\ \xi_k & \eta_k & \zeta_k \\ \xi_l & \eta_l & \zeta_l \end{vmatrix}. \tag{49}$$

They satisfy the anti-symmetry conditions, in that they change sign when any two of their indices are interchanged. The number of independent components is four. In contrast to the "surface" tensor, (49) represents already the most general "volume" tensor. In other words, every tensor of rank 3 for which the components satisfy the above symmetry conditions, can be represented in the form (49).

Four vectors $x^{(1)}$, $x^{(2)}$, $x^{(3)}$, $x^{(4)}$ can be said to span a four-dimensional volume element. In a Cartesian coordinate system, its magnitude is simply equal to the value of the determinant formed by the 4×4 components of the vectors x. In an oblique coordinate system, its value can be expressed in terms of the corresponding components, using (34) and (39),

$$S = \det|x^{(i)k}| \cdot \det|e_i| = \det|x^{(i)}{}_k| \cdot \det|e^*{}_i|, \tag{50}$$

with the help of the multiplication rules for determinants. Here, $\det|e_i|$ and $\det|e'_i|$ are the determinants of the 4×4 components of the vectors

---

[60a] In this connection we should like to mention the Plücker line coordinates. If $x_1 \ldots x_4$ and $y_1 \ldots y_4$ are the homogeneous coordinates of two points on a straight line in three dimensions (so that x_1/x_4, x_2/x_4, x_3/x_4 and y_1/y_4, y_2/y_4, y_3/y_4 are their ordinary coordinates), then the straight line can be defined by the six quantities $p_{ik} = x_i y_k - x_k y_i$ whose *ratios* are independent of the particular choice of the two points on the line. These quantities satisfy (46). The formal analogy with the skew-symmetric tensor of rank 2 in four-dimensional space is a complete one.

If, next, ξ_{ik} is a special "surface" tensor of type (44a), then, since $dx^i = \xi^{ik} x_k$, there corresponds to each vector x^i an infinitesimal displacement. Since dx^i lies in the plane of the "surface" tensor ξ^{ik} and is perpendicular to x_k, we are dealing here with an infinitesimal rotation in R_4 of the same magnitude and sense as ξ_{ik}. If ξ_{ik} is a "surface" tensor of the general kind, then the corresponding displacement is obtained by adding two mutually orthogonal rotations and can be described as a *screw*. Minkowski himself (Cf. footnote 54, p. 21, III) stresses this analogy between a "surface" tensor and a screw. The corresponding analogy in three-dimensional space is applied extensively by Sir Robert Ball, *A Treatise on the Theory of Screws* (Cambridge 1900). See also the paper by F. Klein, *Z. Math. Phys.*, **47** (1902) 237 and *Math. Ann.* **62** (1906), 419.

It should further be noted that the independent meaning of skew-symmetric tensors of second rank, as opposed to vectors, in multi-dimensional manifolds was already recognized and described by H. Grassmann (Cf. footnote 56, p. 24, *ibid.*).

e_i and $e^*{}_i$, respectively, in the Cartesian coordinate system. Their magnitudes are obtained by squaring and further application of the multiplication rules. Using (37), (41), and (26 a),

$$(\det|e_i|)^2 = \det|e_i \cdot e_k| = \det|g_{ik}| = g;$$

$$(\det|e^*{}_i|)^2 = \det|e^*{}_i \cdot e^*{}_k| = \det|g^{ik}| = \frac{1}{g}.$$

Thus, finally, the *invariant* volume is

$$S = \det|x^{(i)k}| \cdot \sqrt{g} = \det|x^{(i)}{}_k| \cdot \frac{1}{\sqrt{g}}. \tag{51}$$

Since the quadruple integral

$$\int dx^1 \, dx^2 \, dx^3 \, dx^4,$$

or $\int dx$ for short, transforms like the determinant $\det |x^{(i)k}|$, the volume of an arbitrary region is given by

$$\varSigma = \int \sqrt{g} \, dx \tag{52}$$

with the help of (51). If the integral

$$\int \mathfrak{W} \, dx$$

is an invariant, then $\mathfrak{W}$ is called a *scalar density*, following Weyl's[61] terminology. Such a scalar density is formed by multiplying an ordinary scalar quantity by $\sqrt{g}$.

A vector density with components $\mathfrak{w}^i$ is defined correspondingly by the condition that the integrals (over an infinitesimally small region)

$$\int \mathfrak{w}^i \, dx$$

form a vector. Generalizing, tensor densities can be defined in an analogous way. They are obtained by multiplying ordinary tensors by $\sqrt{g}$.

When the classification of tensors was discussed in § 9, no account was taken of the symmetry relations between tensor components. We saw, however, that for instance the skew-symmetrical tensors of second rank were, from a geometrical point of view, completely different from the symmetrical ones. This difference will again be brought out in our discussion of tensor analysis (see §§ 19 and 20). It would therefore seem advisable to follow Weyl's[62] treatment and introduce (adhering closely to the terminology of Grassmann's "expansion theory") a new classification of tensors, side by side with the old one. As in (44) and (49), let us form the

[61] H. Weyl, *Math. Z.*, **2** (1918) 384; *Raum-Zeit-Materie* (3rd edn., Berlin 1920) p. 92 *et seq*. [*Space-Time-Matter* (London 1922) p. 109 *et seq*.].
[62] H. Weyl, *Raum-Zeit-Materie* (1st edn., Berlin 1918), pp. 45–51.

sequence ξ^i, ξ^{ik}, ξ^{ikl},.... . Tensors of the first "grade" ("line" tensors)[†] and of first, second, third, etc., rank are then produced by the linear, bilinear, trilinear, etc., forms of the *single* displacement ξ^i,

$$a_i \xi^i, \quad a_{ik} \xi^i \xi^k, \quad a_{ikl} \xi^i \xi^k \xi^l, ... ;$$

and similarly, for tensors of the second "grade" ("surface" tensors),

$$b_{ik} \xi^{ik}, \quad b_{iklm} \xi^{ik} \xi^{lm}, ...$$

Certain normalization conditions have to be satisfied by the coefficients, in order that they can be determined unambiguously from the forms. Thus, for instance, the a_{ik}, a_{ikl} have to remain unaltered by an interchange of any two of their indices, the b_{ik} have to be skew-symmetric, and the components b_{iklm} of the "surface" tensor of rank 2[†] have to satisfy the conditions

$$b_{iklm} = -b_{kilm} = -b_{ikml} = b_{lmik} \tag{53a}$$

and

$$b_{iklm} + b_{ilmk} + b_{imkl} = 0. \tag{53 b}$$

(53 b) follows from the relations (46). For instance, the curvature tensor (see § 16) is such a "surface" tensor of rank 2.[†] The number of independent components of such a tensor in n-dimensional space reduces to $n^2(n^2 - 1)/12$, because of (53 a) and (53 b). This classification by no means comprises all those entities which fall under the definition of a tensor as given in § 9. But only tensors that can be classified as above have any importance in physical applications.

12. Dual tensors

With every surface element

$$\xi^{ik} = x^i y^k - x^k y^i \tag{54}$$

in a four-dimensional manifold can be associated another, normal to it, which has the property that all straight lines in the one are perpendicular to all straight lines in the other. Such a surface element is called dual to ξ^{ik}, if, in addition, it is of the same magnitude. First of all, it is determined by

$$\xi^{*ik} = x^{*i} y^{*k} - x^{*k} y^{*i},$$

where the vectors x^{*i}, y^{*i} are normal to x^i and y^i, i.e.

$$x_i^* x^i = 0, \quad x_i^* y^i = 0, \quad y_i^* x^i = 0, \quad y_i^* y^i = 0.$$

A simple calculation then shows that the components ξ^{*ik} are directly obtained from the components ξ_{ik} by an even permutation of the indices, but with the addition of a factor $\sqrt{g}$ or $1/\sqrt{g}$ respectively.

$$\xi^{*14} = \frac{1}{\sqrt{g}} \xi_{23}, \qquad \xi^{*24} = \frac{1}{\sqrt{g}} \xi_{31}, \qquad \xi^{*34} = \frac{1}{\sqrt{g}} \xi_{12},$$

$$\tag{54 a}$$

$$\xi^{*23} = \frac{1}{\sqrt{g}} \xi_{14}, \qquad \xi^{*31} = \frac{1}{\sqrt{g}} \xi_{24}, \qquad \xi^{*12} = \frac{1}{\sqrt{g}} \xi_{34}.$$

[†] See suppl. note 5 a.

Corresponding results are obtained by interchanging ξ^{ik} and ξ^{*ik},

$$\xi^{*}_{14} = \sqrt{g}\,\xi^{23}, \qquad \xi^{*}_{24} = \sqrt{g}\,\xi^{31}, \qquad \xi^{*}_{34} = \sqrt{g}\,\xi^{12},$$
$$\xi^{*}_{23} = \sqrt{g}\,\xi^{14}, \qquad \xi^{*}_{31} = \sqrt{g}\,\xi^{24}, \qquad \xi^{*}_{12} = \sqrt{g}\,\xi^{34}. \qquad (54\,\mathrm{b})$$

These same relations are used for associating with ξ^{ik} a dual tensor, even when the former is not of the special form (44). Scalar multiplication of the "surface" tensor ξ_{ik} with its dual tensor ξ^{*ik} according to (48) results in an invariant of especially simple structure,

$$J = \tfrac{1}{2}\xi_{ik}\xi^{*ik} = \frac{1}{\sqrt{g}}(\xi_{12}\xi_{34} + \xi_{13}\xi_{42} + \xi_{14}\xi_{23}). \qquad (46\,\mathrm{a}$$

In a similar way a dual vector ξ^{*i} can be associated with a "volume" tensor ξ^{ikl}. The dual vector is then that line which is perpendicular to all straight lines of the space element and whose length is equal to the volume of the space element. For any even permutation $iklm$ we have

$$\xi^{*m} = \frac{1}{\sqrt{g}}\xi_{ikl}, \qquad \xi^{*}_{m} = \sqrt{g}\,\xi^{ikl}. \qquad (55)$$

13. Transition to Riemannian geometry

We now go on to discuss the theory of invariants of the group of all point transformations. For this it will be necessary to consider first of all the determination of length and the theorems of general Riemannian geometry. The older geometries of Bolyai and Lobachevski, in which the Euclidean axiom of parallelism had been abandoned, had all retained the axiom of the free mobility of rigid point systems (axiom of congruency) and had therefore only arrived at the special case of spaces of constant curvature. Starting from projective geometry one would still not obtain a more general metric. It was Riemann[63] who first envisaged such a possibility. The modification of the concept of a rigid body in the special and general theory of relativity has meant that the axioms of congruency, which had for so long been considered as self-evident, have to be discarded nowadays and that it is the general Riemannian geometry which must form the basis of our considerations of space and time.

We shall assume that a given, finite, neighbourhood of every point of the manifold can be characterized in a unique and continuous manner by the coordinates $x^1, x^2, ..., x^n$. This neighbourhood we shall, for the present, simply call "space". Such an assumption need by no means be assumed to hold for the *entire* manifold. The number of dimensions, n, of the manifold shall be arbitrary. The basis for the metric is then the length s of a given curve

$$x^k = x^k(t) \qquad (k = 1, 2, ..., n)$$

where t is an arbitrary parameter. Not until this length has been defined in

[63] B. Riemann, 'Über die Hypothesen, welche der Geometrie zugrunde liegen', Inaugural Lecture (1854). Posthumously published in *Nachr. Ges. Wiss. Göttingen*, **13** (1868) 133 (edited by Dedekind), also in Riemann's *Collected Works*, p. 254. Recently published separately in pamphlet form, edited by H. Weyl (Berlin 1920).

some physical way, can we apply the results of our mathematical investigations to the manifold which exists in reality. In R_3, the rigid measuring rod will have to be thought of as replaced by a measuring thread.

We now have to see what plausible assumptions we can make about the function $s(t)$. Since such assumptions are only made about the derivative ds/dt, Riemannian geometry is characterized as a differential geometry in contrast to Euclidean, "finite", geometry. Our first axiom will be:

Axiom I. The differential coefficient ds/dt at a given point on the curve shall only depend on the derivatives dx^k/dt at this point and not on higher derivatives or on the behaviour of the curve elsewhere.

Since the arc length s is independent of the choice of the parameter t, it then follows that ds/dt has to be a homogeneous function of first degree of the quantities dx^k/dt. The distance between two points will be denoted by the arc length of the shortest line connecting them. Such a line is said to be perpendicular to a second line if the distance of an arbitrary point P of line 1 from the point of intersection, S, of the two lines is less than the distance of P from any other point Q on line 2. According to axiom I the position of the point P on line 1 is quite immaterial and we are only concerned with the differential coefficients $(dx^k/dt)_1$ and $(dx^k/dt)_2$ at S. One can therefore also say that the direction 1 is orthogonal to direction 2. In general it does not follow from this that direction 2 is also orthogonal to direction 1. But we shall limit the character of the function ds/dt by a second axiom:

Axiom II. ds/dt is to be the square root of a quadratic form of the dx^k/dt:

$$\frac{ds}{dt} = \sqrt{\left(g_{ik}\frac{dx^i}{dt}\frac{dx^k}{dt} \right)}$$

or more concisely

$$ds^2 = g_{ik}\,dx^i\,dx^k. \qquad (19)$$

This is the equation which we wrote down in § 8. Axiom II can be looked upon as Pythagoras' theorem for two infinitesimally near points. It is just this limitation of its region of validity which characterizes the transition from a "finite" to a differential geometry. According to axiom II the orthogonality of two directions is a reciprocal relation. The corollary to this is that if such a reciprocal relation always holds, then the line element has to be of the form (19).[64] For this reason, axiom II can also be replaced by another:

Axiom II'. When direction 1 at P is orthogonal to direction 2, then 2 is also orthogonal to 1.

Taking $n = 2$, with axiom II as basis, one gets back to the Gaussian geometry on arbitrarily curved surfaces. Just as one can think of each such surface as being in Euclidean R_3-space, so every Riemannian space R_n can be embedded in a Euclidean space $R_{n(n+1)/2}$ (here $n(n+1)/2$ corresponds to the number of the components g_{ik}). However, all those geo-

---

[64] D. Hilbert, 'Grundlagen der Physik, 2. Mitt.', *Nachr. Ges. Wiss. Göttingen* (1917) 53; W. Blaschke, *S.B. naturf. Ges., Lpz., math.-phys.Kl.*, **68** (1916) 50.

metrical theorems which are of importance for the theory of relativity can also be derived without making use of such a possibility. The angle $(1, 2)$ between two directions dx^i and δx^i at a point P can be defined in exactly the same way as in Euclidean space, as long as the straight lines are replaced by infinitesimally small shortest lines. Analogously to (32),

$$\cos(1, 2) = \frac{g_{ik}\, dx^i\, \delta x^k}{\sqrt{(g_{ik}\, dx^i\, dx^k)}\sqrt{(g_{ik}\, \delta x^i\, \delta x^k)}}. \tag{56}$$

The g_{ik} can be obtained at every point by determining the line element in $n(n+1)/2$ independent directions (i.e. in $n(n+1)/2$ directions for which the determinant of $n(n+1)/2$ rows of the corresponding quantities $dx^i\, \delta x^k$ does not vanish).

For an arbitrary point transformation

$$x'^i = x'^i(x^1, ..., x^n) \qquad (i = 1, 2, ..., n) \tag{57}$$

the differentials dx^k obey linear homogeneous transformation laws, just as the coordinates in (22),

$$dx'^i = \alpha_k{}^i\, dx^k \tag{58}$$

$$\alpha_k{}^i = \frac{\partial x'^i}{\partial x^k}, \tag{59}$$

with the corresponding inverse relations

$$dx^k = \bar{\alpha}_i{}^k\, dx'^i \tag{60}$$

$$\bar{\alpha}_i{}^k = \frac{\partial x^k}{\partial x'^i}. \tag{61}$$

This is the connection of the general transformation group with the affine group. An essential point to watch, however, is that the $\alpha_k{}^i$ cannot be arbitrary functions of the coordinates, but have to satisfy the integrability conditions

$$\frac{\partial \alpha_k{}^i}{\partial x^l} = \frac{\partial \alpha_l{}^i}{\partial x^k} \tag{62}$$

or the inverse conditions

$$\frac{\partial \bar{\alpha}_i{}^k}{\partial x'^l} = \frac{\partial \bar{\alpha}_l{}^k}{\partial x'^i}. \tag{63}$$

At any given point P_0, however, the $\alpha_i{}^k$ can take on arbitrary values. As long as we are dealing with relations between tensors at one and the same point, and not with the differentiation or integration of a tensor field, we can apply all the tensor operations of the affine group. This can also be expressed in a different way. In tensor algebra, the Riemannian space at a given point P_0 can be replaced by the "tangential" space, which is obtained by giving the g_{ik} everywhere the same constant values $g_{ik}(P_0)$ which, in Riemannian space, they only have at P_0. The form ds^2 is by definition invariant, the g_{ik} are the covariant components of a tensor of rank 2. Also, the rules of tensor algebra can be used for passing over to the contravariant components g^{ik} and for forming the volume element $d\Sigma$.

14. Parallel displacement of a vector

The concept of the parallel displacement of a vector has turned out to be more and more fundamental to the geometrical basis of the tensor calculus in Riemannian space. It was first formulated by Levi-Civita[65] in connection with the embedding of a Riemannian space R_n in a Euclidean space $R_{n(n+1)/2}$ (see § 13) and was later derived in a direct manner by Weyl[66]. Afterwards Weyl defined it axiomatically also for manifolds for which the line element is not even yet defined (cf. Part V).[67]

Let us again consider the curve

$$x^k = x^k(t),$$

and at each of its points P the totality of all vectors which originate from P. We then have to single out from all mappings

$$\xi^i = f^i(\overset{\circ}{\xi}{}^k, t)$$

of the totality of vectors of $P_0(t_0)$ on to the totality of vectors of $P(t)$ a special group in an invariant manner, and we have to distinguish it as one of parallel displacements or translations. Now it is impossible simply to postulate that two parallel vectors at two points, a finite distance apart, should have the same components. If this happens to be the case in *one* coordinate system, it will in general not be true for another. Therefore the above property of translation will have to be described in the following way:

(a) At each point P there exists a coordinate system such that the change in the components of a vector vanishes for an *infinitesimal* translation along all curves originating from P, i.e. at P

$$\frac{d\xi^i}{dt} = 0.$$

By stipulating that the infinitesimal change in the vector components should be transformed away simultaneously for all curves starting from P, one links the parallel displacements along different curves. It is easy to see that, because of postulate (a), the change $d\xi^i/dt$ of the vector components is given by

$$\frac{d\xi^i}{dt} = -\Gamma^i_{rs} \frac{dx^s}{dt} \xi^r \tag{64}$$

in an arbitrary coordinate system, where the Γ^i_{rs} only depend on the coordinates and not on their derivatives. They satisfy the symmetry relation

$$'\Gamma^i_{rs} = \Gamma^i_{sr}. \tag{65}$$

[65] T. Levi-Civita, 'Nozione di parallelismo, etc.', *R.C. Circ. mat. Palermo*, **42** (1917) 173.
[66] H. Weyl, *Raum-Zeit-Materie* (1st edn., Berlin 1918) pp. 97–101.
[67] H. Weyl, *Math. Z.*, **2** (1918) 384; *Raum-Zeit-Materie* (3rd edn., Berlin 1920), pp. 100–102.

Conversely it can be shown that postulate (a) is satisfied when (64) and (65) are true. The $\Gamma^i{}_{rs}$ behave like tensor components under linear coordinate transformations, but not under the general group of transformations. This can already be seen from the fact that the $\Gamma^i{}_{rs}$ can always be made to vanish (in one coordinate system), whereas the components of a tensor would vanish in all coordinate systems if they vanish in *one* system, since they transform homogeneously. We may as well define here another set of quantities, $\Gamma_{i,rs}$, by means of

$$\Gamma_{i,rs} = g_{ik}\,\Gamma^k{}_{rs}, \quad \Gamma^i{}_{rs} = g^{ik}\,\Gamma_{k,rs}. \tag{66}$$

The definition of parallel displacement is completed by the second postulate.†

(b) Translation is a congruent transformation, i.e. it leaves the lengths of the vectors unchanged,

$$\frac{d}{dt}(g_{ik}\,\xi^i\xi^k) = \frac{d}{dt}(\xi_i\,\xi^i) = 0. \tag{67}$$

In this way the geodesic components are linked with the fundamental tensor. A simple consequence of postulate (b) is the fact that angles are also conserved for a parallel displacement. Since relations (64) and (67) have to be valid for arbitrary ξ^i, one obtains immediately

$$\frac{\partial g_{ir}}{\partial x^s} = \Gamma_{i,rs} + \Gamma_{r,is}, \tag{68}$$

$$\frac{1}{2}\left(\frac{\partial g_{ir}}{\partial x^s} + \frac{\partial g_{is}}{\partial x^r} - \frac{\partial g_{rs}}{dx^i}\right) = \Gamma_{i,rs}. \tag{69}$$

The quantities $\Gamma^i{}_{rs}$ then follow from (66). Christoffel, in whose paper[68] the quantities defined by (69) and (66) appeared in print for the first time, used the notation $\left[{rs\atop i}\right]$ and $\left\{{rs\atop i}\right\}$ instead of $\Gamma_{i,rs}$ and $\Gamma^i{}_{rs}$. They are often called Christoffel symbols (of the first and second kind, respectively) Weyl[69] denoted them as components of the affine connection, since the infinitesimal translation is, according to (64), an affine mapping of the vectors. We shall simply call them the *geodesic components* of the reference system in question. A coordinate system in which they vanish at a point P is called geodesic, at the point P.

Similarly, the transformation formula for the covariant components ξ_i can be obtained from the invariance requirement of $\xi_i\,\eta^i$, for arbitrary η^i, with the help of (64),

$$\frac{d\xi_i}{dt} = \Gamma^r{}_{is}\frac{dx^s}{dt}\xi_r = \Gamma_{r,is}\frac{dx^s}{dt}\xi^r; \tag{!(70)}$$

---

† See suppl. note 6.

[68] E. B. Christoffel, *J. reine angew. Math.*, **70** (1869) 46. Cf. also R. Lipschitz, *J. reine angew. Math.*, **70** (1869) 71. The relevant derivations by Riemann were written down in 1861 in a Prize Essay (Paris), and were not published until 1876, in the first edition of Riemann's *Collected Works*, p. 370.

[69] H. Weyl, *Raum-Zeit-Materie* (3rd edn., Berlin 1920), p. 101.

and from

$$\frac{d}{dt}(g^{ik}\xi_i\xi_k) = 0$$

follows the identity

$$\frac{\partial g^{ik}}{\partial x^s} + g^{ir}\Gamma^k_{rs} + g^{kr}\Gamma^i_{rs} = 0. \tag{71}$$

The following equations follow from (26) and (27) by differentiation,

$$dg^{ik} = -g^{ir}g^{ks}dg_{rs}, \qquad dg_{ik} = -g_{ir}g_{ks}dg^{rs}, \tag{72}$$

$$\frac{\partial g^{ik}}{\partial x^l} = -g^{ir}g^{ks}\frac{\partial g_{rs}}{\partial x^l}, \qquad \frac{\partial g_{ik}}{\partial x^l} = -g_{ir}g_{ks}\frac{\partial g^{rs}}{\partial x^l}, \tag{72a}$$

and

$$dg = gg^{ik}dg_{ik} = -gg_{ik}dg^{ik} \tag{73}$$

$$\frac{\partial g}{\partial x^l} = gg^{ik}\frac{\partial g_{ik}}{\partial x^l} = -gg_{ik}\frac{\partial g^{ik}}{\partial x^l}. \tag{73a}$$

By contraction, one can also obtain

$$\Gamma^r_{ir} = g^{rs}\Gamma_{r,is} = \tfrac{1}{2}g^{rs}\frac{\partial g_{rs}}{\partial x^i} = \frac{1}{\sqrt{g}}\frac{\partial\sqrt{g}}{\partial x^i} = \frac{\partial\log\sqrt{g}}{\partial x^i} \tag{74}$$

from (69) and then, using (71),

$$\frac{1}{\sqrt{g}}\frac{\partial\sqrt{g}\,g^{ik}}{\partial x^k} + g^{rs}\Gamma^i_{rs} = 0. \tag{75}$$

15. Geodesic lines

The direction of the curve $x^k = x^k(t)$ at any of its points P is characterized by the vector u^i,

$$u^i = \frac{dx^i}{ds}, \qquad (s = \text{arc length}) \tag{76}$$

which has the direction of the tangent to the curve at P and is of unit length. In fact,

$$u_i u^i = g_{ik}\frac{dx^i}{ds}\frac{dx^k}{ds} = 1. \tag{77}$$

The geodesic line, then, is a curve which *always maintains its direction*[70]. That is to say, if one constructs the appropriate direction vector u^i at an arbitrary point P_0 of the geodesic line, one obtains the direction vectors at other points by a parallel displacement of u^i along the geodesic line.

[70] H. Weyl, *Raum-Zeit-Materie* (1st edn., Berlin 1918), p. 102.

According to (64) and (70) this can be expressed analytically by means of the (completely equivalent) relations

$$\frac{du_i}{ds} = \Gamma_{r,is} u^r u^s = \frac{1}{2} \frac{\partial g_{rs}}{\partial x^i} u^r u^s \tag{78}$$

and

$$\frac{du^i}{ds} = -\Gamma^i_{rs} u^r u^s; \tag{79}$$

the latter can also be written

$$\frac{d^2 x^i}{ds^2} + \Gamma^i_{rs} \frac{dx^r}{ds} \frac{dx^s}{ds} = 0. \tag{80}$$

These are the differential equations of the geodesic line. Since the length of a vector is invariant for a parallel displacement, it follows from (80) that

$$g_{ik} \frac{dx^i}{ds} \frac{dx^k}{ds} = \text{const.} \tag{77a}$$

Thus (80) *only* holds for such parameters s which, apart from a constant factor, are equal to the arc length.

The geodesic lines can also be characterized by means of a variational principle. For they are, at the same time, the "shortest" lines already mentioned in § 13, or, put more precisely, they are the "extremals"[70a] for which the variation of the curve length vanishes (the latter need not necessarily be a minimum). Let A and B be the fixed initial and end points, s the arc length, λ an arbitrary parameter. It will then have to be shown that for the geodesic lines

$$\delta \int_A^B ds = \delta \int_A^B \sqrt{\left(g_{ik} \frac{dx^i}{d\lambda} \frac{dx^k}{d\lambda} \right)} d\lambda = 0. \tag{81}$$

Here, the g_{ik} are given functions of the coordinates x^i, and it is the functions $x^k = x^k(\lambda)$ which are varied.

Relation (81) will now be transformed in a manner well known from mechanics[70b]. For this purpose we choose the parameter λ in such a way that it coincides with the arc length s on the extremal and always covers the same range of values. In the resulting differential equations, then, λ can be replaced by s. If we now put

$$L = \tfrac{1}{2} g_{ik} \frac{dx^i}{d\lambda} \frac{dx^k}{d\lambda}, \tag{82}$$

then

$$\delta \int_A^B ds = \int_A^B \frac{\delta L}{\sqrt{[g_{ik}(dx^i/d\lambda)(dx^k/d\lambda)]}} d\lambda,$$

[70a] Cf. A. Kneser, *Encykl. math. Wiss.*, II, 8, pp. 597 and 600.

[70b] There, it is a question of going over to Hamilton's Principle from Jacobi's form of the Principle of Least Action. Cf. A. Voss, *Encykl. math. Wiss.*, IV, 1, p. 96.

and since the square root becomes equal to 1 for the extremal, we can simply write, instead of (81),

$$\int_A^B \delta L \, d\lambda = \delta \int_A^B L \, d\lambda = 0. \tag{83}$$

We have thus obtained a complete analogy to Hamilton's Principle in mechanics, if L is looked upon as the Lagrangian. Let us therefore write, for the moment, $\dot{x}^i$ instead of $dx^i/d\lambda = dx^i/ds$. Then the differential equations resulting from (83) will be[70c]

$$\frac{d}{ds} \frac{\partial L}{\partial \dot{x}^i} - \frac{\partial L}{\partial x^i} = 0. \tag{84}$$

Since, according to (20),

$$\frac{\partial L}{\partial \dot{x}^i} = g_{ik} \frac{dx^k}{ds},$$

we do obtain in fact

$$\frac{d}{ds}\left(g_{ik} \frac{dx^k}{ds} \right) = \frac{1}{2} \frac{\partial g_{rs}}{\partial x^i} \frac{dx^r}{ds} \frac{dx^s}{ds},$$

which agrees with (78). (In a manifold in which ds^2 is an indefinite form, the derivation given here does not apply to those curves on which $ds = 0$ throughout. Cf. § 22 for the exceptional case of such "null lines".)

16. Space curvature

The concept of the curvature of space was first put forward by Riemann[71] as a generalization of the Gaussian curvature of a surface to n-dimensional manifolds (see § 17). His analytical treatment of the problem, however, remained unknown until the publication of his Paris Prize Essay[72]; it contains his treatment in its entirety, using both the elimination method and the variational method. Before this, however, Christoffel[73] and Lipschitz[74] had obtained the same results, by formulating the condition for a given quadratic form

$$g_{ik} \, dx^i \, dx^k \qquad\qquad (g_{ik} \text{ functions of the } x)$$

to be transformed into

$$\sum_i (dx^i)^2.$$

[70c] Whereas in ordinary mechanics the Lagrange equations are obtained by permitting all possible point transformations for the space coordinates, the above shows that the equations are of the same form, even though the time coordinate is also subjected to arbitrary transformations; the independent variable now is of course not t, but s. Cf. T. Levi-Civita, *Enseign. math.*, **21** (1920) 5.

[71] Cf. footnote 63, p. 34, *ibid.*

[72] Cf. footnote 68, p. 38, *ibid.*

[73] Cf. footnote 68, p. 38, *ibid.*

[74] R. Lipschitz, *J. reine angew. Math.*, **70** (1869) 71, **71** (1870) 244 and 288, and **72** (1870) 1, as well as **82** (1877) 316. The last paper of this series was published only after the publication of Riemann's Prize Essay.

This is itself again a special case of the problem of the equivalence of quadratic differential forms, which Christoffel had also studied, namely under what conditions the two forms

$$g_{ik}\,dx^i\,dx^k \quad \text{and} \quad g'_{ik}\,dx'^i\,dx'^k$$

can be transformed into each other. This general question of equivalence has however not turned out to be of importance for physics, so far. Ricci and Levi-Civita[75], whose presentation was followed closely by Einstein[76], managed to derive the curvature tensor in a purely formal, but very concise, way, in contrast to Christoffel's rather lengthy calculations. Finally, Hessenberg[77], and Levi-Civita[78] found a geometrical, intuitive, interpretation of the curvature tensor by starting from the concept of the parallel displacement of a vector.

In § 14 it was always a question of the parallel displacement of a vector along a given curve, and never of the parallel displacement from a point P to just any other point P'. In fact, only in Euclidean geometry is this independent of the path chosen. If, however, a vector ξ^i undergoes parallel displacement along a closed curve, one obtains a vector ξ^{*i} which is different from the initial vector ξ^i. This fact can be used for the definition of the curvature tensor. Let the two-parameter set of curves

$$x^k = x^k(u, v)$$

be given and displace an arbitrary vector ξ^h from the point $P_{00}(u, v)$ via $P_{10}(u+\Delta u,\ v)$, $P_{11}(u+\Delta u,\ v+\Delta v)$, $P_{01}(u,\ v+\Delta v)$ back again to $P_{00}(u, v)$, alternately along curves with constant v and constant u. The difference $\xi^{*h} - \xi^h = \Delta\xi^h$ then must obviously be of order $\Delta u\,\Delta v$, since it becomes zero as soon as *one* of the quantities Δu or Δv vanishes. The limit

$$\lim_{\substack{\Delta u \to 0 \\ \Delta v \to 0}} \frac{\Delta\xi^h}{\Delta u \Delta v},$$

which alone is of importance here, can be worked out immediately with the help of (64) and is found to be

$$\lim \frac{\Delta\xi^h}{\Delta u \Delta v} = R^h{}_{ijk}\,\xi^i\,\frac{\partial x^j}{\partial u}\,\frac{\partial x^k}{\partial v} \tag{85}$$

where

$$R^h{}_{ijk} = \frac{\partial\Gamma^h{}_{ij}}{\partial x^k} - \frac{\partial\Gamma^h{}_{ik}}{\partial x^j} + \Gamma^h{}_{k\alpha}\Gamma^\alpha{}_{ij} - \Gamma^h{}_{j\alpha}\Gamma^\alpha{}_{ik}. \tag{86}$$

It follows from the vector character of the left-hand side of (85) that the right-hand side, too, is a vector (it is to be noted that $\Delta\xi^h$ is the difference of two vectors *at the same point*). Hence, the quantities $R^h{}_{ijk}$ are the components of a tensor. This is the [mixed] curvature tensor, which is called

---

[75] Cf. footnote 56, p. 24, *ibid.*

[76] Cf. footnote 56, p. 24, *Ann. Phys., Lpz., loc. cit.*

[77] Cf. footnote 58 a, p. 27, *ibid.*

[78] Cf. footnote 65, p. 37, *ibid.* Cf. also Weyl's treatment in *Raum-Zeit-Materie* (1st and 3rd edns.)

the Riemann-Christoffel tensor, after its discoverers. The meaning of formula (85) becomes somewhat more apparent if one writes it in terms of differentials, rather than differential coefficients. Setting dx^j for $(\partial x^j/\partial u)\,du$ and δx^k for $(\partial x^k/\partial v)\,dv$, and introducing the "surface" tensor†

$$d\sigma^{jk} =_. dx^j\,\delta x^k - dx^k\,\delta x^j$$

(since $R^h{}_{ijk}$ is antisymmetric in j and k), formula (85) can then be written[78a]

$$\Delta\xi^h = \tfrac{1}{2}R^h{}_{ijk}\,\xi^i\,d\sigma^{jk}. \tag{87}$$

The same procedure which leads to formula (86) can be used to obtain the change in the *covariant* components for a parallel displacement along the above closed contour. With the help of (70),

$$\Delta\xi_h = \tfrac{1}{2}R_{hijk}\,\xi^i\,d\sigma^{jk}, \tag{88}$$

where

$$
R_{hijk} = \frac{\partial \Gamma_{i,hk}}{\partial x^j} - \frac{\partial \Gamma_{i,hj}}{\partial x^k} + g^{\alpha\beta}(\Gamma_{\alpha,hj}\Gamma_{\beta,ik} - \Gamma_{\alpha,hk}\Gamma_{\beta,ij})
$$

$$
= \frac{1}{2}\left(\frac{\partial^2 g_{hj}}{\partial x^i\,\partial x^k} + \frac{\partial^2 g_{ik}}{\partial x^h\,\partial x^j} - \frac{\partial^2 g_{hk}}{\partial x^i\,\partial x^j} - \frac{\partial^2 g_{ij}}{\partial x^h\,\partial x^k} \right) + \tag{89}
$$

$$
+ g^{\alpha\beta}(\Gamma_{\alpha,hj}\Gamma_{\beta,ik} - \Gamma_{\alpha,hk}\Gamma_{\beta\,ij}).
$$

Also, it is easy to show that

$$\Delta\xi_h = g_{h\alpha}\Delta\xi^\alpha; \tag{90}$$

hence the R_{hijk} are the covariant components associated with $R^h{}_{ijk}$,

$$R_{hijk} = g_{h\alpha}R^\alpha{}_{ijk}. \tag{91}$$

From (89) it follows that R_{hijk} satisfies the symmetry conditions

$$R_{hijk} = -R_{hikj} = -R_{ihjk} = R_{jkhi}, \tag{92}$$

$$R_{hijk} + R_{hjki} + R_{hkij} = 0.$$

According to § 11 the [covariant] curvature tensor can thus be looked upon as a "surface" tensor of rank 2.† As Hessenberg[79] has shown, the relations (92) can also be obtained directly from the definition (87) of the curvature tensor. Since Riemann writes $(hijk)$ in place of R_{hijk}, these quantities are sometimes called four-index symbols. In Euclidean space they are zero, for they certainly vanish in those coordinate systems in which the g_{ik} are constant, and it then follows from their tensor character that they must vanish in *all* coordinate systems. This vanishing is therefore the necessary condition for $g_{ik}dx^i dx^k$ to be transformable into $\Sigma(dx'^i)^2$.

---

† See suppl. note 5 a.

[78a] For a two-dimensional manifold, this method leads to the well-known relation between the Gaussian curvature and the excess (defect) of the sum of the angles in a geodesic triangle. This relation was already demonstrated by Gauss.

[79] Cf. footnote 58a, p. 27, *ibid.*

From the second-rank "surface" tensor $R^h{}_{ijk}$ one obtains by contraction a "line" tensor of second rank†, R_{ik},

$$R_{ik} = R^\alpha{}_{i\alpha k} = g^{\alpha\beta} R_{\alpha i\beta k} = g^{\alpha\beta} R_{i\alpha k\beta}. \tag{93}$$

Its symmetry properties follow from

$$g^{\alpha\beta} R_{\alpha i\beta k} = g^{\alpha\beta} R_{\beta k\alpha i} = g^{\alpha\beta} R_{\alpha k\beta i}.$$

With the help of (86) its components are given by

$$R_{ik} = \frac{\partial \Gamma^\alpha{}_{i\alpha}}{\partial x^k} - \frac{\partial \Gamma^\alpha{}_{ik}}{\partial x^\alpha} + \Gamma^\beta{}_{i\alpha} \Gamma^\alpha{}_{k\beta} - \Gamma^\alpha{}_{ik} \Gamma^\beta{}_{\alpha\beta}. \tag{94}$$

A further contraction then leads to the curvature invariant[79a]

$$R = g^{ik} R_{ik}. \tag{95}$$

It should be noted that Herglotz[80], and Weyl in his more recent papers[81], define the curvature tensors with opposite sign from that used here and by the other authors.

17. Riemannian coordinates and their applications

For many purposes it is useful to introduce the following coordinate system, which is due to Riemann (Inaugural Lecture). Let an arbitrary coordinate system x^i be given, and let all geodesic lines originating from an arbitrary point P_0 be drawn. Their directions are characterized by the tangential vectors at P_0, having components $(dx^k/ds)_0$. In a certain neighbourhood of P_0 only *one* geodesic line will exist which passes through a given point P as well as P_0. If s is the geodesic arc length PP_0, then the point P is unambiguously determined by means of the quantities

$$y^k = \left(\frac{dx^k}{ds}\right)_0 s. \tag{96}$$

The y^k are called Riemannian coordinates. Evidently the y-coordinate system is tangential to the x-coordinate system at P_0, so that at this point the g_{ik} (and, for that matter, the components of any arbitrary tensor) are equal in the two systems. We distinguish them by a superscript zero, e.g. $\overset{\circ}{g}_{ik}$. To an arbitrary transformation of the x-system corresponds an affine transformation of the y-system. Let us now leave the x-system aside and investigate the form taken by the line element in the Riemannian coordinate system. First of all, at P_0 the $\Gamma^i{}_{rs}$ have to vanish, according to (80), since *all* geodesic lines originating from P_0 have linear equations,

$$\overset{\circ}{\Gamma}{}^i{}_{rs} = 0. \tag{97}$$

In other words, the Riemannian coordinate system is geodesic at P_0. At an

† See suppl. note 5 a.
[79a] It is first mentioned by R. Lipschitz, cf. footnote 74, p. 41, *ibid.*, **72**.
[80] G. Herglotz 'Zur Einsteinschen Gravitationstheorie', *S. B. naturf. Ges., Lpz., math.-phys. Kl.*, **68** (1916) 199.
[81] Cf. footnote 67, p. 37, *ibid.*

arbitrary point P, *no other* geodesic lines originating from it have linear equations, *except for the one* which also passes through P_0. This is expressed by

$$\Gamma^i{}_{rs}(y)\,y^r\,y^s = 0. \tag{98}$$

where $\Gamma^i{}_{rs}(y)$ are the values of the geodesic components at the point with coordinates y. This equation must hold for all y. If, conversely, relations (97) and (98) are satisfied for a given coordinate system, then it is Riemannian. It can be proved[81a] that, as a consequence of these relations, the line element ds^2 has to be of the form

$$ds^2 = \mathring{g}_{ik}dy^i\,dy^k + \sum_{(hi)(jk)} p_{hijk}(y)(y^h\,dy^i - y^i\,dy^h)(y^j\,dy^k - y^k\,dy^j). \tag{99}$$

The summation is carried out over the $n(n-1)/2$ possible combinations of each of the index pairs (hi) and (jk). Conversely, (97) and (98) can be obtained from (99), so that this form of the line element is a necessary and sufficient condition for the y-coordinate system to be Riemannian. The $p_{hijk}(y)$ are regular functions of the coordinates and behave under linear transformations of the y like the components of a "surface" tensor of the second rank†; they can always be determined[81b] in such a way that they satisfy the symmetry conditions (53) (cf. § 11). The curvature tensor at the origin ["pole"] is connected with the values of the p_{hijk} there in a very simple way,

$$\mathring{R}_{hijk} = 3\mathring{p}_{hijk}. \tag{100}$$

Thus the R_{hijk} measure directly, in this representation, the deviation of the geometry from Euclidean geometry. Moreover, Riemann observed that for the case of a two-dimensional manifold (of which the line element is given by

$$ds^2 = \gamma_{11}\,du^2 + 2\gamma_{12}\,du\,dv + \gamma_{22}\,dv^2, \quad \text{say}),$$

R_{1212}, the only independent component of the curvature tensor, determines the Gaussian curvature K of the surface according to the formula

$$K = -\frac{R_{1212}}{\gamma_{11}\gamma_{22} - \gamma_{12}{}^2}. \tag{101}$$

This is shown by direct comparison of (89) with the Gaussian formulae. If, in particular, u, v are the Riemannian coordinates of the surface, so that the line element is of the form

$$ds^2 = \mathring{\gamma}_{11}\,du^2 + 2\mathring{\gamma}_{12}\,du\,dv + \mathring{\gamma}_{22}\,dv^2 + \pi(u,v)(u\,dv - v\,du)^2, \tag{102}$$

then the Gaussian curvature at P_0 can also be written

$$\mathring{K} = -\frac{3\mathring{\pi}}{\mathring{\gamma}_{11}\mathring{\gamma}_{22} - \mathring{\gamma}_{12}{}^2}, \tag{103}$$

[81a] Cf. comments by H. Weber in *Riemann's Collected Works*, (2nd edn.), p. 405, and F. Schur, *Math. Ann.* **27** (1886) 537.

† See suppl. note 5a.

[81b] H. Vermeil, *Math. Ann.*, **79** (1918) 289.

because of (100) and (101). There is a historic reason for the choice of sign of K, because of its relation to the Euclidean R_3 in which it is embedded, and it has nothing to do with the metrical properties of the surface itself. In view of the form of the line element (99) it would seem more natural to choose the opposite sign, so that, for example, the curvature of a sphere would be negative.

With the help of the Riemannian coordinates the concept of a curvature of R_n-space can now be associated with that of the curvature of a surface. This is, in fact, the way in which Riemann first came to think of it. Let there be two directions which are characterized by the vectors ξ^i and η^i. The lengths of these vectors is immaterial. They, in turn, define the linear pencil

$$\xi^i u + \eta^i v$$

and the surface direction

$$\xi^{ik} = \xi^i \eta^k - \xi^k \eta^i.$$

Along each direction contained in the pencil we construct the geodesic line originating from P_0. The totality of these geodesic lines then form a surface, of which we want to determine the curvature. The line element of the surface is obtained by substituting

$$y^i = \xi^i u + \eta^i v$$

in (99). It will be of the form (102), where the quantities $\overset{\circ}{\gamma}_{ik}$ and π take on the following values,

$$\overset{\circ}{\gamma}_{11} = \overset{\circ}{g}_{ik} \xi^i \xi^k = \xi_i \xi^i, \qquad \overset{\circ}{\gamma}_{12} = \tfrac{1}{2} \overset{\circ}{g}_{ik}(\xi^i \eta^k + \xi^k \eta^i) = \xi_i \eta^i,$$

$$\overset{\circ}{\gamma}_{22} = \overset{\circ}{g}_{ik} \eta^i \eta^k = \eta_i \eta^i, \qquad \pi = \sum_{(hi)(jk)} p_{hijk} \xi^{hi} \xi^{jk}.$$

Formulae (100) and (103) then lead immediately to an expression for the curvature (dropping the superscript 0)

$$-K = \frac{\sum\limits_{(hi)(jk)} R_{hijk} \xi^{hi} \xi^{jk}}{\tfrac{1}{2} \xi_{ik} \xi^{ik}} = \frac{\sum\limits_{(hi)(jk)} R_{hijk} \xi^{hi} \xi^{jk}}{\sum\limits_{(hi)(jk)} (g_{hj} g_{ik} - g_{hk} g_{ij}) \xi^{hi} \xi^{jk}}. \tag{104}$$

This result has no longer anything to do with the Riemannian coordinates. The *magnitude* of ξ^{ik} obviously cancels, and what we have is an invariant Gaussian curvature associated with every surface direction. This is called, after Riemann, the [sectional] curvature of R_n-space in the given surface direction (after giving it the opposite sign). It also becomes evident in this connection that the quantities R_{hijk} are the components of a "surface" tensor of second rank†.

In connection with formula (104), which was derived by Riemann,

† See suppl. note 5 a.

Herglotz[82] showed that the contracted curvature tensor and the curvature invariant could also be interpreted geometrically. His conclusions are as follows. Let n orthogonal directions be given, which determine $\binom{n}{2}$ surface directions. If $K(rs)$ is the [sectional] curvature with respect to the surface spanned by the rth and sth vectors, then the curvature invariant R is equal to the double sum

$$R = 2\sum_{(rs)} K(rs) \tag{105}$$

(summed over all index combinations (rs)). This is independent of the choice of the n directions $1, 2, ..., n$ and can be described as the mean curvature of R_n at the particular point. If, now, a further direction o is determined by the vector ξ^i, then

$$\sum_{(or)} K(or) \sin^2(o,r) = \frac{R_{ik}\,\xi^i\,\xi^k}{\xi_i\,\xi^i} \tag{106}$$

determines the contracted curvature tensor (this sum, too, turns out to be independent of the choice of the n directions). This is the geometrical proof for the tensor character of R_{ik} and for the invariance of R, both of which had previously only been established algebraically. If, in particular, one lets one of the orthogonal directions, say 1, coincide with direction o, then

$$\frac{R_{ik}\,\xi^i\,\xi^k}{\xi_i\,\xi^i} = \sum_{r=2}^{n} K(1r). \tag{107}$$

Finally, from (105) and (107) follows the expression for the mean curvature of R_{n-1}, which is perpendicular to direction 1, and which is characterized by the vector ξ^i,

$$\sum_{\substack{(rs) \\ r \neq 1, s \neq 1}} K(rs) = \tfrac{1}{2}R - \frac{R_{ik}\,\xi^i\,\xi^k}{\xi_i\,\xi^i} = -\frac{G_{ik}\,\xi^i\,\xi^k}{\xi_i\,\xi^i}, \tag{108}$$

where

$$G_{ik} = R_{ik} - \tfrac{1}{2}g_{ik}R. \tag{109}$$

This tensor is of great importance in the general theory of relativity [and is called the Einstein tensor].

In addition, let us mention a simple theorem by Vermeil[83], which is based on expression (99) for the line element. The volume V_n of a [hyper-] sphere of radius r in Euclidean R_n-space has the simple value

$$V_n = C_n r^n$$

where C_n is a numerical factor whose value is of no importance here. In an arbitrary Riemannian manifold, V becomes a complicated function

---

[82] Cf. footnote 80, p. 44. An interpretation of the curvature invariant was already given before Herglotz by H. A. Lorentz, *Versl. gewone Vergad. Akad. Amst.* **24** (1916) 1389.

[83] H. Vermeil, 'Notiz über das mittlere Krümmungsmass einer n-fach ausgedehnten Riemannschen Mannigfaltigkeit', *Nachr. Ges. Wiss. Göttingen* (1917) 334.

of r. We can imagine it to be expanded in a power series in r and retain only the term following $C_n r^n$. This gives,

$$V_n = C_n r^n \left\{ 1 + \frac{R}{6} \frac{r^2}{n+2} + \ldots \right\}, \tag{110}$$

where R is the curvature invariant at the centre of the sphere. Differentiating, one obtains from (110) the formula for the surface S_n of the sphere,

$$S_n = n C_n r^{n-1} \left\{ 1 + \frac{R}{6} \frac{r^2}{n} + \ldots \right\}. \tag{111}$$

These relations can be used for a new geometrical definition of the curvature invariant,

$$R = \lim_{r \to 0} \left(\frac{V_n}{C_n r^n} - 1 \right) \frac{6(n+2)}{r^2} = \lim_{r \to 0} \left(\frac{S_n}{n C_n r^{n-1}} - 1 \right) \frac{6n}{r^2}. \tag{112}$$

The introduction of Riemannian coordinates reduces the problem of invariants under general transformations to one of invariants under linear transformations.[84] It can thus be proved that R (apart from an unimportant constant factor) is the only invariant which contains only the g_{ik} themselves as well as their first and second derivatives and which is linear in the latter.[84a] Also, all "line" tensors of rank 2 which have this property with respect to the g_{ik} are of the form

$$c_1 R_{ik} + c_2 R g_{ik} + c_3 g_{ik} \tag{113}$$

(c_1, c_2, c_3 are constants).[84a]

18. The special cases of Euclidean geometry and of constant curvature

It is easy enough to see that in Euclidean space the curvature tensor R_{hijk} vanishes (see § 16). But it was already pointed out by Riemann in his Inaugural Lecture that the converse of this theorem is also true: If the curvature tensor vanishes, the space is Euclidean, i.e. one can then always find a coordinate system in which the g_{ik} are constants. This was first shown by Lipschitz[85], though his proof is very long-winded. Weyl[86] indicated a way of reasoning which is very clear and easy to visualize. In the general case the result of a parallel displacement of a vector depends essentially on the path along which it is carried out. This will only *not* be so when the vector components are not just functions of s, but can be determined as functions of the coordinates in such a way that everywhere and for all directions of the curve Eq. (64) is satisfied. This means,

---

[84]   Cf. general remarks in E. Noether, 'Invarianten beliebiger Differentialausdrücke', *Nachr. Ges. Wiss. Göttingen* (1918) 37 and also H. Vermeil (Cf. footnote 81 b, p. 45, *ibid.*).
[84a]   H. Vermeil (cf. footnote 83, *ibid.*), and H. Weyl, *Raum-Zeit-Materie* (4th edn.), appendix.
[85]   Cf. footnote 74, p. 41, *ibid.*, **70**.
[86]   H. Weyl, *Raum-Zeit-Materie* (1st edn., Berlin 1918), p. 111.

however, that the ξ^i have to satisfy the differential equations

$$\frac{\partial \xi^i}{\partial x^s} = -\Gamma^i_{rs}\xi^r. \tag{114}$$

It then follows that the conditions of integrability amount to the requirement that $R_{hijk} = 0$. In other words, if the curvature tensor vanishes, then the system of equations (114) can always be solved, the direction transfer is independent of the path taken, i.e. it is integrable. All that is needed now is to introduce, in place of the given coordinate system K with base vectors e_k, a new coordinate system K' with base vectors e'_i, having the following properties. The e'_i at an arbitrary point P_1 are to be parallel to the e'_i at a second arbitrary point P_2. Their components $\alpha_i{}^k$ in system K (see § 10) will, because of (114), satisfy the equations

$$\frac{\partial \bar{\alpha}_i{}^k}{\partial x^s} = -\Gamma^k_{rs}\bar{\alpha}_i{}^r. \tag{115}$$

Such a choice of coordinates becomes possible because, by reason of (115), the conditions of integrability (63) are automatically fulfilled. In fact,

$$\frac{\partial \bar{\alpha}_i{}^k}{\partial x'^l} = \frac{\partial \alpha_i{}^k}{\partial x^s}\bar{\alpha}^s = -\Gamma^k_{rs}\bar{\alpha}_i{}^r\bar{\alpha}_l{}^s$$

is symmetrical in i and l. There are now at every point n vectors (the n base vectors e'_i) whose components in K' remain constant for all infinitesimal translations. Since an arbitrary vector x can be linearly built up from the e'_i and since the infinitesimal translation is an affine one according to § 14, the translation will not change the components of x in K'. This, however, is only possible when the geodesic components of K' vanish everywhere, i.e. when the g'_{ik} are constants. This is easily verified by direct calculation of $\partial g'_{ik}/\partial x'^l$. The proof is thus completed.

A wider class of Riemannian manifolds are those whose curvature is independent both of the surface direction and the position. From (104) it is seen that they are characterized by the relation

$$R_{hijk} + \alpha(g_{hj}g_{ik} - g_{hk}g_{ij}) = 0, \tag{116}$$

where α is a (positive or negative) constant. On contraction it follows that

$$R_{ik} + (n-1)\alpha g_{ik} = 0 \tag{117}$$

and

$$R = -n(n-1)\alpha. \tag{118}$$

For future applications let us also write down the expression for the [Einstein] tensor G_{ik}, defined by (109),

$$G_{ik} = \frac{(n-1)(n-2)}{2}\alpha g_{ik}. \tag{119}$$

For $\alpha = 0$ we get back to the case of zero curvature.

An example of a space of constant curvature is the n-dimensional sphere, which can be thought of as embedded in a Euclidean R_{n+1}-space.

It is better to speak here of a spherical space R_n, if we are to concern ourselves only with the intrinsic metrical relationships of the sphere. We then have

$$ds^2 = \sum_i (dx^i)^2 + (dx^{n+1})^2, \tag{120}$$

$$\sum_i (x^i)^2 + (x^{n+1})^2 = a^2, \tag{121}$$

where the summation indices always run from 1 to n. Let us, first of all, introduce as coordinates on the sphere the x^i ($i = 1, ..., n$), which corresponds to a parallel projection on to the equatorial plane $x^{n+1} = 0$. Then, eliminating x^{n+1} from (120) by means of (121),

$$ds^2 = \sum_i (dx^i)^2 + \frac{(x^i\,dx^i)^2}{a^2 - r^2}, \qquad \left(r^2 = \sum_i (x^i)^2\right). \tag{122}$$

The equator $x^{n+1} = 0$ is a singular line of the coordinate system, and to each set of values of the coordinates there correspond *two* points in the spherical space R_n. One can also project the points on the sphere from the centre on to the plane $x^{n+1} = -a$. This corresponds to a coordinate transformation

$$x^i = \frac{r}{r'}x'^i, \qquad \left(r'^2 = \sum_i (x'^i)^2\right); \qquad \frac{r}{r'} = \frac{|x^{n+1}|}{a} = \frac{a}{\sqrt{(a^2 + r'^2)}}. \tag{123}$$

Dropping primes in the final result, the line element then is of the form

$$ds^2 = \frac{a^2}{(a^2 + r^2)^2}\left[(a^2 + r^2) \sum_i (dx^i)^2 - (x^i\,dx^i)^2\right]. \tag{124}$$

The coordinate system only includes one hemisphere, the equator is removed to infinity ($r = \infty$).

In the same way one obtains from a stereographic projection,

$$x^i = \frac{r}{r'}x'^i, \qquad \frac{r}{r'} = \frac{a - x^{n+1}}{2a} = \frac{1}{1 + (r'^2/4a^2)}, \tag{125}$$

$$ds^2 = \frac{\sum (dx^i)^2}{[1 + (r^2/4a^2)]^2}, \tag{126}$$

where again the primes have been dropped in the last equation. This coordinate system is only singular at the pole $x^{n+1} = a$, where $r = \infty$.

A fourth form of the line element is obtained by introducing Riemannian coordinates, and substituting

$$x^i = \frac{r}{\rho}y^i, \qquad \left(\rho = \sum_i (y^i)^2\right); \qquad \frac{r}{\rho} = \frac{a}{\rho}\sin\frac{\rho}{a}; \qquad x^{n+1} = a\cos\frac{\rho}{a} \tag{127}$$

into (122). The result is

$$ds^2 = \frac{a^2}{\rho^2} \sin^2\frac{\rho}{a} \sum (dx^i)^2 + \frac{1}{\rho^2}\left(1 - \frac{a^2}{\rho^2}\sin^2\frac{\rho}{a}\right)(y^i\, dy^i)^2. \qquad (128)$$

Because

$$\sum_{(ik)} (y^i\, dy^k - y^k\, dy^i)^2 = \rho^2 \sum (dy^i)^2 - (y^i\, dy^i)^2 \qquad (129)$$

(in the sum on the left-hand side each combination (ik) is to be counted only once), this can also be written

$$ds^2 = \sum (dy^i)^2 - \frac{1}{\rho^2}\left(1 - \frac{a^2}{\rho^2}\sin^2\frac{\rho}{a}\right) \sum_{(ik)} (y^i\, dy^k - y^k\, dy^i)^2 \qquad (128\,\text{a})$$

The y^i are thus in fact Riemannian coordinates. These expressions can also be derived by way of polar coordinates. The pole $x^{n+1} = a$ corresponds to the origin of the y^i-system; the system becomes singular for $\rho = a\pi$ because the same point, viz. the pole $x^{n+1} = -a$, corresponds to all values of y^i which satisfy the condition $\rho = a\pi$. All points on the sphere can already be obtained if one restricts ρ by the condition $\rho < a\pi$.

Because of (99) and (100), it follows from (128a) that at the point $y^i = 0$ the space curvature is independent of the surface direction, so that a relation of the form (116) holds at that point. Since

$$\frac{1}{\rho^2}\left(1 - \frac{a^2}{\rho^2}\sin^2\frac{\rho}{a}\right)\Bigg|_{\rho = 0} = \frac{1}{3a^2},$$

and because of formula (100), the coefficient α has the value

$$\alpha = \frac{1}{a^2}. \qquad (130)$$

Relations (116), with the same value of α, hold at all points of the spherical space. This follows from the existence of a transformation group $G_{n(n+1)/2}$, which allows us to transform a given point and an associated n-dimensional subspace into another arbitrary point and another n-dimensional subspace. It is essential that this should be done in such a way that the lengths of all curves remain unchanged. For, let S be the transformation (127) of the system $x^1, ..., x^{n+1}$ of Euclidean R_{n+1} into the Riemannian coordinate system in a spherical space R_n, and T be the $\frac{1}{2}n(n+1)$-parameter group of the orthogonal transformations of the former system, then

$$G_{n(n+1)/2} = S^{-1}\,TS$$

is the required transformation group. It shows that the line element has the same form in all Riemannian coordinate systems, whatever the position of their origin in spherical space R_n. From this it immediately follows that relations (116) and (130) are generally valid in spherical space. This can, of course, be verified by direct calculation.

The curvature is therefore always a constant when R_n has the following property. In a certain (finite) neighbourhood of every point of R_n a coordinate system can be determined such that the line element at this

point is one of the four equivalent forms (120), (122), (124) and (128);
α need not necessarily be positive. If α is negative, then a^2 has to be replaced by $-a^2$ in the relevant formulae, and

$$\alpha = -\frac{1}{a^2}. \tag{130a}$$

That, conversely, R_n always has this property when (116) is valid was
pointed out by Riemann in his Inaugural Lecture and was first proved by
Lipschitz[88]. Vermeil[89], using a power expansion for the line element in
Riemannian coordinates, gave a simpler proof of the general theorem
that for a given curvature tensor the form of the line element in Riemannian coordinates is already unambiguously determined. This, too, had
been hinted at by Riemann. In physics this corollary has not, so far,
found any applications.

For cosmological problems (cf. Part IV), however, the following
circumstance is of importance. The large-scale metrical relationships
of R_n are not uniquely defined by the form of the line element, and it is
here that the projective point of view has to supplement that of differential geometry. The former enables one immediately to determine the
relationships of the *whole* space for spaces of constant curvature. In this
way, there exist two alternatives for spaces of constant *positive* curvature,
as was first shown by Klein[90]. In the coordinate system represented by
(122), there correspond to each set of values of the coordinates either
two or only one space point. In the first case the space is called spherical,
in the second case elliptic, following the projective viewpoint. Both kinds
of spaces are, although unbounded, nevertheless finite in the Riemannian
sense. The total volume of the elliptic space is evidently half that of the
spherical space having the same curvature. The same is true of the ratio
of the total lengths of the (closed) geodesic lines in the two spaces. For
spaces of constant *negative* curvature, the number of alternatives is much
greater. Especially worth noting is the Clifford surface, which demonstrates the possibility of having a *finite* manifold with *zero curvature*. The
whole problem of the large-scale metrical relationships of manifolds of
constant curvature was given the name of "the problem of the Clifford–
Klein space forms" by Killing.

19. The integral theorems of Gauss and Stokes in a four-dimensional Riemannian manifold

The complications of the tensor analysis of the general transformation
group, as opposed to that of the affine group, arise from the fact that
the components of two tensors which are associated with different points,
can now no longer be simply added. In order to use the differentiation

[88] Cf. footnote 74, p. 41, *ibid.*
[89] Cf. footnote 81 b, p. 45, *ibid.*
[90] F. Klein, *Math. Ann.* 4 (1871) 573, 6 (1872) 112, and in particular *Math. Ann.* 37 (1890)
544, where the problem is solved in all its generality. See also *Programm zum Eintritt in
die philosophische Fakultät in Erlangen* ['*Erlanger Programm*'] (1872), reprinted in *Math.
Ann.* 43 (1893) 63.

of tensors for deriving new tensors, one will therefore have to take recourse to the concept of parallel displacement, developed in § 14. The rules for this were first derived in a purely formal way by Christoffel[91] and later systematized by Ricci and Levi-Civita[56]. Simplifications and geometrical interpretations were introduced in the papers by Weyl[92], Hessenberg[58 a] and Lang[56].

For certain operations, the geodesic components do not enter the final result. Such operations will now be considered and refer to tensors of the first rank (see § 11). It would therefore seem a natural requirement not to make use of the concept of parallel displacement in their derivation. First of all, a vector Grad ϕ can be derived from a scalar ϕ by differentiation. This follows immediately from the invariance of

$$d\phi = \frac{\partial \phi}{\partial x^i} dx^i.$$

It should be observed that the $\partial \phi / \partial x^i$ are *covariant* components,

$$\text{Grad}_i \phi = \frac{\partial \phi}{\partial x^i}. \tag{131}$$

To find further relations, we have to apply the integral theorems of Gauss and Stokes to our case, but shall confine ourselves to four-dimensional manifolds. The corresponding generalizations of Gauss' and Stokes' theorems to manifolds of arbitrary dimensions can be found, in their most general form, in the papers by Poincaré[93a] and Goursat[93b]. For the case of the special theory of relativity (Euclidean geometry and orthogonal coordinates), the formulae were also derived by Sommerfeld[55b].

Let

$$f^i, \quad F^{ik}, \quad A^{ikl} \tag{132}$$

be the components of a "line", "surface" and "volume" tensor, respectively, and let

$$ds^i, \quad d\sigma^{ik}, \quad dS^{ikl}, \quad d\Sigma \tag{133}$$

be the components of a line-, surface-, space-, and world-element, having absolute magnitudes

$$ds, \quad d\sigma, \quad dS, \quad |d\Sigma|. \tag{133a}$$

The components (133) can be expressed in terms of the coordinates in the following way. The ds^i are equal to the coordinate differentials

$$ds^i = dx^i; \tag{134a}$$

moreover, if $dx^i, \delta x^i$ and $dx^i, \delta x^i, \mathfrak{d}x^i$ are the components of two and three

[91]   Cf. footnote 68, p. 38, *ibid.*
[92]   H. Weyl, *Raum-Zeit-Materie* (1st edn., 1918), pp. 103–107.
[93 a]   H. Poincaré, *Acta math., Stockh.*, 9 (1887) 321.
[93 b]   E. Goursat, *J. Math. pures appl.* (6) 4 (1908) 331.
[55 b]   A. Sommerfeld (Cf. footnote 55, p. 22, *ibid.*). His derivation of the divergence of a "surface" tensor is different from ours.

line elements in independent directions on the surface- and space-element respectively, then

$$d\sigma^{ik} = \left| \begin{array}{cc} dx^i & \delta x^i \\ dx^k & \delta x^k \end{array} \right| \tag{134 b}$$

$$dS^{ikl} = \left| \begin{array}{ccc} dx^i & \delta x^i & \eth x^i \\ dx^k & \delta x^k & \eth x^k \\ dx^l & \delta x^l & \eth x^l \end{array} \right|. \tag{134 c}$$

Substituting these expressions into surface- or space-integrals of the type $\int \phi(x)\, d\sigma^{ik}$ or $\int \phi(x)\, dS^{ikl}$, respectively, this corresponds to the way of writing multiple integrals which Klein[94] introduced and called the Grassmann notation. This is a natural notation, since it immediately allows one to read off the behaviour of multiple integrals under coordinate transformations and Klein[94] therefore preferred it to the ordinary notation. While the latter has the advantage of greater simplicity, it has, on the other hand, the disadvantage that the behaviour of the integrand under coordinate transformations is not shown up directly. The ordinary notation is arrived at by assuming the independent directions d, δ $(d, \delta, \eth)$ for the separate components of the surface (space) element to be parallel to the corresponding coordinates. In that case,

$$d\sigma^{ik} = dx^i\,\delta x^k, \qquad dS^{ikl} = dx^i\,\delta x^k\,\eth x^l$$

or, more simply,

$$d\sigma^{ik} = dx^i\,dx^k, \qquad dS^{ikl} = dx^i\,dx^k\,dx^i. \tag{135}$$

It is however to be noted that these expressions behave, under transformations of the coordinates, like the components of a "surface" and "volume" tensor, respectively.

We can now form two kinds of invariants from tensors (132) and (133):

(a) The orthogonal projections of f, **F**, **A** on ds, $d\sigma$, dS, multiplied by the magnitude of the latter tensors

$$f_s\,ds = f_i\,dx^i \tag{136a}$$

$$F_\sigma\,d\sigma = F_{ik}\,d\sigma^{ik} \tag{136 b}$$

$$A_S\,dS = A_{ikl}\,dS^{ikl}. \tag{136 c}$$

(b) The orthogonal projection of the vector f on the direction normal to dS, that of **F** on the direction normal to $d\sigma$, and that of **A** on the direction normal of ds, in each case multiplied by the magnitude of the second tensor. The values of these expressions are obtained with the help of the tensors dual to ds, $d\sigma$, dS (see § 12, (54b) and (55)),

$$f_n\,dS = f^i\,dS^*_i = \sum_{(iklm)} \sqrt{g}\,f^i\,dS^{klm} = \sum_{(ikm)} \mathfrak{f}^i\,dS^{klm} \tag{137a}$$

$$F_N\,d\sigma = F^{ik}\,d\sigma^*_{ik} = \sum_{(iklm)} \sqrt{g}\,F^{ik}\,d\sigma^{lm} = \sum_{(iklm)} \mathfrak{F}^{ik}\,d\sigma^{lm} \tag{137 b}$$

[94] F. Klein, 'Über die Integralform der Erhaltungssätze und die Theorie der räumlich geschlossenen Welt', *Nachr. Ges. Wiss. Göttingen* (1918) 394.

$$A_n ds = A^{ikl} ds^*_{ikl} = \sum_{(iklm)} \sqrt{g}\, A^{ikl} ds^m = \sum_{(iklm)} \mathfrak{A}^{ikl} ds^m. \qquad (137\,\mathrm{c})$$

The sums $\sum\limits_{(iklm)}$ are to be taken over even permutations, and $\mathfrak{f}^i$, $\mathfrak{F}^{ik}$, $\mathfrak{A}^{ikl}$ are the tensor densities corresponding to f, $\mathbf{F}$, $\mathbf{A}$ (§ 11).

The generalizations of Gauss' and Stokes' theorems can now be formulated. Let us integrate (136 a) over a closed curve, (136 b) and (137 b) over a closed surface, and (137 a) over a closed space region. (We shall leave out the analogous theorems for (136 c) and (137 c), since they have not, as yet, had any applications in physics.) These integrals can be transformed into integrals over the surface-, space-, and world-region, respectively, which are enclosed by them.

$$\int f_s\, ds = \int \mathrm{Curl}_N\, f \cdot d\sigma = \int \mathrm{Curl}_{ik}\, f \cdot d\sigma^{ik} \qquad (138\,\mathrm{a})$$

$$\int F_\sigma\, d\sigma = \int \mathrm{Curl}_n\, \mathbf{F} \cdot dS = \int \mathrm{Curl}_{ikl}\, \mathbf{F} \cdot dS^{ikl} \qquad (138\,\mathrm{b})$$

$$\int f_n\, dS = \int \mathrm{Div}\, f \cdot d\Sigma = \int \mathfrak{Div}\, f \cdot dx \qquad (139\,\mathrm{a})$$

$$\int F_N\, d\sigma = \int \mathrm{Div}_n\, \mathbf{F} \cdot dS = \int \sum_{(iklm)} \mathfrak{Div}^i\, \mathbf{F} \cdot dS^{klm}, \qquad (139\,\mathrm{b})$$

where

$$\mathrm{Curl}_{ik}\, f = \frac{\partial f_k}{\partial x^i} - \frac{\partial f_i}{\partial x^k} \qquad (140\,\mathrm{a})$$

$$\mathrm{Curl}_{ikl}\, \mathbf{F} = \frac{\partial F_{ik}}{\partial x^l} + \frac{\partial F_{li}}{\partial x^k} + \frac{\partial F_{kl}}{\partial x^i} \qquad (140\,\mathrm{b})$$

and

$$\mathfrak{Div}\, f = \frac{\partial \mathfrak{f}^i}{\partial x^i} \qquad \left(\mathrm{Div}\, f = \frac{1}{\sqrt{g}} \frac{\partial \sqrt{g}\, f^i}{\partial x^i} \right) \qquad (141\,\mathrm{a})$$

$$\mathfrak{Div}^i\, \mathbf{F} = \frac{\partial \mathfrak{F}^{ik}}{\partial x^k} \qquad \left(\mathrm{Div}^i\, \mathbf{F} = \frac{1}{\sqrt{g}} \frac{\partial \sqrt{g}\, F^{ik}}{\partial x^k} \right). \qquad (141\,\mathrm{b})$$

The important point here is that the invariance of the initial integral also implies the invariance of the final integral. But this can only be the case if the integrand itself is invariant everywhere, since the region of integration can always be chosen arbitrarily small. It follows that Curl_{ik} and $\mathrm{Curl}_{ikl}\, \mathbf{F}$ are the covariant components of a "surface" and "volume" tensor, respectively, that $\mathfrak{Div}\, f$ is a scalar density, and that $\mathfrak{Div}^i\, \mathbf{F}$ are the contravariant components of a vector density. These properties of the operations Curl and $\mathfrak{Div}$ can be summarized by the following rules:

(a) The operation Curl raises the "grade" of a tensor (cf. § 11), the operation $\mathfrak{Div}$ lowers it.

(b) In operation Curl the covariant components of the tensor, in operation $\mathfrak{Div}$ the contravariant components of the tensor density, are differentiated. We can add a third rule:

(c) The operations Curl and $\mathfrak{Div}$ are dual to each other. This follows from relations (137). For example

$$\mathrm{Curl}_{ikl}\ \mathbf{F} = \mathfrak{Div}^m\ \mathbf{F^*}, \tag{142}$$

as can easily be verified.

As in ordinary vector analysis, the operations Grad, Curl, and $\mathfrak{Div}$ can also be combined, and one obtains

$$\mathrm{Curl\ Grad}\,\phi = \mathfrak{Div}\ \mathfrak{Div}\ \mathbf{F} = \mathrm{Curl\ Curl}\,f = 0. \tag{143}$$

By operating successively with Div and Grad on a scalar ϕ, one obtains the generalization of the Laplace operator Δ. Following a suggestion of Cauchy's, it is denoted by $\square$. In the theory of invariants of n-dimensional manifolds it had already been introduced by Beltrami[94a]; its first application to the case of the special theory of relativity occurs in Poincaré's work. It is to be noted that after forming the Grad one has to go over to the contravariant components of the vector density, because of (141 a),

$$\square\phi = \mathrm{Div\ Grad}\,\phi = \frac{1}{\sqrt{g}}\ \frac{\partial}{\partial x^k}\left(\sqrt{g}\,g^{ik}\frac{\partial\phi}{\partial x^i}\right). \tag{144}$$

For constant g_{ik} this becomes

$$\square\phi = g^{ik}\frac{\partial^2\phi}{\partial x^i\,\partial x^k}. \tag{144a}$$

In this special case one can also derive a new vector from the vector f_i by means of the operation $\square$. For, just as in ordinary vector analysis,

$$\mathrm{Div}_i\ \mathrm{Curl}\,f = \mathrm{Grad}_i\ \mathrm{Div}\,f - \square f_i. \tag{145}$$

This relation cannot, however, be generalized for the case where the g_{ik} are not constant.

Finally it should be remarked that the scheme of tensors, introduced in § 11 for geometrical reasons, has turned out to be well justified for our calculations here. Tensors of first rank are distinguished analytically from those of higher rank, because new tensors can be formed from them by differentiation, without making use of the geodesic components of the reference system.

20. Derivation of invariant differential operations, using geodesic components

We now come to the second group of differential operations, for which the concept of parallel displacement plays an important rôle. For physical

[94a] E. Beltrami, 'Sulla teoria generale dei parametri differenziali', *Mem. R. Accad. Bologna* (2) **8** (1869) 549.

applications, only two of these operations are of importance, namely those which correspond to the operations

$$a_{ik} = \frac{\partial a_i}{\partial x^k}$$

and

$$\overset{\cdot}{t_i} = \frac{\partial t_i{}^k}{\partial x^k} \quad \text{(Div of a tensor of second rank)}$$

of the affine group. To obtain such expressions for the general transformation group, we have to make the following construction. First of all, let a vector with components a^i be given at every point of the curve $x^k = x^k(t)$. If P is an arbitrary point on the curve, we can displace the vector $a^i(P)$ parallel to itself along the curve and thus construct a second set of vectors $\bar{a}^i(P')$ (P' arbitrary). Then $\bar{a}^i$ and a^i coincide at P,

$$\bar{a}^i(P) \equiv a^i(P).$$

A vector can now be defined in an invariant manner by setting

$$A^i = \lim_{P' \to P} \frac{a^i(P') - \bar{a}^i(P')}{\Delta t},$$

since the numerator is the difference of two vectors *at the same point*. From (64) and (70) we have immediately,

$$A^i = \frac{da^i}{dt} + \Gamma^i{}_{rk} a^r \frac{dx^k}{dt} \tag{146a}$$

and

$$A_i = \frac{da_i}{dt} - \Gamma^r{}_{ik} a_r \frac{dx^k}{dt} \tag{146b}$$

Setting the arc length s in place of t, and the tangential vector $u^i = dx^i/ds$ in place of a^i, one obtains in this way the "acceleration" vector whose components B^i agree with the left-hand side of equation (80),

$$B^i = \frac{d^2 x^i}{ds^2} + \Gamma^i{}_{rs} \frac{dx^r}{ds} \frac{dx^s}{ds}. \tag{147}$$

If a^i is not only given along a curve, but as a vector field, then $da^i/dt = (\partial a^i/\partial x^k)(dx^k/dt)$, and it follows from (146) that a vector

$$A_i = a_{ik} \frac{dx^k}{dt}, \qquad A^i = a^i{}_k \frac{dx^k}{dt}$$

is associated with each direction dx^k/dt. Thus

$$a^i{}_k = \frac{\partial a^i}{\partial x^k} + \Gamma^i{}_{rk} a^r \tag{148a}$$

$$a_{ik} = \frac{\partial a_i}{\partial x^k} - \Gamma^r{}_{ik} a_r \tag{148b}$$

form the components of a tensor. This tensor is the required generalization of the tensor $\partial a_i/\partial x^k$ of the affine group.

A vector field a^i for which the associated tensor a_{ik} vanishes at a point P is called *stationary* at that point. According to §§ 16 and 18 there exist in Euclidean space—and *only* there—vector fields which are stationary at all points of a finite region.

Since the set of quantities a_{ik} are neither symmetrical nor skew-symmetrical, we are not dealing here with a tensor in the geometrical sense of § 11, but only with a tensor in the wider sense of § 9. We can split a_{ik} into a skew-symmetrical part

$$\frac{1}{2}\left(\frac{\partial a_i}{\partial x^k} - \frac{\partial a_k}{\partial x^i}\right)$$

and a symmetrical part

$$\hat{a}_{ik} = \frac{1}{2}\left(\frac{\partial a_i}{\partial x^k} + \frac{\partial a_k}{\partial x^i}\right) - \Gamma^r_{ik}\,a_r. \tag{148c}$$

With the help of stationary vector fields we can now derive the divergence of a tensor of second rank, T^{ik}, following Weyl's[95] procedure. Let ξ^i be a vector field stationary at P, so that at this point

$$\frac{\partial \xi^i}{\partial x^k} = -\,\Gamma^i_{rk}\,\xi^r$$

and

$$\frac{\partial \xi_i}{\partial x^k} = \Gamma^r_{ik}\,\xi_r.$$

We then form the divergence of the vector

$$f^i = T^{ik}\xi_k = T^i_k\,\xi^k,$$

using (141a). If we now substitute the above derivatives of the ξ_i, we obtain

$$\mathfrak{Div}\, f = \frac{\partial \mathfrak{f}^i}{\partial x^i} = \mathfrak{Div}_i \mathfrak{T}\cdot \xi^i = \mathfrak{Div}^i \mathfrak{T}\cdot \xi_i\,, \tag{149}$$

where

$$\mathfrak{Div}_i \mathfrak{T} = \frac{\partial \mathfrak{T}_i{}^k}{\partial x^k} - \mathfrak{T}_r{}^s\,\Gamma^r_{is} = \frac{\partial \mathfrak{T}_i{}^k}{dx^k} - \frac{1}{2}\frac{\partial g_{rs}}{\partial x^i}\mathfrak{T}^{rs}\,, \tag{150a}$$

$$\mathfrak{Div}^i \mathfrak{T} = \frac{\partial \mathfrak{T}^{ik}}{\partial x^k} + \mathfrak{T}^{rs}\,\Gamma^i_{rs}. \tag{150b}$$

$\mathfrak{T}$ is the tensor density corresponding to **T**, and it follows from the invariance of (149) that (150a) and (150b) are the covariant and contravariant components, respectively, of a vector density.

In Euclidean space the divergence of a tensor of rank 2 can also be interpreted in a different way. Let r^i and s^i be two unit vectors and let $T_{(rs)} = T_{ik}\,r^i s^k$ be the component of the tensor in these two directions. If r^i is arbitrarily given at P, then, in Euclidean space, one can associate

[95] H. Weyl, *Raum-Zeit-Materie* (3rd edn., 1920), p. 104.

with this direction a parallel direction $\bar{r}^i$ at P' in a unique and invariant manner. The vector field $\bar{r}^i$ is obviously stationary everywhere and can be used for the ξ^i in (149), so that

$$\mathfrak{Div}(\mathfrak{T}\bar{r}) = \mathfrak{Div}_{\bar{r}}\mathfrak{T}.$$

Putting now $\mathfrak{f} = (\mathfrak{T}\bar{r})$ in (139 a) we obtain directly

$$\int T_{(\bar{r}n)}dS = \int \mathfrak{Div}_{\bar{r}}\mathfrak{T}\,dx = \int \mathrm{Div}_{\bar{r}}\mathfrak{T}\,d\Sigma, \tag{151}$$

$$\mathrm{Div}_{\bar{r}}\mathfrak{T} = \lim_{S\to 0}\frac{\int T_{(\bar{r}n)}dS}{\int d\Sigma}, \tag{151a}$$

a formula which was derived by Lang[96]. For this purpose, every non-Euclidean space can be replaced by a Euclidean space tangential to it. This is possible because the second derivatives of the g_{ik} do not appear in the final result (150) and because the first derivatives of the g_{ik} can always be brought to coincidence in the two spaces, by a suitable choice of the coordinates. Therefore the result of the limiting process (151 a), i.e. the vector character of $\mathfrak{Div}_i\,\mathfrak{T}$, can be taken to be generally valid, although the integral on the right-hand side has a meaning only in Euclidean space.

For completeness we shall also quote a general formula, which is not of importance for physics, though. From a tensor $a^{ikl\cdots}{}_{rst\cdots}$ one obtains, by differentiation, a tensor of higher rank,

$$a^{ikl\cdots}{}_{rst\cdots,p} = \frac{\partial a^{ikl\cdots}{}_{rst\cdots}}{\partial x^p} + \Gamma^i{}_{\rho p}a^{\rho kl\cdots}{}_{rst\cdots} + \Gamma^k{}_{\rho p}a^{i\rho l\cdots}{}_{rst\cdots} +$$
$$+ \ldots - \Gamma^\rho{}_{pr}a^{ikl\cdots}{}_{\rho st\cdots} - \Gamma^\rho{}_{ps}a^{ikl\cdots}{}_{r\rho t\cdots} - \ldots. \tag{152}$$

This operation, which is already found in Christoffel's work, is called covariant differentiation by Ricci and Levi-Civita.

Formerly it was used to derive the divergence of a tensor of second rank. With the help of (152), the tensor $T^{ik}{}_l$ was formed by differentiating T^{ik} and then contracted,

$$\mathrm{Div}^i T = T^{ik}{}_k.$$

It should also be mentioned here how Ricci and Levi-Civita[56a] obtained an expression for the curvature tensor. Starting with an arbitrary vector a_i, one forms a_{ik} using (148 b), and then $a_{ik,l}$ using (152). On the right-hand side, then, there are terms which contain the a_i only, and also terms which contain their first and second derivatives. The latter cancel, though, if one forms the difference $a_{ik,l} - a_{il,k}$ and one is then just left with

$$a_{ik,l} - a_{il,k} = -R^h{}_{ikl}a_h.$$

With this, the tensor character of the set of quantities $R^h{}_{ikl}$ is proved.

[96] H. Lang, dissertation (Munich 1919).
[56a] Cf. footnote 56, p. 24, *ibid.*, See also the derivation by Einstein (footnote 56, *Ann. Phys., Lpz., loc. cit.*).

This method, however, does not give us any insight into its inherent geometrical significance.†

21. Affine tensors and free vectors

Although the general theory of relativity only deals with equations which are covariant with respect to arbitrary transformations of the coordinates, nevertheless certain sets of quantities are of importance in it which behave like tensors only under linear (affine) coordinate transformations. We shall call them *affine tensors*. Such affine tensors, for example, are the geodesic components. But there also exist affine tensors $U_i{}^k$, in particular, whose corresponding tensor densities $\mathfrak{U}_i{}^k = U_i{}^k \sqrt{g}$ satisfy the equations

$$\frac{\partial \mathfrak{U}_i{}^k}{\partial x^k} = 0 \tag{153}$$

in every reference system. It is clear that the $U_i{}^k$ cannot transform in a linear-homogeneous manner for general coordinate transformations. But a second set of quantities J_k can be derived from the $U_i{}^k$ by integration, which behaves like a vector under a transformation group which is much more general than the affine group.

To show this, it will be of help to consider first the following case. Let a four-vector s^k, with its associated vector density $\mathfrak{s}^k$, be given, whose Div vanishes everywhere,

$$\frac{\partial \mathfrak{s}^k}{\partial x^k} = 0. \tag{154}$$

Also, $\mathfrak{s}^k$ is supposed to have non-zero values only inside a "world-canal", or at least to decrease outside it sufficiently rapidly, so that integrals over regions outside the world-canal, and sufficiently far removed, will vanish. Moreover, we shall only consider those coordinate systems in which spaces of constant time $x^4 = \text{const.}$ intersect the world-canal in simply-connected regions. We now make use of the fact that, according to (139 a) and (154), the integral $\int s_n \, dS$ always vanishes when it is taken over a closed space region. We choose, first of all, as region of integration two hyperplanes $x^4 = \text{const.}$ which we can think of as connected by space elements lying outside the world-canal. Because of (137 a), the integral

$$J = \int \mathfrak{s}^4 \, dx^1 \, dx^2 \, dx^3. \tag{155}$$

has then the same value for both hyperplanes, in other words, it is independent of x^4. Next, we introduce a second coordinate system K'. Inside the world-canal, this has only to satisfy the condition that surfaces $x'^4 = \text{const.}$ intersect the world-canal in simply-connected regions. Outside, it has to have constant g_{ik}. We now take as region of integration the hyperplanes $x^4 = \text{const.}$ and $x'^4 = \text{const.}$ These can always be chosen in such

† See suppl. note 7.

a way that they do not intersect each other. Then,

$$\int \mathfrak{s}^4 dx^1 dx^2 dx^3 = \int \mathfrak{s}'^4 dx'^1 dx'^2 dx'^3,$$

i.e. the integral J is invariant under all coordinate transformations considered here.

The problem of an integral over the components of an affine tensor can now be reduced to the above case. We multiply such an affine tensor with a vector, p^k, whose components inside a world-canal are constants,

$$U^k = U_i{}^k p^i.$$

Under all linear transformations, U^k behaves like a vector. In all coordinate systems K' which are produced from the original system K by such transformations, the components p'^i, too, are constant inside the world-canal. Thus the equation

$$\frac{\partial \mathfrak{U}'^k}{\partial x'^k} = 0$$

also holds in K'. It follows from (155) that the integral

$$J = \int \mathfrak{U}^4 dx^1 dx^2 dx^3$$

is invariant with respect to linear transformations and has the same value for each cross section. But since

$$J = J_k p^k,$$

where

$$J_k = \int \mathfrak{U}_k{}^4 dx^1 dx^2 dx^3, \tag{156}$$

and where the vector p^k has been quite arbitrary, the quantities J_k exhibit a vector character under linear transformations.[97]

We are now going to show, following Einstein's[98] procedure, that they retain this vector character also if we go over from K to an arbitrary coordinate system K' which coincides with K outside the world-canal. For this purpose we need only construct a coordinate system which coincides with K on *one* hyperplane $x''^4 = c_1$, and with K' on another hyperplane $x''^4 = c_2$. Since it has already been proved that for two different hyperplanes $x^4 = $ const. of the same coordinate system the J_k have the same value, we have thus shown that the J_k have the same value in K and K'. Thus, *the J_k are quite independent of the choice of the coordinate system inside the world-canal*. It is interesting that, starting with the affine tensor $U_i{}^k$ which behaves covariantly only under *linear* coordinate transformations, one should obtain by integration a set of quantities J_k which behave as a vector under a *much more general* transformation group. The vector J_k is distinguished from ordinary vectors by not being related to a given

[97] This was first proved by F. Klein (Cf. footnote 94, p. 54, *ibid*, where free vectors are discussed in detail). The derivation given here is due to H. Weyl, *Raum-Zeit-Materie* (3rd edn., 1920), p. 234.
[98] A. Einstein, *S.B. preuss. Akad. Wiss.* (1918) 448.

point. We shall follow Klein in calling it a *free vector* [non-localized vector], in analogy to the terminology used in mechanics.

22. Reality relations

Up to now we had tacitly assumed that ds^2 was a definite form. This is by no means the case in the actual space-time world, where the normal form of ds^2 has three positive and one negative sign. Formally, all results previously obtained hold also for this case, since one can reduce ds^2 to a definite form by introducing an imaginary coordinate (see § 7). The geometrical interpretation of the formulae, however, will have to be changed.

Let us, first of all, consider the case where the special theory of relativity is valid, and introduce, as fourth coordinate, $x^4 = ct$. Then, for a given origin of the coordinate system, the world can be divided in a Lorentz-invariant way into two parts, which are characterized by

$$x_1{}^2 + x_2{}^2 + x_3{}^2 - x_4{}^2 < 0 \quad \text{(past and future)} \quad (A)$$

and $\qquad\qquad x_1{}^2 + x_2{}^2 + x_3{}^2 - x_4{}^2 > 0 \text{ (intermediate region) } (B)$

They are separated by the cone

$$x_1{}^2 + x_2{}^2 + x_3{}^2 - x_4{}^2 = 0, \quad \text{(light cone)} \quad (C)$$

on which the world lines of the light rays lie.

If one lets the starting point of a vector coincide with the origin of the coordinate system, the vector is called space-like if its end point lies in world region (B), time-like, if it is in (A). It is called a null vector (vector of magnitude zero) if it lies on the cone (C). Actually, because of the changed sign of the fourth coordinate, the Lorentz transformation cannot be looked upon as a rotation of the coordinate system, but as a transformation of one system of conjugate diameters of the hyperboloid

$$x_1{}^2 + x_2{}^2 + x_3{}^2 - x_4{}^2 = 1$$

into another. (This interpretation of the Lorentz transformation, and also the terminology employed here, occur first in Minkowski's work.) By means of a simple geometrical consideration, or by a straightforward application of formula (I) for the Lorentz transformation, it can be shown that for a suitable choice of the coordinates one can always obtain spatial coincidence for points in region (A), time coincidence (simultaneity) for those in (B), with the origin. This amounts to essentially the same thing as saying that, with a suitable choice of the coordinates, the time component of a space-like vector, or all the space components of a time-like vector can always be made to vanish. From the results of § 6 it also follows that only world points of the type (A) can have a causal connection with the origin. We shall introduce Klein's and Hilbert's terminology by calling the geometry which was discussed here, and which is determined by the line element

$$ds^2 = (dx^1)^2 + (dx^2)^2 + (dx^3)^2 - (dx^4)^2,$$

a *pseudo-Euclidean* geometry.

Analogous distinctions between the geometries of positive-definite and of indefinite line elements apply in the case of general Riemannian geometry. Let us construct all geodesic lines which start from a point P_0 and which satisfy at P_0 the conditions

$$g_{ik} \frac{dx^i}{dt} \frac{dx^k}{dt} < 0, \qquad (A')$$

or

$$g_{ik} \frac{dx^i}{dt} \frac{dx^k}{dt} > 0, \qquad (B')$$

or

$$g_{ik} \frac{dx^i}{dt} \frac{dx^k}{dt} = 0 \qquad (C')$$

(t = curve parameter). They span certain world regions, or the surface of the light cone (C), in a continuous manner. The associated directions (vectors) at P_0 are again called time-like, space-like and null-directions (null vectors), respectively.

This division of the space–time world carries with it a restriction on the permissible point transformations, as was stressed by Hilbert[99]. The reason for this is that, in permissible coordinate systems, the first three coordinate axes have always to be in space-like directions, the fourth in a time-like direction. This condition is fulfilled, first, when the quadratic form obtained from ds^2 by setting dx^4 equal to zero is positive-definite, for which the conditions are

$$g_{11} > 0, \qquad \begin{vmatrix} g_{11} & g_{12} \\ g_{21} & g_{22} \end{vmatrix} > 0, \qquad \begin{vmatrix} g_{11} & g_{12} & g_{13} \\ g_{21} & g_{22} & g_{23} \\ g_{31} & g_{32} & g_{33} \end{vmatrix} > 0$$

and, secondly, when

$$g_{44} < 0.$$

These inequalities must not be violated by the permissible coordinate transformations. Since the determinant g of the g_{ik} is always negative (this follows from the inequalities), we shall always have to replace $\sqrt{g}$ by $\sqrt{(-g)}$ in the tensor formulae which were developed for the definite case[99 a].

According to (A'), the arc length of a world line can also become imaginary. In fact, this is always so for the world line of a material body. For

[99] D. Hilbert, 'Grundlagen der Physik', 2. Mitt., *Nachr. Ges. Wiss. Göttingen* (1917) 53.
[99a] Minkowski (Cf. footnote 54, II, p. 21) and Klein (Cf. footnote 55 a, p. 23, *Phys. Z.*, *loc. cit.*) impose a further restriction on the permissible point transformations. $\partial x'^4/\partial x^4$ is to be always > 0, in other words, an interchange of past and future is to be excluded, in order that one should be dealing with a really continuous group. But the covariance with respect to this more restricted group entails already in a purely formal way the covariance with respect to time reversal, as long as the equations do not contain quite artificial irrationalities (on this latter point, see Part V). Moreover, the covariance of all physical laws under time reversal seems to be required also for physical reasons, according to our present-day view. For this reason we shall not make use of the above-mentioned restriction.

such a case it is therefore practicable to introduce, in place of the arc length, s, the *proper time* τ, given by

$$s = ic\tau. \tag{157}$$

This gives the time which is shown by a clock moving along this world line. For, in a coordinate system in which the clock is momentarily at rest, $d\tau = dt$. We also introduce in place of

$$u^i = \frac{dx^i}{ds}$$

the vector
$$u^i = \frac{dx^i}{d\tau}, \tag{158}$$

for which
$$g_{ik}u^i u^k = u_i u^i = -c^2. \tag{159}$$

Among the geodesic lines the geodesic *null lines*, which lie on the light cone (C), form an exception. While the variational principle (83) and the differential equations (80) apply to them, the variational principle (81) does not do so. For, first, the coordinates can no longer be represented as functions of the arc length in this case, because it vanishes. Another curve parameter, determined only to within an arbitrary factor, will therefore have to be chosen in (80), as well. Secondly, since the term $\sqrt{[g_{ik}(dx^i/d\lambda)(dx^k/d\lambda)]}$ vanishes, it can no longer be taken into the denominator, when deriving (83) from (81). Eq. (83) can therefore no longer be deduced from (81), and the geodesic null lines will have to be defined in a different way. The geodesic null lines are distinguished from all other curves lying on the light cone (C) by the existence of a curve parameter for which the differential equations

$$\frac{d^2x^i}{d\lambda^2} + \Gamma^i_{rs}\frac{dx^r}{d\lambda}\frac{dx^s}{d\lambda} = 0$$

as well as the variational principle (83) hold. On the other hand, for those geodesic lines which are not null lines, the derivations of § 15 remain unchanged.

Vermeil's result, on the connection of the volume of a sphere in Riemannian space with the curvature invariant (§ 17), too, cannot be immediately applied to the indefinite case, because here the sphere corresponds to a hyperboloid of infinite extent.

It should also be mentioned that usually in the special theory of relativity the normal form of the line element contains, by definition, three positive signs and one negative sign, whereas in the general theory of relativity one assumes three negative signs and one positive sign. We shall use the former convention throughout.

23. Infinitesimal coordinate transformations and variational theorems

If, in general, a quantity is invariant under coördinate transformations, then, in particular, it will also be invariant under *infinitesimal* coordinate transformations. The usefulness of a study of the latter lies in the fact that from its invariance under such transformations certain differential equations can be derived, which the quantity will have to satisfy. We now define such an infinitesimal coordinate transformation by

$$x'^i = x^i + \epsilon \xi^i(x) \tag{160}$$

where ϵ is an infinitely small quantity. The dependence of the ξ^i on the coordinates can be quite arbitrary. All differences between primed and unprimed functions will have to be thought of as developed in powers of ϵ, in what is to follow. Eventually, we shall only be interested in the first-order term, which is called the variation of the function in question. To obtain the variation of an arbitrary tensor for the transition from the unprimed to the primed coordinate system, we shall have to substitute the values

$$\alpha_k{}^i = \frac{\partial x'^i}{\partial x^k} = \delta_k{}^i + \epsilon \frac{\partial \xi^i}{\partial x^k}; \qquad \bar\alpha_i{}^k = \frac{\partial x^k}{\partial x'^i} = \delta_i{}^k - \epsilon \frac{\partial \xi^k}{\partial x^i} \tag{161}$$

into the general transformation formula (25). These values follow from $(\partial x'^i / \partial x^k)(\partial x^k / \partial x'^\alpha) = \delta_\alpha{}^i$ and are of course only correct up to first order in ϵ. Let us also write down the value of the transformation determinant,

$$\det\left|\frac{\partial x'^i}{\partial x^k}\right| = 1 + \epsilon \frac{\partial \xi^i}{\partial x^i}, \qquad \det\left|\frac{\partial x^k}{\partial x'^i}\right| = 1 - \epsilon \frac{\partial \xi^i}{\partial x^i}. \tag{162}$$

In this way one obtains for the variation of a vector,

$$\delta a^i = \epsilon \frac{\partial \xi^i}{\partial x^r} a^r, \qquad \delta a_i = -\epsilon \frac{\partial \xi^r}{\partial x^i} a_r, \tag{163}$$

and for that of a tensor of second rank,

$$\delta a^{ik} = \epsilon \left(\frac{\partial \xi^i}{\partial x^r} a^{rk} + \frac{\partial \xi^k}{\partial x^r} a^{ir} \right),$$

$$\delta a_k{}^i = \epsilon \left(\frac{\partial \xi^i}{\partial x^r} a_k{}^r - \frac{\partial \xi^r}{\partial x^k} a_r{}^i \right), \tag{164}$$

$$\delta a_{ik} = -\epsilon \left(\frac{\partial \xi^r}{\partial x^i} a_{rk} + \frac{\partial \xi^r}{\partial x^k} a_{ir} \right).$$

Corresponding formulae hold, in particular, for the variation of g_{ik}. It should also be noted that, for an arbitrary symmetrical system of numbers t_{ik}, it follows from (72) that

$$t_{ik} \delta g^{ik} = - t^{ik} \delta g_{ik} \tag{165}$$

(As usual, we have put here $t^{ik} = g^{i\alpha} g^{k\beta} t_{\alpha\beta}$). Equally, it follows from (73) that

$$\delta \sqrt{(-g)} = \tfrac{1}{2} \sqrt{(-g)}\, g^{ik} \delta g_{ik} = -\tfrac{1}{2} \sqrt{(-g)}\, g_{ik} \delta g^{ik}. \tag{73 b}$$

In (163) and (164) we are always dealing with a variation of the type

$$\delta a^i = a'^i(x') - a^i(x), ... \qquad \delta a^{ik} = a'^{ik}(x') - a^{ik}(x), ... \text{ etc.} \qquad (166)$$

Essentially different from this is the variation

$$\delta^* a^i = a'^i(x) - a^i(x), ... \qquad \delta^* a^{ik} = a'^{ik}(x) - a^{ik}(x), ... \text{ etc.} \qquad (167)$$

Evidently, this is connected with (166) by means of the symbolic relation

$$\delta^* = \delta - \epsilon \frac{\partial}{\partial x^r} \xi^r, \qquad (168)$$

from which the expressions $\delta^* a^i$, $\delta^* a_i$, etc., are immediately obtained. From (164) and (167) follows the important formula

$$\frac{1}{2} \int \mathfrak{T}^{ik} \delta^* g_{ik} \, dx = \epsilon \int \left[-\mathfrak{T}_i{}^k \frac{\partial \xi^i}{\partial x^k} - \frac{1}{2} \frac{\partial g_{rs}}{\partial x^i} \mathfrak{T}^{rs} \xi^i dx \right]$$

or, using (150a),

$$\frac{1}{2} \int \mathfrak{T}^{ik} \delta^* g_{ik} \, dx = \epsilon \left[\int \mathfrak{Div}_i \mathfrak{T} \cdot \xi^i \, dx - \int \frac{\partial}{\partial x^k} (\mathfrak{T}_i{}^k \xi^i) dx \right]. \qquad (169)$$

Finally, let us consider the variation of the integral

$$J = \int \mathfrak{W}(x) \, dx.$$

It is

$$\delta J = \int_{X'} \mathfrak{W}'(x') \, dx' - \int_{X} \mathfrak{W}(x) \, dx = \int_{X} \mathfrak{W}'(x') \det \left| \frac{\partial x'^i}{\partial x^k} \right| dx - \int_{X} \mathfrak{W}(x) \, dx,$$

or, because of (162), and because $\mathfrak{W}'(x') = \mathfrak{W}'(x) + \epsilon (\partial \mathfrak{W}/\partial x^i) \xi^i$,

$$\delta \int \mathfrak{W} \, dx = \int \delta^* \mathfrak{W} \, dx + \epsilon \int \frac{\partial (\mathfrak{W} \xi^i)}{\partial x^i} \, dx. \qquad (170)$$

Here, $\delta^* \mathfrak{W} = \mathfrak{W}'(x) - \mathfrak{W}(x)$. If ξ^i vanishes at the boundary of the region of integration, the second term of the right-hand side of (170) does not contribute to $\delta \int \mathfrak{W} dx$, since it can be transformed into a surface integral over the boundary, according to (139a). *If now J is an invariant, in other words $\mathfrak{W}$ a scalar density, then the variation (170) has to vanish for arbitrary ξ^i.* One can first set up a general expression for $\delta \mathfrak{W}$ for an arbitrary variation of the field tensors from which $\mathfrak{W}$ is built up, and then perform this latter variation specifically by means of an infinitesimal change in the coordinate system, using (164). In this way, one can obtain certain identities from formula (170). In certain cases the ξ^i can in addition be assumed to vanish at the boundary, which simplifies the calculation. This will now be illustrated by means of the following examples. They will be worked through in full, because of their physical applications, later on.

(a) Let us form the Curl of a vector ϕ_i

$$F_{ik} = \frac{\partial \phi_k}{\partial x^i} - \frac{\partial \phi_i}{\partial x^k}, \qquad (171)$$

and from this derive the invariant

$$L = \tfrac{1}{2}F_{ik}F^{ik}.\tag{172}$$

If $\mathfrak{L}$ is the scalar density associated with L,

$$\mathfrak{L} = L\sqrt{(-g)},$$

one can derive from the invariant integral

$$\int \mathfrak{L}\,dx$$

a transformation which is of importance for the ponderomotive force in electrodynamics. We shall limit ourselves to those variations of the fields and coordinates which vanish at the boundary of the region of integration. We consider, first of all, ϕ_i and g_{ik} as independent variables. If they are varied in the manner given above, a simple calculation, using (165), leads to

$$\delta\mathfrak{L} = \mathfrak{F}^{ik}\,\delta F_{ik} - \mathfrak{S}^{ik}\,\delta g_{ik},$$

where

$$S_{ik} = \frac{1}{\sqrt{(-g)}}\,\mathfrak{S}_{ik} = F_{ri}F_{sk}\,g^{rs} - \tfrac{1}{4}F_{rs}F^{rs}g_{ik}.\tag{173}$$

An integration by parts then gives

$$\delta\int \mathfrak{L}\,dx = \int (2\mathfrak{s}^i\,\delta\phi_i - \mathfrak{S}^{ik}\,\delta g_{ik})\,dx,\tag{174}$$

with

$$\mathfrak{s}^i = \frac{\partial\mathfrak{F}^{ik}}{\partial x^k},\tag{175}$$

from which also follows

$$\frac{\partial\mathfrak{s}^i}{\partial x^i} = 0.\tag{175a}$$

We now generate the variations $\delta\phi_i$ and δg_{ik} specifically from an infinitesimal coordinate transformation. This can be done by replacing, in (174), $\delta\phi_i$ and δg_{ik} by $\delta^*\phi_i$ and δ^*g_{ik}, since $\delta\phi_i$ and δg_{ik} vanish at the boundary, and because of (170). From (163) and (168) we first of all obtain

$$\mathfrak{s}^i\,\delta^*\phi_i = -\,\epsilon\!\left(\mathfrak{s}^k\frac{\partial\phi_k}{\partial x^i}\xi^i + \mathfrak{s}^k\phi_i\frac{\partial\xi^i}{\partial x^k}\right)$$

and, after integrating by parts, because of (169) and (175 a),

$$0 = \int (2\mathfrak{s}^i\delta^*\phi_i - \mathfrak{S}^{ik}\,\delta^*g_{ik})\,dx = -\,2\epsilon\int (F_{ik}\mathfrak{s}^k + \mathfrak{Div}_i\,\mathfrak{S})\xi^i\,dx.$$

Since the last expression has to vanish for arbitrary ξ^i, we have

$$F_{ik}\mathfrak{s}^k = -\,\mathfrak{Div}_i\,\mathfrak{S}$$

or, written out in full,

$$F_{ik}\mathfrak{s}^k = -\left(\frac{\partial\mathfrak{S}_i{}^k}{\partial x^k} - \frac{1}{2}\frac{\partial g_{rs}}{\partial x^i}\mathfrak{S}^{rs}\right).\tag{176}$$

This identity will be used in §§ 30 and 54.

(b) The variation of the invariant integral

$$\int \mathfrak{R}\, dx$$

corresponding to the curvature invariant R, becomes of particular interest because of the investigations by Lorentz[100], Hilbert[101], Einstein[102], Weyl[103], and Klein[104] on the rôle of Hamilton's Principle in the general theory of relativity, the physical significance of which will be discussed in Part IV.

Let us, first of all, split the invariant $\int \mathfrak{R}\, dx$ into a volume integral which only contains *first* derivatives of the g_{ik}, and a surface integral, i.e.

$$\int \mathfrak{R}\, dx = - \int \mathfrak{G}\, dx + \int_{\text{surface}} (\ldots), \tag{177}$$

where

$$\mathfrak{G} = \sqrt{(-g)}\, G, \qquad G = g^{ik}(\Gamma^r_{is}\Gamma^s_{kr} - \Gamma^r_{ik}\Gamma^s_{rs}). \tag{178}$$

G is evidently only invariant under linear transformations, i.e. it is an affine scalar. Beyond this, however, the integral $\int \mathfrak{G}\, dx$ is, according to (177), invariant under all transformations which apply only to the interior of the region of integration and leave the boundary values of the coordinates and of the g_{ik} and their derivatives unchanged. These two invariance properties of $\int \mathfrak{G}\, dx$ will now be used to derive some mathematical identities which are of importance for the theory. Since now the integrand of the integral to be varied no longer contains second derivatives of the g_{ik}, this simplifies matters considerably, though it is not essential for the method to be described below.

For an arbitrary variation of the g_{ik}-field we obtain first of all (using the abbreviation $g^{ik}{}_\sigma = (\partial g^{ik}/\partial x^\sigma)$),

$$- \int \delta\mathfrak{G}\, dx = - \int \left(\frac{\partial \mathfrak{G}}{\partial g^{ik}}\delta g^{ik} + \frac{\partial \mathfrak{G}}{\partial g^{ik}{}_\sigma}\delta g^{ik}{}_\sigma \right) dx$$

$$= \int \left(\frac{\partial}{\partial x^\sigma}\frac{\partial \mathfrak{G}}{\partial g^{ik}{}_\sigma} - \frac{\partial \mathfrak{G}}{\partial g^{ik}} \right)\delta g^{ik}\, dx - \int \frac{\partial}{\partial x^\sigma}\left(\frac{\partial \mathfrak{G}}{\partial g^{ik}{}_\sigma}\delta g^{ik} \right) dx.$$

[100] H. A. Lorentz, *Versl. gewone Vergad. Akad. Amst.*, **23** (1915) 1073; **24** (1916) 1389 and 1759; **25** (1916) 468 and 1380.

[101] D. Hilbert, 'Grundlagen der Physik', 1. Mitt., *Nachr. Ges. Wiss. Göttingen* (1915) 395.

[102] A. Einstein, *S.B. preuss. Akad. Wiss.* (1916) 1115 (Also reprinted in the collection, Lorentz, Einstein, Minkowski, *Das Relativitätsprinzip* (3rd edn., Leipzig 1920)).

[103] H. Weyl, *Ann. Phys., Lpz.*, **54** (1917) 117; *Raum-Zeit-Materie*, (1st edn., 1918 and 3rd edn., 1920).

[104] F. Klein, 'Zu Hilberts erster Note über die Grundlagen der Physik', *Nachr. Ges. Wiss. Göttingen* (1917) 469; 'Über die Differentialgesetze von Impuls und Energie in der Einstein-schen Gravitationstheorie', *ibid.* (1918) 235. The simplification of Klein's (as compared with Hilbert's) approach is due to his using also variations of the coordinates which do *not* vanish at the boundary of the region of integration. Since Lagrange, this has frequently been done in classical mechanics, too. In this way, many of the relations become more transparent. Lorentz (Cf. footnote 100, *ibid.*) already uses a similar procedure, though not so systematically.

An explicit evaluation now shows[105] that

$$\frac{\partial}{\partial x^{\sigma}} \frac{\partial \mathfrak{G}}{\partial g^{ik}_{\sigma}} - \frac{\partial \mathfrak{G}}{\partial g^{ik}} = \mathfrak{G}_{ik} = \sqrt{(-g)}\, G_{ik},$$

where G_{ik} is the tensor defined in (109).† We thus obtain

$$-\int \delta\mathfrak{G}\, dx = \int \mathfrak{G}_{ik} \delta g^{ik}\, dx - \int \frac{\partial}{\partial x^{\sigma}} \left(\frac{\partial \mathfrak{G}}{\partial g^{ik}_{\sigma}} \delta g^{ik} \right) dx. \tag{179}$$

Since the last integral can obviously be written as a surface integral, we also have, from (177),

$$\delta \int \mathfrak{R}\, dx = \int \mathfrak{G}_{ik} \delta g^{ik}\, dx + \int\limits_{\text{surface}} (\ldots). \tag{180}$$

We now produce the variation of the g_{ik} specifically from a variation δ^* of the coordinate system. Because of (169) and (170), we then obtain from (179),

$$\delta^* \int \mathfrak{G}\, dx = 2\epsilon \int \mathfrak{Div}_i\, \mathfrak{G} \cdot \xi^i\, dx + \epsilon \int \frac{\partial}{\partial x^k} \left(\frac{\partial \mathfrak{G}}{\partial g^{rs}_k} \delta^* g^{rs} - 2\mathfrak{G}_i{}^k \xi^i + \mathfrak{G}\xi^k \right) dx. \tag{181}$$

Next, we specialize the infinitesimal coordinate transformation still further in such a way that it leaves $\int \mathfrak{G}\, dx$ invariant.

(i) *The ξ^i are to vanish at the boundary.* It then follows that

$$\mathfrak{Div}_i\, \mathfrak{G} = \frac{\partial \mathfrak{G}_i{}^k}{\partial x^k} - \frac{1}{2} \frac{\partial g_{rs}}{\partial x^i} \mathfrak{G}^{rs} \equiv 0 \tag{182a}$$

$$\mathfrak{Div}^i\, \mathfrak{G} = \frac{\partial \mathfrak{G}^{ik}}{\partial x^k} + \mathfrak{G}^{rs} \Gamma^i_{rs} \equiv 0. \tag{182b}$$

If we had only wanted to derive this identity, we could have shortened the calculation quite considerably.‡ Herglotz[82] pointed out that the identity leads very simply to an interesting theorem, which had previously been derived by Schur[105a] in a different way. If, analogously to (116),

$$R_{hijk} = -\alpha(g_{hj}\, g_{ik} - g_{ij}\, g_{hk}),$$

where α can however still be a function of the coordinates, it follows by substitution of (119) in (182a) immediately that, for $n > 2$,

$$\frac{\partial \alpha}{\partial x^i} = 0, \qquad \alpha = \text{const.}$$

This means that *if the curvature of a Riemannian space R_n $(n > 2)$ is independent of the surface direction at each of its points, it is also independent of the position.*

[105] For details, see H. Weyl, *Raum-Zeit-Materie* (1st edn., 1918), p. 191, (3rd edn., 1920), pp. 205, 206; also A. Palatini, *R. C. Circ. mat. Palermo* **43** (1919) 203.

† See suppl. note 8.

‡ See suppl. note 7.

[105a] F. Schur, *Math. Ann.* **27** (1886), 537.

(ii) *The ξ^i are to be constants.* We could even more generally take the ξ^i to be linear functions of the coordinates, but the further identities which would result from this are not of importance. Since now the first integral in (181) can be omitted because of (182), the second integral must also vanish identically for constant ξ^i. But this is only possible provided the integrand vanishes identically for this case, since the region of integration can be assumed arbitrarily small. For constant ξ^i we now have to put $\delta^* g^{rs} = -g^{rs}{}_i\,\xi^i$, because of (164) and (168). The integrand then takes the form

$$\xi^i\frac{\partial}{\partial x^k}\left(-\frac{\partial \mathfrak{G}}{\partial g^{rs}{}_k}g^{rs}{}_i - 2\mathfrak{G}_i{}^k + \mathfrak{G}\,\delta_i{}^k\right).$$

We can finally put

$$\mathfrak{U}_i{}^k = \frac{1}{2}\left(\frac{\partial \mathfrak{G}}{\partial g^{rs}{}_k}g^{rs}{}_i - \mathfrak{G}\,\delta_i{}^k\right) \tag{183}$$

so that

$$\frac{\partial(\mathfrak{U}_i{}^k + \mathfrak{G}_i{}^k)}{\partial x^k} \equiv 0. \tag{184}$$

An evaluation of the expression (183) for $\mathfrak{U}_i{}^k$, using the value (178) for $\mathfrak{G}$, then gives

$$\mathfrak{U}_i{}^k = \frac{1}{2}\left\{\Gamma^r{}_{\alpha r}\frac{\partial[g^{\alpha k}\sqrt{(-g)}]}{\partial x^i} - \Gamma^k{}_{rs}\frac{\partial[g^{rs}\sqrt{(-g)}]}{\partial x^i} - \mathfrak{G}\,\delta_i{}^k\right\}. \tag{185}$$

For the case $\sqrt{(-g)} = \text{const.}$, this can also be written[106]

$$\mathfrak{U}_i{}^k = \sqrt{(-g)}U_i{}^k, \qquad U_i{}^k = \Gamma^k{}_{rs}\Gamma^r{}_{\alpha i}\,g^{\alpha s} - \tfrac{1}{2}G\delta_i{}^k. \tag{185a}$$

Evidently, we are dealing here with an affine tensor of the kind discussed in § 21. On its physical significance, see Part IV, §§ 57 and 61.†

[106] For details of calculation see A. Einstein, *Ann. Phys., Lpz.*, **49** (1916) 806, Eq. (50), for the case $\sqrt{(-g)} = \text{const.}$; W. Pauli, jr., *Phys. Z.* **20** (1919) 25, for the general case.
† See suppl. note 8 a.

PART III. SPECIAL THEORY OF RELATIVITY. FURTHER ELABORATIONS.

(a) KINEMATICS

24. Four-dimensional representation of the Lorentz transformation

The kinematic results of the theory of relativity, which we had discussed in Part I, can be presented much more clearly by making use of the four-dimensional space–time world. Two different representations can be employed side by side, the one imaginary,

$$x^1 = x, \quad x^2 = y, \quad x^3 = z, \quad x^4 = ict,$$

the other real,

$$x^1 = x, \quad x^2 = y, \quad x^3 = z, \quad x^4 = ct.$$

The first is historically the older and was already used by Poincaré[107], the second was used by Minkowski in his lecture 'Space and Time' ('Raum und Zeit'). The special Lorentz transformation (I), in which x^2 and x^3 remain unchanged, is given by the formulae

$$\left.\begin{aligned}
x'^1 &= x^1 \cos\phi + x^4 \sin\phi & x'^1 &= x^1 \cosh\psi - x^4 \sinh\psi \\
x'^4 &= -x^1 \sin\phi + x^4 \cos\phi & x'^4 &= -x^1 \sinh\psi + x^4 \cosh\psi \\
& & (\phi &= i\psi),
\end{aligned}\right\} \quad (186)$$

in complete analogy to a rotation of the coordinate system in R_3. The former first occur explicitly in Minkowski II (Eq. (1))—he writes $i\psi$ in place of ϕ. Since for $x'^1 = 0$ we must have $x = vt$, ϕ and ψ are determined by

$$\tan\phi = i\beta, \qquad \tanh\psi = \beta,$$

and hence

$$\left.\begin{aligned}
\cos\phi &= \frac{1}{\sqrt{(1-\beta^2)}}, & \cosh\psi &= \frac{1}{\sqrt{(1-\beta^2)}}, \\
\sin\phi &= \frac{i\beta}{\sqrt{(1-\beta^2)}}, & \sinh\psi &= \frac{\beta}{\sqrt{(1-\beta^2)}}.
\end{aligned}\right\} \quad (187)$$

In an imaginary coordinate system the special Lorentz transformation is a rotation, in a real one it stands for the transformation of one pair of conjugate diameters of the invariant hyperbola

$$(x^1)^2 - (x^4)^2 = 1$$

into another. In the former, there is no difference between the covariant and contravariant components of a vector. In the latter, $a_4 = -a^4$.

[107] Cf. footnote 11, p. 2, *R.C. Circ. mat. Palermo, loc. cit.*, p. 168.

and in general, for an arbitrary tensor, the raising or lowering of an index
4 results in a change of sign.

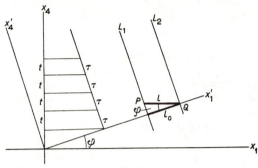

Fig. 2

The Lorentz contraction is illustrated in the right-hand part of Fig. 2,
in which $x^1 = x$ is the abscissa and $x^4 = ict$ the ordinate. The figure is
drawn as if x^4 were real.

L_1 and L_2 are the world lines of a rod at rest in system K', the distance
l_0 between them is equal to its rest length. In the moving system K,
the length l of the rod is to be looked upon as the distance between the
points of intersection P and Q of L_1 and L_2 with a line parallel to the
x^1-axis. Obviously

$$l = \frac{l_0}{\cos \phi}, \tag{188}$$

which agrees with (7), using (187).[108] In an analogous way, Einstein's
time dilatation is illustrated by the left-hand part of Fig. 2. Any periodic
occurrence can be used for a clock, assumed to be at rest in system K'.
The world points corresponding to a succession of periodic times lie on a
line which is parallel to the x'^4-axis. The distance between any two of them,
τ, is the length of the period normally measured. (For simplicity, the unit
of time is chosen such as to make the velocity of light equal to unity.)
The length t of the period, as measured in K, is then given by the projections of the lines of length τ on the x^4-axis. Thus

$$\tau = \frac{t}{\cos \phi}, \tag{189}$$

which is identical with (8), because of (187).

A simple generalization of this argument leads to an illustration of the
clock paradox (cf. § 5).[109] In Fig. 3, L_1 and L_2 are the world lines of the
clocks C_1 and C_2 which were discussed in § 5.

[108] The corresponding figure for the real coordinate system can be found in Minkowski's
lecture 'Raum und Zeit'.
[109] Cf. footnote 4 in Minkowski's lecture 'Raum und Zeit', in the collection of papers,
Das Relativitätsprinzip (Leipzig 1913), and M. v. Laue, *Phys. Z.* **13** (1912) 118.

The time τ shown by the clock C_2 at the world point Q (which coincides spatially with P when viewed in system K), is, apart from a factor $1/c$, equal to the length s of the curve L. We may generalize this for arbitrarily

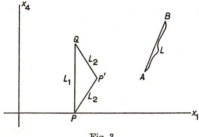

Fig. 3

moving clocks, provided the acceleration is not too great, and assume that they indicate the time τ, where

$$\tau = \int \sqrt{(1 - \beta^2)}\, dt = \frac{s}{ic} \qquad (157\,\text{a})$$

and where s is again the length of the corresponding world line. Evidently, τ is the *proper time* (defined by (157)) of the clock in question, i.e. it is the time measured by an observer moving with the clock. Of two clocks which are moved from world point A to world point B, the one which is moving uniformly will indicate the shortest time interval (cf. Fig. 3).

25. The addition theorem for velocities

The transformation formulae for velocities for the transition to a moving coordinate system K' can be written down in a simple and clear way by introducing, in place of the three-dimensional vector $\mathbf{u}$, the four-dimensional vector u^i (defined in (158) and (159)). Its components have, for our case, the values

$$(u^1, u^2, u^3) = \frac{\mathbf{u}}{\sqrt{[1 - (u^2/c^2)]}}, \qquad u^4 = \frac{ic}{\sqrt{[1 - (u^2/c^2)]}}. \qquad (190)$$

The transformation formulae for going over to the system K' will then read, according to (186) and (187),

$$\left.\begin{array}{l} u'^1 = \dfrac{u^1 + i(v/c)u^4}{\sqrt{[1 - (v^2/c^2)]}}, \\[2mm] u'^4 = \dfrac{-i(v/c)u^1 + u^4}{\sqrt{[1 - (v^2/c^2)]}}, \\[2mm] u'^2 = u^2, \qquad u'^3 = u^3. \end{array}\right\} \qquad (191)$$

From these, formulae (10) to (12) of § 6 can easily be obtained; in particular, (11 a) is identical with the transformation formula for u^4. The corresponding formulae for real coordinates are obtained by using the prescription at the beginning of § 24.

Another interpretation of the composition law for velocities was first given by Sommerfeld[110]. It follows from the observation that the angles ϕ_1 and ϕ_2 for two successive rotations are simply additive, if the addition theorem for two velocities in the *same direction* is first considered. Sommerfeld's interpretation is then a consequence of the compound angle formula for the tangent

$$\tan\phi = \tan(\phi_1 + \phi_2) = \frac{\tan\phi_1 + \tan\phi_2}{1 - \tan\phi_1 \tan\phi_2},$$

because of (187). Analogous interpretations for the more general case, where the velocities are in different directions, are possible. In particular, one can demonstrate formula (11 a), which gives the *magnitude* of the resulting velocity, and the invalidity of the commutative law for the velocity *directions*. This can be done by considering the spherical geometry on a sphere[110] of radius i. Varičak[111] points out the analogy between the composition of velocities in the theory of relativity and the addition of lines of given lengths in the Bolyai–Lobachevski plane.

26. Transformation law for acceleration. Hyperbolic motion

Similarly to the case of velocity one introduces in the theory of relativity in place of the three-dimensional vector $\mathbf{\dot{u}}$ the four-dimensional vector B, defined by (147) and specified for the line element of special relativity. It will have components

$$B^i = \frac{du^i}{d\tau} = \frac{d^2x^i}{d\tau^2}. \tag{147a}$$

In the special theory of relativity

$$u_i \frac{du^i}{d\tau} = u^i \frac{du_i}{d\tau},$$

so that it follows from $u_i u^i = -c^2$ by differentiation that

$$u_i B^i = u_i \frac{du^i}{d\tau} = 0. \tag{192}$$

[110] A. Sommerfeld, *Phys. Z.* **10** (1909) 826.

[111] Varičak establishes a formal connection between the Lorentz transformation, as well as the relativistic formulae for the Doppler effect, aberration of light, and reflection at a moving mirror, with the Bolyai–Lobachevski geometry. Cf. V. Varičak, *Phys. Z.* **11** (1910) 93, 287, 586; *Glas srpsk. kralj. Akad.* **88** (1911); a summarizing report in *Jber. dtsch. Mat.Ver.* **21** (1912) 103; *Rad. Jug. Akad. Znan Umj.* (1914) 46; (1915) 86 and 101; (1916) 79; (1918) 1; (1919) 100.

This connection with the Bolyai–Lobachevski geometry can be briefly described in the following way (this had not been noticed by Varičak): If one interprets dx^1, dx^2, dx^3, dx^4 as homogeneous coordinates in a three-dimensional projective space, then the invariance of the equation $(dx^1)^2 + (dx^2)^2 + (dx^3)^2 - (dx^4)^2 = 0$ amounts to introducing a Cayley system of measurement, based on a *real* conic section. The rest follows from the well-known arguments by Klein (*Math. Ann.* **4**(1871) 112).

The connection of B with the vector $\dot{\mathbf{u}}$ of the three-dimensional space is given by

$$(B^1, B^2, B^3) = \dot{\mathbf{u}}\frac{1}{1-\beta^2} + \mathbf{u}\frac{(\mathbf{u}\cdot\dot{\mathbf{u}})}{c^2}\cdot\frac{1}{(1-\beta^2)^2},$$

$$B^4 = i\frac{(\dot{\mathbf{u}}\cdot\dot{\mathbf{u}})}{c}\cdot\frac{1}{(1-\beta^2)^2}.$$

$$\tag{193}$$

Of particular interest are the formulae for transforming the acceleration of matter from a system K', instantaneously moving with the medium, to a system K relative to which matter moves with a velocity $\mathbf{u}$. If we take the x-axis along the direction of motion, then in this case

$$(B'^1, B'^2, B'^3) = \dot{\mathbf{u}}', \qquad B'^4 = 0,$$

$$B^1 = \frac{\dot{u}_x}{1-\beta^2} + \frac{\beta}{i}B^4, \qquad B^2 = \frac{\dot{u}_y}{1-\beta^2}, \qquad B^3 = \frac{\dot{u}_z}{1-\beta^2}.$$

The transformation formulae for the components of B are obtained from the relations inverse to (186),

$$B^1 = \frac{B'^1}{\sqrt{(1-\beta^2)}}, \qquad B^2 = B'^2, \qquad B^3 = B'^3, \qquad B^4 = \frac{i\beta B'^1}{\sqrt{(1-\beta^2)}}.$$

From these follow the relations

$$\dot{u}_x = \dot{u}'_x(1-\beta^2)^{3/2}, \qquad \dot{u}_y = \dot{u}'_y(1-\beta^2), \qquad \dot{u}_z = \dot{u}'_z(1-\beta^2), \quad (194)$$

which are to be found already in Einstein's first paper[112].

In relativistic kinematics one will naturally describe as "uniformly accelerated" a motion for which the acceleration in a system K', moving with the medium or particle, is always of the same magnitude b. The system K' is a different one at each instant; for one and the same Galilean system K, the acceleration of such a motion is not constant in time. All this applies to *rectilinear* uniformly accelerated motion. We can restrict ourselves to a discussion of this case, since the more general case can be reduced to it by means of a Lorentz transformation. From (194) one easily obtains by integration

$$(x-x_0)^2 - c^2(t-t_0)^2 = \frac{c^4}{b^2} = \text{const} = a^2,$$

and if the origin of the space- and time-coordinates is so chosen that

$$t = 0, \qquad \dot{x} = 0, \qquad x = \frac{c^2}{b},$$

then the equation of the path taken is of the form

$$x^2 - c^2t^2 = \frac{c^4}{b^2} = \text{const} = a^2. \tag{195a}$$

---

[112] A very simple and elementary derivation of them can be found in Sommerfeld's *Atombau und Spektrallinien* (1st edn., Braunschweig 1919) pp. 320 and 321; (2nd edn., 1920), pp. 317 and 318.

The velocity does not increase indefinitely, but approaches the velocity of light asymptotically. The associated world line is a hyperbola, for which reason uniformly accelerated motion in relativity is also called hyperbolic motion, in contrast to the "parabolic motion" of classical mechanics. In an imaginary coordinate system the world line is a circle of radius a,

$$(x^1)^2 + (x^4)^2 = a^2. \tag{195 b}$$

The coordinates x^1 and x^4 can be expressed in terms of the imaginary arc length s of the world line

$$x^1 = a \cos\frac{s}{a}, \qquad x^4 = a \sin\frac{s}{a} \tag{196 a}$$

or, respectively, in the real coordinate system,

$$x^1 = a \cosh\frac{c\tau}{a}, \qquad x^4 = a \sinh\frac{c\tau}{a}. \tag{196 b}$$

It follows from this that the vector B is in the direction of the radius and has the value $c^2/a = b$. Since one can construct a radius of curvature at each of the points of an arbitrary path in the (x, ict) plane, one can associate with every motion of a particle at each instant an osculating hyperbolic motion.

Hyperbolic motion was first recognized by Minkowski[113] as a particularly simple type of motion, and was later discussed more fully by Born[114] and Sommerfeld[115]. On its significance in dynamics and electrodynamics, cf. §§ 37 and 32 (γ).

(b) ELECTRODYNAMICS

27. Conservation of charge. Four-current density

In Lorentz's electron theory the density ρ and velocity $\mathbf{u}$ of an electric charge satisfy the continuity equation[115a]

$$\frac{\partial \rho}{\partial t} + \operatorname{div}(\rho \mathbf{u}) = 0. \tag{A}$$

This suggests our trying to write the equation in the form of a four-dimensional Div,

$$\frac{\partial s^i}{\partial x^i} = 0 \qquad (\operatorname{Div} s = 0), \tag{197}$$

where the quantities s^i are defined by

$$(s^1, s^2, s^3) = \rho \frac{\mathbf{u}}{c}, \qquad s^4 = i\rho. \tag{198}$$

---

[113] Cf. footnote 54, p. 21, Minkowski III.
[114] M. Born, *Ann. Phys., Lpz.*, **30** (1909) 1.
[115] Cf. footnote 55, p. 22, *ibid.*, **33**, 670.
[115a] Cf. footnote 4, p. 1, H. A. Lorentz, *ibid.*, § 2, Eq. (II).

It now has to be required that (A), and therefore also (197), should be valid for every Galilean reference system. From this we can deduce that the s^i are the components of a four-vector, which is called the four-current density. In its essentials this is already found in Poincaré's work. From the invariance of (197) we can derive the transformation formulae for the s^i. These, it is true, originally contain an undetermined factor which might depend in some way on the velocity in the Lorentz transformation. It can be shown, however, that such a factor must be equal to unity, following the same kind of argument which was used in § 5 for the case of the factor κ in the transformation formulae for the coordinates. Apart from the transformation formulae for the velocity, already mentioned several times before, the vector character of s^i also leads to the transformation formula for the charge density,

$$\rho' = \frac{\rho[1 - (v/c^2)u_x]}{\sqrt{[1 - (v^2/c^2)]}}. \tag{199}$$

Its physical meaning can be made more apparent by choosing the coordinate system K' in such a way that the charge density is at rest in K'. Then $u_x = u = v$ and if we also write for this case ρ_0 instead of ρ', we obtain

$$\rho_0 = \rho\sqrt{[1 - (u^2/c^2)]}, \qquad \rho = \frac{\rho_0}{\sqrt{[1 - (u^2/c^2)]}}. \tag{199a}$$

Conversely, (199) follows from (199 a), if one uses the addition theorem for velocities. Now, because of the Lorentz contraction of a material volume element dV (see (7a)),

$$dV = dV_0\sqrt{[1 - (u^2/c^2)]},$$

we have

$$\rho dV = \rho_0 dV_0 \tag{200a}$$

or

$$de = de_0. \tag{200b}$$

The charge contained in a given material volume element is an invariant. That the *total* charge of a particle is not changed by setting it in motion, is of course a direct consequence of (A) and can reasonably safely be assumed to be in keeping with experiment, since otherwise the neutral character of an atom would be upset by the mere motion of the electrons it contains. Relation (200 b) states beyond this that the charge of each material volume remains invariant.

Sommerfeld[116], on the other hand, starts from (200 a) and arrives at the vector character of s^i in the following way. The four-dimensional volume

$$dV \cdot dx^4 \qquad (x^4 = ict)$$

swept out by the space-volume element dV in time dt is, as such, an invariant. Because of assumption (200 a), the same is true for the product

$$i\rho\, dV.$$

[116] Cf. footnote 55, p. 22, *ibid.*, **32**.

The set of quantities s^i, given in (198), is now obtained by taking the (equally invariant) quotient $i\rho/dx^4$ and multiplying it with the vector components $dx^1, ..., dx^4$. Therefore, s^i is a four-vector.

In terms of the vector u^i, and with the help of (190) and (199 a), s^i can be written simply

$$s^i = \frac{1}{c}\rho_0 u^i, \tag{201}$$

and the equation of continuity becomes

$$\frac{\partial(\rho_0 u^i)}{\partial x^i} = 0. \tag{197a}$$

See § 28 for the proof of the vector character of s^i, based on Maxwell's equations, and Part V, § 65(δ) on the interpretation of the conservation law (197) in Weyl's theory.

28. Covariance of the basic equations of electron theory

As was already stressed in § 1, the non-covariance of Maxwell's equations under the Galilean transformation was one of the main factors which gave an impetus to the development of the theory of relativity. In his paper of 1904[117], Lorentz came very near to proving the covariance of these equations under the relativistic transformation group. The complete proof was given, independently, by Poincaré[118] and Einstein[119]. The four-dimensional formulation is due to Minkowski[120], who first stressed the concept of a "surface" tensor, as we would call it now.

To write down the field equations in a four-dimensionally invariant way, one can take first the four equations which do not contain the charge density, i.e.[120a]

$$\text{curl}\,\mathbf{E} + \frac{1}{c}\dot{\mathbf{H}} = 0,$$
$$\text{div}\,\mathbf{H} = 0. \tag{B}$$

Putting, in the real coordinate system,

$$(F_{41}, F_{42}, F_{43}) = i\mathbf{E}, \qquad (F_{23}, F_{31}, F_{12}) = \mathbf{H}$$
$$(F_{ik} = -F_{ki}), \tag{202}$$

(B) can be written in the form

$$\frac{\partial F_{ik}}{\partial x^l} + \frac{\partial F_{li}}{\partial x^k} + \frac{\partial F_{kl}}{\partial x^i} = 0 \qquad (\text{Curl }\mathbf{F} = 0) \tag{203}$$

(cf. (140 b)).

From the invariance of (203) under the Lorentz transformation it then follows that the F_{ik} are the components of a "surface" tensor.

[117] Cf. footnote 10, p. 2, *ibid.*
[118] Cf. footnote 11, p. 2, *ibid.*
[119] Cf. footnote 12, p. 2, *ibid.*
[120] Cf. footnote 54, p. 21, Minkowski I.
[120] Cf. footnote 4, p. 1, H. A. Lorentz, *ibid.*, Eqs. (IV) and (V).

Once again, the undetermined factor remaining in the transformation formulae is eliminated as before. If instead of F_{ik} the dual tensor F^{*ik}, defined by (54 a) and (54 b), is introduced,

$$(F^{*41}, F^{*42}, F^{*43}) = -\mathbf{H}, \qquad (F^{*23}, F^{*31}, F^{*12}) = -i\mathbf{E}, \qquad (202\,\text{a})$$

the system of equation (203) can also be written, because of (142) and (141 b),

$$\frac{\partial F^{*ik}}{\partial x^k} = 0 \qquad (\text{Div } \mathbf{F}^* = 0). \qquad (203\,\text{a})$$

But it is well known that, in ordinary space, $\mathbf{E}$ is a polar, and $\mathbf{H}$ an axial vector ("surface" tensor), not the reverse. We therefore consider the "surface" tensor (202) to be the *natural* representation of the electro-magnetic field and the dual tensor (202 a) to be an artificiality. Minkowski[121] uses both conventions for writing down the field equations. The former, which in many cases is the more transparent and convenient (in particular in the general theory of relativity), has later tended to be forgotten and Sommerfeld[122] does not mention it. It was only in the year 1916 that Einstein[123] revived an interest in it.

From the tensor character of F_{ik} follow the transformation formulae for the field intensities, for going over to a moving system of reference. Let us assume that the velocity $\mathbf{v}$ of the Lorentz transformation has an arbitrary direction with respect to the x-axis of the coordinate system. Then,

$$\mathbf{E}'_{\parallel} = \mathbf{E}, \qquad \mathbf{E}'_{\perp} = \frac{[\mathbf{E} + (1/c)(\mathbf{v} \wedge \mathbf{H})]_{\perp}}{\sqrt{[1 - (v^2/c^2)]}},$$

$$\mathbf{H}'_{\parallel} = \mathbf{H}, \qquad \mathbf{H}'_{\perp} = \frac{[\mathbf{H} - (1/c)(\mathbf{v} \wedge \mathbf{E})]_{\perp}}{\sqrt{[1 - (v^2/c^2)]}}. \qquad (204)$$

The splitting of the field into an electric and a magnetic field is only of relative importance. If e.g. in system K only an electric field is present, then in a system K' moving relative to K, a magnetic field, too, will exist. This remark removes certain conceptual difficulties in understanding the induction due to a moving magnet on the one hand, and the motion of a conductor in which the current is induced, on the other.

Also, the electromagnetic potentials, the scalar potential φ and the vector potential $\mathbf{A}$ of Lorentz's theory, admit of a simple four-dimensional interpretation. As was first noticed by Minkowski[54a], they can be combined to form a vector in the four-dimensional world, the four-vector potential,

$$(\phi_1, \phi_2, \phi_3) = \mathbf{A}, \qquad \phi_4 = i\varphi. \qquad (205)$$

[121] Cf. footnote 54, p. 21, Minkowski I.
[122] Cf. footnote 55, p. 22, *ibid.*
[123] A. Einstein, 'Eine neue formale Deutung der Maxwellschen Gleichungen', *S.B. preuss. Akad. Wiss.* (1916) 184.
[54a] Cf. footnote 54, p. 21, Minkowski I.

The expressions for the field intensities[124]

$$\mathbf{H} = \operatorname{curl}\mathbf{A}, \qquad \mathbf{E} = -\operatorname{grad}\varphi - \frac{1}{c}\dot{\mathbf{A}}$$

then take the form

$$F_{ik} = \frac{\partial\phi_k}{\partial x^i} - \frac{\partial\phi_i}{\partial x^k} \qquad (\mathbf{F} = \operatorname{Curl}\boldsymbol{\varphi}) \qquad (206)$$

(cf. (140 a)).

The four-vector potential is a mathematical auxiliary function which is often of great use, but it has no immediate physical significance in Lorentz's theory. Eqs. (203) follow from (206), and conversely, when (203) applies, a vector field ϕ_i can always be so determined that (206) is satisfied. This relation, however, does not determine ϕ_i unambiguously. For if ϕ_i is a solution of (206) for given F_{ik}, then (206) is also satisfied by $\phi_i + \partial\psi/\partial x^i$, where ψ is an arbitrary scalar function of the space-time coordinates. For an unambiguous definition of ϕ_i, therefore, the condition[125]

$$\operatorname{div}\mathbf{A} + \frac{1}{c}\frac{\partial\varphi}{\partial t} = 0$$

is added in the Lorentz theory, which can be written in its four-dimensional form.

$$\frac{\partial\phi^i}{\partial x^i} = 0. \qquad (\operatorname{Div}\boldsymbol{\varphi} = 0) \qquad (207)$$

Up to now, a four-dimensional meaning has not been assigned to the Hertz vector $\mathbf{Z}$.

The second system of Maxwell's equations[125a], containing the charge density,

$$\operatorname{curl}\mathbf{H} - \frac{1}{c}\dot{\mathbf{E}} = \rho\frac{\mathbf{u}}{c}, \qquad \operatorname{div}\mathbf{E} = \rho, \qquad (C)$$

can be dealt with analogously to (B). It follows immediately from (198) and (202) that

$$\frac{\partial F^{ik}}{\partial x^k} = s^i \qquad (\operatorname{Div}\mathbf{F} = s) \qquad (208)$$

(cf. (141b)). If the charge density is defined by (C), this establishes at once the vector character of s^i, which was previously derived in a different way. One can express the field intensities in (208) in terms of the four-vector potential, using (206), and thus obtain (cf. (145))

$$\operatorname{Div}_i\operatorname{Curl}\boldsymbol{\varphi} = \operatorname{Grad}_i\operatorname{Div}\boldsymbol{\varphi} - \square\phi_i = s_i$$

and, because of (207),

$$\square\phi_i = -s_i. \qquad (209)$$

[124] Cf. footnote 4, p. 1, H. A. Lorentz, *ibid.* Eqs. (IX) and (X).
[125] Cf. footnote 4, p. 1, H. A. Lorentz, *ibid.*, § 4, Eq. (2).
[125a] Cf. footnote 4, p. 1, H. A. Lorentz, *ibid.*, § 2, Eqs. (I), (Ia) and (IV).

Because of the covariance of the electromagnetic field equations under the Lorentz group it would seem natural to enquire whether there exist wider groups for which this covariance applies. This question was answered by Cunningham and Bateman[126]. The most general such group is formed by the affine transformations which transform the equation of the light cone

$$s^2 = 0$$

into itself (§ 8, (B')). Apart from the transformations of the Lorentz group, it also contains the inversion with respect to a four-dimensional sphere, or a hyperboloid in the real coordinate system, respectively. New light is thrown on Bateman's theorem by Weyl's theory (cf. Part V). P. Frank[127] gives a simple proof for the fact that the Lorentz group, together with the group of the ordinary affine transformations, is the only *linear* group with respect to which Maxwell's equations are covariant.

29. Ponderomotive forces. Dynamics of the electron

Even in his first paper Einstein had shown that the theory of relativity could enable us to make quite definite statements about the motion of a point charge moving with arbitrary velocity in an electromagnetic field, provided it was known for infinitely small velocities. What we mean by a point charge, here, is a charge whose dimensions are so small that the external field can be assumed homogeneous throughout the region filled by the charge. The "point charge", therefore, need not be an electron. Let $\mathbf{E}$ be the intensity of the external electric field, e and m_0 the charge and mass of such a "point charge" in a coordinate system K', in which the point charge is momentarily at rest. Then, in K',

$$m_0 \frac{d^2 \mathbf{r}'}{dt'^2} = e\mathbf{E}'. \tag{210}$$

With the help of formulae (194) and (204) we can immediately derive the equations of motion holding in a system K, in which the charge (and the system K') moves with velocity $\mathbf{u}$ in the positive x-direction. We obtain

$$
\left.
\begin{aligned}
\frac{m_0}{(1-\beta^2)^{3/2}} \cdot \frac{d^2 x}{dt^2} &= eE_x = e\left[\mathbf{E} + \frac{1}{c}(\mathbf{u} \wedge \mathbf{H})\right]_x \\
\frac{m_0}{\sqrt{(1-\beta^2)}} \cdot \frac{d^2 y}{dt^2} &= e\left[\mathbf{E} + \frac{1}{c}(\mathbf{u} \wedge \mathbf{H})\right]_y \\
\frac{m_0}{\sqrt{(1-\beta^2)}} \cdot \frac{d^2 z}{dt^2} &= e\left[\mathbf{E} + \frac{1}{c}(\mathbf{u} \wedge \mathbf{H})\right]_z .
\end{aligned}
\right\} \tag{211}
$$

First of all, it is seen that the right-hand side is precisely the expression for the Lorentz force[128]. *The Lorentz force had, in earlier treatments, been*

[126] E. Cunningham, *Proc. Lond. math. Soc.*, **8** (1910) 77; H. Bateman, *Proc. Lond. math. Soc.*, 8 (1910) 223.
[127] P. Frank, *Ann. Phys., Lpz.*, **35** (1911) 599.
[128] Cf. H. A. Lorentz, footnote 4, p. 1, *ibid.*, § 3, Eq. (VI).

introduced as a new axiom, whereas here it is a consequence of the relativity principle. It should be mentioned, however, that this statement contains not a physical law but a definition of the force, as far as terms of second and higher order in u/c are concerned. In fact, our choice of what to put on the right-hand or left-hand sides of the equations (211) would at first seem quite arbitrary. We could, e.g., multiply through by $(1-\beta^2)^{3/2}$ or $(1-\beta^2)^{1/2}$ and then denote the corresponding expressions on the right-hand side as components of a force. Einstein had, initially, considered eE' to be a force also in the moving system K. But relativistic mechanics shows that the most convenient and in fact the only natural definition is the one given above, which is due to Planck[129]. According to this, the Lorentz expression for an arbitrarily moving charge

$$\mathbf{K} = e\left[\mathbf{E} + \frac{1}{c}(\mathbf{u} \wedge \mathbf{H})\right],\qquad(212)$$

is defined as force. For it is seen that only with this definition can the force be considered as the time derivative of a momentum, which remains constant in closed systems (cf. § 37). From (212) and (207) we obtain the transformation formulae for the force,

$$K_x = K'_x, \qquad K_y = K'_y \sqrt{(1-\beta^2)}, \qquad K_z = K'_z \sqrt{(1-\beta^2)}, \quad(213)$$

where the assumption has been made that the medium on which the force acts is, at the given moment, at rest in the coordinate system K'.

In the earlier literature, $m_0/(1-\beta^2)^{3/2}$ has frequently been called the longitudinal, $m_0/(1-\beta^2)^{1/2}$ the transversal mass, because of (211); but it is more convenient to write (211) in the form

$$\frac{d}{dt}(m\dot{\mathbf{r}}) = \mathbf{K}\qquad(214)$$

where now

$$m = \frac{m_0}{\sqrt{(1-\beta^2)}}\qquad(215)$$

appears throughout as the mass.[130] This expression for the velocity-dependence of the mass was derived for the first time by Lorentz[131] specifically for the electron mass, on the assumption that electrons, too, suffer a "Lorentz contraction" during their motion. Abraham's theory of the rigid electron had resulted in a more complicated formula for the change in mass.[132] It constituted a definite progress that Lorentz's law of the variability of mass could be derived from the theory of relativity without

---

[129] M. Planck, *Verh. dtsch. phys. Ges.*, **4** (1906) 136.
[130] This result is already contained implicitly in Planck's work (cf. footnote 129, *ibid.*) and was later stressed in particular by R. C. Tolman, *Phil. Mag.*, **21** (1911) 296.
[131] Cf. footnote 10, p. 2, *Proc. Acad. Sci. Amst., loc. cit.*
[132] Cf. H. A. Lorentz, footnote 4, *ibid.*, § 21, Eqs. (77) and (78). Because of its historical interest, Bucherer's (deformable) electron of constant volume should be mentioned: A. H. Bucherer, *Mathematische Einleitung in die Elektronentheorie* (1904), p. 58. See also M. Abraham, *Theorie der Elektrizität*, Vol. 2. (3rd edn., Leipzig 1914) p. 188.

making any specific assumptions on the electron shape or charge distribution. Also, nothing need be assumed about the nature of the mass: formula (215) is valid for every kind of ponderable mass, as was shown here for the case of electromagnetic forces and will be generalized for arbitrary forces in relativistic mechanics (cf. § 37). The old idea that one could distinguish between the constant "true" mass and the "apparent" electromagnetic mass[133], by means of deflection experiments on cathode rays, can therefore not be maintained.

Formula (215), or rather the equations of motion (211), provide an opportunity for testing the theory of relativity by means of experiments on the deflection of fast cathode-rays or β-rays in the presence of electric and magnetic fields. The older experiments of Kaufmann[134] seem to bear out Abraham's formula, but Kaufmann over-estimated the accuracy of his measurements. After the experiments by Bucherer[135], Hupka[136], and Ratnowsky[137] the relativistic formula seemed the more likely one, and the more recent results by Neumann[138] (with a supplementary result by Schäfer[139]) and Guye and Lavanchy[140] are quiteun ambiguously in favour of the latter. The theory of spectra however gives us today, with the fine structure of the hydrogen lines, a much more accurate means for determining the velocity dependence of the electron mass[141]. This leads to a complete confirmation of the relativistic formula, which can thus be considered as experimentally verified. It has not been possible up till now to establish this variability for masses other than that of the electron experimentally, because of the smallness of the effect, even for fast α-particles.†

Equation (211) can be written in an invariant form if, instead of the force on the total charge, one considers the force per unit volume (force density),

$$\mathbf{f} = \rho\left[\mathbf{E} + \frac{1}{c}(\mathbf{u} \wedge \mathbf{H})\right]. \tag{212a}$$

---

[133] See, e.g., footnote 4, p. 1, H. A. Lorentz, *ibid.*, § 65.

[134] W. Kaufmann, *Nachr. Ges. Wiss. Göttingen, math.-nat. Kl.* (1901) 143; (1902) 291; (1903) 90; *Ann. Phys., Lpz.,* **19** (1906) 487 and **20** (1906) 639.

[135] A. H. Bucherer, *Verh. dtsch. phys. Ges.* **6** (1908) 688; *Phys. Z.* **9** (1908) 755; *Ann. Phys., Lpz.,* **28** (1909) 513 and **29** (1909) 1063. See also subsequent experiments by K. Wolz, *Ann. Phys., Lpz.,* **30** (1909) 373; and the discussion between Bucherer and Bestelmeyer: A. Bestelmeyer, *Ann. Phys., Lpz.,* **30** (1909) 166; A. H. Bucherer, *ibid.,* **30** (1909) 974; A. Bestelmeyer, *ibid.,* **32** (1910) 231.

[136] E. Hupka, *Ann. Phys., Lpz.,* **31** (1910) 169; cf. also discussion by W. Heil, *ibid.,* **31** (1910) 519.

[137] S. Ratnowsky, *Dissertation* (Geneva 1911).

[138] G. Neumann, *Dissertation* (Breslau 1914). Résumé in *Ann. Phys., Lpz.,* **45** (1914) 529; report on Neumann's experiments by C. Schäfer, *Verh. dtsch. phys. Ges.,* **15** (1913) 935; *Phys. Z.* **14** (1913) 1117.

[139] C. Schäfer, *Ann. Phys., Lpz.,* **49** (1916) 934.

[140] Ch. E. Guye and Ch. Lavanchy, *Arch. Sci. phys. nat.,* **41** (1916) 286, 353 and 441.

[141] K. Glitscher, *Dissertation* (Munich 1917), résumé in *Ann. phys., Lpz.,* **52** (1917) 608. Cf. also A. Sommerfeld, *Atombau und Spektrallinien,* (1st edn., Braunschweig 1919, p. 373 *et seq.*; 2nd edn., 1920, p. 370). Sommerfeld points to W. Lenz as the originator of this method.

† See suppl. note 9.

This expression would suggest the formation of the product of the skew-symmetric tensor F_{ik} with the four-current density s^k,

$$f_i = F_{ik}s^k. \tag{216}$$

The resulting vector f_i has components

$$(f_1, f_2, f_3) = \mathbf{f}, \qquad f_4 = i\rho\left(\frac{\mathbf{E}\cdot\mathbf{u}}{c}\right). \tag{217}$$

The force per unit volume (force density) produces the three space-components of a four-vector whose time-component is the work done per unit time per unit volume (power per unit volume), divided by c. This important fact had, in its essentials, already been recognized by Poincaré[11] and was subsequently stated clearly by Minkowski[54a]. From (201) and (216) we see that the four-vector f_i is perpendicular to the velocity vector u^i,

$$f_i u^i = 0. \tag{218}$$

We can now write down the equation of motion (214) in an invariant form; in fact, this can be done in two ways. The first is to introduce the four-vector K_i with components

$$(K_1, K_2, K_3) = \frac{\mathbf{K}}{\sqrt{(1-\beta^2)}}, \qquad K_4 = i\frac{(\mathbf{K}\cdot\mathbf{u})/c}{\sqrt{(1-\beta^2)}}. \tag{219}$$

That these quantities actually form a four-vector, can be seen from the transformation formulae (213) for the force. One calls $\mathbf{K}/\sqrt{(1-\beta^2)}$ the Minkowski force, in contrast to the Newtonian force $\mathbf{K}$. The equations of motion are simply

$$m_0\frac{d^2x^i}{d\tau^2} = K^i \quad \text{or} \quad m_0\frac{du_i}{d\tau} = K_i. \tag{220}$$

The second way is to apply the equations of motion to a unit volume. If μ_0 is the rest-mass density m_0/V_0, then

$$\mu_0\frac{d^2x^i}{d\tau^2} = f^i, \qquad \mu_0\frac{du_i}{d\tau} = f_i. \tag{221}$$

It should be noted, however, that the physical meaning of these latter equations, when referring to an electron, is not quite clear (cf. Part V, § 63), at least so long as one attributes to f_i the meaning of (216); in that case they are only valid when the self-field of the particle is not included on the right-hand side.

For $i = 4$, (220) and (221) result in the energy equation, which is a consequence of the equations of motion. In fact, the four equations (220) (or (221)) are not independent, for by scalar multiplication with u^i one obtains the identity $0 = 0$, using (192) and (218).

On the generalization of the definition of the force vector and of the equations of motion, see § 37.

[54a] Cf. footnote 54, p. 21, Minkowski I.

30. Momentum and energy of the electromagnetic field. Differential and integral forms of the conservation laws

In electrodynamics it is shown that the Lorentz force density $\mathbf{f}$ can be represented as the resultant of the surface forces produced by the Maxwell stresses and the (negative) time derivative of the momentum density of the aether.[142] The stress tensor is defined by

$$T_{ik} = (E_i E_k - \tfrac{1}{2} \mathbf{E}^2 \delta_{ik}) + (H_i H_k - \tfrac{1}{2} \mathbf{H}^2 \delta_{ik}), \qquad (i, k = 1, 2, 3),$$

and the electromagnetic momentum by[142]

$$\dot{\mathbf{g}} = \frac{1}{c^2} \mathbf{S}, \qquad \mathbf{S} = c(\mathbf{E} \wedge \mathbf{H}).$$

Then
$$\mathbf{f} = \operatorname{div} \mathbf{T} - \dot{\mathbf{g}}. \tag{D}$$

It will now be seen that this vector equation can be combined with the (scalar) energy equation

$$\frac{\partial W}{\partial t} + \operatorname{div} \mathbf{S} = -\mathbf{f} \cdot \mathbf{u}, \qquad W = \tfrac{1}{2}(\mathbf{E}^2 + \mathbf{H}^2) \tag{E}$$

into a four-vector equation.[143] Let us, first of all, form the symmetrical tensor of rank 2,

$$S_i{}^k = \tfrac{1}{2}(F_{ir} F^{kr} - F^*{}_{ir} F^{*kr}) = F_{ir} F^{kr} - \tfrac{1}{4} F_{rs} F^{rs} \delta_i{}^k, \tag{222}$$

from the surface tensor F_{ik}. It should be observed that its scalar vanishes,

$$S_i{}^i = 0. \tag{223}$$

Its components are

$$S_i{}^k = -T_{ik} \qquad (i, k = 1, 2, 3)$$

$$\left.\begin{array}{c} (S_1{}^4, S_2{}^4, S_3{}^4) = (S_{14}, S_{24}, S_{34}) = \dfrac{i}{c}\mathbf{S} = ic\mathbf{g} \\[2mm] S_4{}^4 = S_{44} = S^{44} = -W. \end{array}\right\} \tag{224}$$

The space components of $\mathbf{S}$ are thus essentially equal to the components of the electromagnetic stress tensor, the space–time components equal to the Poynting vector and the momentum density, the time component equal to the energy density. Equations (D) and (E) can then be combined with the help of the four-vector f (defined by (216)) to form the system of equations

$$f_i = -\frac{\partial S_i{}^k}{\partial x^k} \qquad (f = -\operatorname{Div} \mathbf{S}). \tag{225}$$

(This was first noticed by Minkówski[144].) For $i = 1, 2, 3$, they represent momentum conservation, for $i = 4$, energy conservation. For this reason

---

[142] Cf. H. A. Lorentz, footnote 4, p. 1, *ibid.*, § 7.
[143] Cf. H. A. Lorentz, footnote 4, p. 1, *ibid.*, § 6.
[144] Cf. footnote 54, p. 21, Minkowski II.

(225) is usually called the energy–momentum conservation law, and **S** is called the energy–momentum tensor of the electromagnetic field.

It is further seen that the derivation of equations (C) and (D) from the field equations (A) and (B) is considerably simplified by the introduction of the four-dimensional notation. Formula (176) is identical with (225) if one identifies ϕ_i with the four-vector potential, F_{ik} with the tensor of the field intensities, and s^i with the four-current density. We need therefore only consider the special case of constant g_{ik} for the derivation given in § 23 (a) (which simplifies it considerably), in order to obtain (225). But the direct calculation, too, is easy enough to carry out.

The relativistic interpretation of the energy–momentum equations contributes something new, not just in a formal respect, but also in its physical content. If the energy equation (4th component of (225)), holds in every coordinate system then the momentum equations follow automatically. Both are perfectly equivalent in their description of the physical processes. Corresponding to the interpretation of **S** in (E) as an energy current, the quantities T_{ik} are to be looked upon as components of the momentum current. Since the momentum is itself already a vector, they form (in ordinary space) a *tensor*, in contrast to the vector **S**. In this way Maxwell's electromagnetic stress tensor, which had previously been regarded as a purely mathematical quantity[145], is given a very real physical meaning. This argument is due to Planck.[146] (On the generalization of this interpretation and of equations (225) for non-electromagnetic momenta, cf. § 42.) At those points at which ponderomotive forces act on matter, electromagnetic momentum (energy) is produced by, or changed into, mechanical momentum (energy), according to (225). (See Part V about attempts to regard every kind of momentum and energy as electromagnetic in origin.) At all other points, however, the momentum and energy of the electromagnetic field have the same flow properties as a fluid which is compressible in general. The special case of stationary fields would correspond to an incompressible fluid consisting of indestructible matter.

The tensor S_{ik} refers to the energy–momentum *density* and the question arises, how the total energy and momentum of a system behave under a transformation to a moving coordinate system. We shall not anticipate the discussion of the general case in § 42. But let us consider here the case where the momentum and energy are purely electromagnetic in nature, where in other words the force density f_i and the electric charge density vanish everywhere, so that, from (225),

$$\frac{\partial S_i{}^k}{\partial x^k} = 0 \qquad (\text{Div } \mathbf{S} = 0). \tag{225a}$$

This is so for the field of a light wave of arbitrary shape, freely travelling

---

[145] Cf. H. A. Lorentz, footnote 4, p. 1, *ibid.*, § 7, p. 163.
[146] M. Planck, *Verh. dtsch. phys. Ges.*, **6** (1908) 728; *Phys. Z.*, **9** (1908) 828. If one accepts this interpretation one must not be unduly perturbed by the following paradox: a momentum *current* may exist even if the momentum *density* vanishes everywhere (as is, for example, the case for a purely electrostatic field). For the energy current a corresponding situation cannot arise.

through space. Let us assume that it fills a finite volume so that also its total energy and momentum are finite. In the world picture, this corresponds to a tube of finite cross-section. We are thus dealing here with precisely the same situation as in § 21 and it follows from the conclusions reached there that we can obtain the components of a four-vector from the quantities $S_k{}^4$ by integrating over the whole volume,

$$J_k = \frac{1}{i} \iiint S_k{}^4 \, dV. \qquad (226)$$

Because of (224), these components are related to the total momentum **G** and the total energy E of the system (light wave) in a very simple way,

$$(J_1, J_2, J_3) = c\mathbf{G}, \qquad J_4 = iE. \qquad (227)$$

We can therefore say that *the total energy and total momentum form, in this case, a four-vector*. Formulae (186) and (187) then immediately give us the transformation formulae

$$G'_x = \frac{G_x - (v/c^2)E}{\sqrt{(1 - \beta^2)}}, \qquad G'_y = G_y, \qquad G'_z = G_z, \left.\vphantom{\frac{G_x - (v/c^2)E}{\sqrt{(1 - \beta^2)}}}\right\}$$
$$E' = \frac{E - vG_x}{\sqrt{(1 - \beta^2)}}. \qquad\qquad (228)$$

It should also be noted that the vector J_k cannot be space-like. If it were, a coordinate system would exist in which $G \neq 0$, but $E = 0$. But this is impossible, since E can only vanish when no field at all exists. We have therefore

$$|J| \leqslant 0, \qquad G \leqslant \frac{E}{c}. \qquad (229)$$

Thus, J can only be a null vector or be time-like. An example for the first alternative would be a laterally bounded plane wave of finite length. For it is well known that for this case $G = E/c$. But since this relation can be written in the form $J_i J^i = 0$, it must hold in every reference system. If α is the angle, measured in K, between the direction of propagation of the light ray and the velocity of K' relative to K, we obtain from (228) Einstein's transformation formula for the energy of a finite plane wave[147]

$$E' = \frac{1 - \beta \cos\alpha}{\sqrt{(1 - \beta^2)}} E. \qquad (228\text{a})$$

If, on the other hand, J is time-like, there always exists a coordinate system K_0 in which the total momentum vanishes. If E_0 is the value of the total energy in K_0, it follows from (228) that for a system K, relative to which K_0 moves with velocity **v**,

$$E = \frac{E_0}{\sqrt{(1 - \beta^2)}}, \qquad \mathbf{G} = \frac{(\mathbf{v}/c^2)E_0}{\sqrt{(1 - \beta^2)}} = \frac{\mathbf{v}}{c^2} E. \qquad (228\,\text{b})$$

[147] Cf. footnote 12, p. 2, *ibid.*, § 8.

An example for this case is a spherical wave of finite width, or a system of two equal and opposite plane waves. See § 42 for the generalization of these relations for non-electromagnetic momenta and energies.

31.   The invariant action principle of electrodynamics

Poincaré[148] had already convinced himself of the invariance of Schwarzschild's action integral[149] with respect to the Lorentz group.† Later, Born[150] formulated the action principle in a very clear way, by writing it in four-dimensional notation.

Schwarzschild[149] first formed the Lagrangian

$$\frac{1}{2} \int (\mathbf{H}^2 - \mathbf{E}^2) dV + \int \rho \Big(\varphi - \frac{1}{c} \mathbf{A} \cdot \mathbf{u} \Big) dV$$

by integration over the three-dimensional volume and obtained from this the action function, by integrating over the time. Naturally, the integrations over space and time can be combined into a quadruple integral.[148] Let L be the invariant

$$L = \tfrac{1}{2} F_{ik} F^{ik} = \mathbf{H}^2 - \mathbf{E}^2, \tag{230}$$

then the (double) action function can simply be written in the form

$$W = \int (L - 2\phi_i s^i) d\Sigma. \tag{231}$$

The action principle in question then states that the variation of W vanishes under certain conditions, i.e.

$$\delta W = 0. \tag{232}$$

These conditions are:

(i) The integral W is to be integrated over a given world region, with the components ϕ_i of the four-vector potential acting as independent variables and having prescribed values at the boundary of the region of integration. The four-current density s^i, i.e. the world lines of the electric charges, and its absolute magnitude are not to be varied. Then, from § 23, Eqs. (174) and (175),

$$\delta W = 2 \int \Big(\frac{\partial F^{ik}}{\partial x^k} - s^i \Big) \delta \phi_i d\Sigma, \tag{233}$$

and (232) produces the second set of Maxwell's equations (208). The first set is already satisfied because of the existence of the four-vector potential and has been assumed from the start.

(ii) The field ϕ^i is a given function of the world coordinates and is not to be varied; on the other hand, the world lines of the matter present are to be varied. The integral over L then does not contribute to the variation,

[148] Cf. footnote 11, p. 2, *R.C. Circ. mat. Palermo, loc. cit.*

[149] K. Schwarzschild, *Nachr. Ges. Wiss. Göttingen*, (1903) 125. See also footnote 4, p. 1, H. A. Lorentz, *loc. cit.*, §9.

† See suppl. note 10.

[150] M. Born, *Ann. Phys., Lpz.*, **28** (1909) 571.

and the second term in the integrand of (231) has first to be transformed. If de is an element of charge, associated with a given element of matter, and τ the proper time of the corresponding world line (taken from some arbitrarily chosen origin), then it follows from the meaning given to s^i in (201) that

$$\int \rho_0 \, d\Sigma = ic \int de \int d\tau, \qquad \int \phi_i s^i \, d\Sigma = i \int de \int \phi_i \frac{dx^i}{d\tau} \, d\tau.$$

We now integrate over the world cylinder which is obtained by measuring off the same length on each world line of the substance. The initial and end points of the world lines are not to be varied. An integration by parts first gives

$$\tfrac{1}{2}\delta W = i \int de \int \left(\frac{d\phi_i}{d\tau} \delta x^i - \frac{\partial \phi_k}{\partial x^i} \delta x^i \frac{dx^k}{d\tau} \right) d\tau$$

$$= -i \int de \int F_{ik} u^k \delta x^i d\tau,$$

or
$$\tfrac{1}{2}\delta W = -\int F_{ik} s^k \delta x^i \, d\Sigma = -\int f_i \delta x^i \, d\Sigma. \tag{234}$$

Born then proceeds in the following way. He adds to the above variation also the auxiliary condition

$$\delta \int ds = 0. \tag{235}$$

For a world line whose direction is always time-like, as here, we have for fixed end points and constant g_{ik}, from § 15,

$$\delta \int d\tau = \frac{1}{c^2} \int \frac{du_i}{d\tau} \delta x^i \, d\tau. \tag{236}$$

In this case, therefore, (232) gives

$$\mu_0 \frac{du_i}{d\tau} = f_i,$$

where μ_0 is a constant Lagrange multiplier. These relations agree with (221) if one interprets μ_0 as the rest-mass density. The self-field of the particular particle is omitted here, as in § 29.

Weyl[151], on the other hand, does not impose the auxiliary condition (235) on the variation, but adds to the action function another term,

$$2 \int \mu_0 c^2 \, d\Sigma = 2ic \int dm \cdot c^2 \int d\tau,$$

so that

$$W_1 = 2 \int \mu_0 c^2 \, d\Sigma + W, \qquad \delta W_1 = 0. \tag{231a}$$

(221) then again follows from (234) and (236).

[151] H. Weyl, *Raum-Zeit-Materie* (1st edn.) § 32, p. 215.

32. Applications to special cases

(α) *Integration of the equations for the potential.* It is well known that the differential equations (207) and (209) have solutions

$$\varphi_{P,t} = \int \frac{\rho_{Q,t-r_{PQ}/c}}{4\pi r_{PQ}} dV_Q, \qquad \mathbf{A}_{P,t} = \int \frac{(\rho\mathbf{u}/c)_{Q,t-r_{PQ}/c}}{4\pi r_{PQ}} dV_Q \qquad (237)$$

when s^i is given as a function of space and time.[152] In these expressions for the potentials, full use is not made of the space-time symmetry of the differential equations. This is done, however, in a method discovered by Herglotz[153] already before the advent of the theory of relativity. The method's starting point is the particular solution

$$\frac{1}{(R_{PQ})^2}$$

of the equations (209), where P and Q are two world points and R their four-dimensional distance. Herglotz obtained from this the usual expression for the potential by multiplying with a suitable function $s(Q)$ and integrating round the line t_P to ∞ in a positive sense in the complex t_Q-plane. The advantage of the method lies in the fact that one can differentiate first and carry out the contour integration afterwards, in order to calculate the field intensities. This makes the calculation much clearer. Sommerfeld[154], influenced by the theory of relativity, later modified and amplified Herglotz's method.

For point charges one obtains from (237) the [Liénard-] Wiechert potentials[155]

$$4\pi\varphi_{P,t} = \frac{e}{r_P - (\mathbf{u}\cdot\mathbf{r}_P/c)}\bigg|_{t-(r/c)}, \qquad 4\pi\mathbf{A}_{P,t} = \frac{e\mathbf{u}/c}{r_P - (\mathbf{u}\cdot\mathbf{r}_P/c)}\bigg|_{t-(r/c)}. \qquad (238)$$

$\mathbf{r}_P$ is the vector joining the position of the charge at time $t-(r_P/c)$ to the field point. According to Minkowski,[156] this admits of an interpretation in terms of four-vectors. Let

$$\xi^i = \xi^i(\tau) \qquad (239)$$

be the world line of the charge as a function of its proper time, and P be the field point, as above. Through P construct a "null cone" into the past. It intersects the world line of the charge at a point Q which will be unique if the direction of the world line is always time-like. If x^i are the coordinates of the field point and

$$X^i = x^i - \xi_Q{}^i, \qquad (240)$$

then the condition

$$X_i X^i = 0 \qquad (241)$$

[152] Cf. H. A. Lorentz, footnote 4, p. 1, *ibid.*, Eqs. (XI) and (XII). We shall not be dealing with the alternative solution by means of advanced potentials [with $t+(r/c)$ instead of $t-(r/c)$ in the formulae], which has been much debated since it was first put forward by Ritz (cf. footnote 21, p. 5, *ibid.*).

[153] G. Herglotz, *Nachr. Ges. Wiss. Göttingen*, (1904) 549.

[154] Cf. footnote 55, p. 22, *ibid.*, **33**, §7, p. 665 *et seq.*

[155] Cf. H. A. Lorentz, footnote 4, p. 1, *ibid.*, § 17, Eq. (70).

[156] Cf. footnote 54, p. 21, Minkowski, III.

determines Q, and hence also the corresponding value of τ as a unique function of the x^i,

$$\tau_Q = f(x^i). \tag{242}$$

Expressions (238) for the potentials can now be combined into

$$4\pi\varphi_i = -\frac{eu_i}{(u_r X^r)}, \tag{238a}$$

using (205) and (190). The introduction of the proper time simplifies the calculation of the field intensities considerably. From (241) one first obtains the derivatives of the function (242) with respect to the co-ordinates x^k of P,

$$X_i\left(\delta_k{}^i - u^i \frac{\partial\tau}{\partial x^k}\right) = 0, \qquad \frac{\partial\tau}{\partial x^k} = \frac{X_k}{(X_r u^r)}. \tag{243}$$

The rest of the calculation is quite elementary and gives

$$4\pi F_{ik} = -\frac{e}{(X_r u^r)^3}\left\{c^2 + \left(X_r \frac{du_r}{d\tau}\right)\right\}(u_i X_k - u_k X_i) + \\ + \frac{e}{(X_r u^r)^2}\left(\frac{du_i}{d\tau}X_k - \frac{du_k}{d\tau}X_i\right), \tag{244}$$

for the field intensities. If now a second charge, $\bar{e}$, is placed at the field point, with velocity vector $\bar{u}^i$, we obtain from (216) the force $\mathbf{K}$ exerted by the first on the second charge. The expression for the Minkowski force, the four-vector K_i defined by (219), then follows,

$$4\pi K_i = 4\pi\bar{e}F_{ik}\bar{u}^k \\ = \frac{e\bar{e}}{(X_r u^r)^3}\left[\left\{c^2 + \left(X_r\frac{du_r}{d\tau}\right)\right\}(u_k\bar{u}^k) - (X_r u^r)\left(\frac{du_k}{d\tau}\bar{u}^k\right)\right]X_i - \\ - \frac{e\bar{e}}{(X_r u^r)^3}\left\{c^2 + \left(X_r\frac{du_r}{d\tau}\right)\right\}(X_k u^k)\bar{u}_i + \frac{e\bar{e}}{(X_r u^r)^3}(X_k\bar{u}^k)\frac{du_i}{d\tau}. \tag{245}$$

Sommerfeld derived not only the [Liénard-] Wiechert potentials (238a), but also expressions (244) and (245), by using the above-mentioned method of complex integration. (245) is in agreement with Schwarzschild's[157] "elementary electrodynamic force".

(β) *The field of a uniformly moving point charge.* Since electron theory is in agreement with relativity, the latter cannot produce results which are not already contained in pre-relativistic Lorentz electron theory, as far as calculations of the electromagnetic field for given electronic motion are concerned. Very often, however, the transformation rules for the field intensities enable us to avoid having to use the differential equations or the general formula (244), provided the field is known for a particular coordinate system. If, e.g. we have to find the field of a point charge

[157] K. Schwarzschild, *Nachr. Ges. Wiss. Göttingen,* (1903) 132; cf. also H. A. Lorentz (footnote 4, p. 1, *ibid.,* § 25).

moving uniformly in a system K, we would first determine the field in a system K' in which the charge is at rest,

$$\mathbf{E}' = \frac{e}{r'^3}\mathbf{r}'.$$

From (207) we then have immediately,

$$E_x = \frac{e}{r'^3}x', \qquad E_y = \frac{e}{r'^3}\frac{y'}{\sqrt{(1-\beta^2)}}, \qquad E_z = \frac{e}{r'^3}\frac{z'}{\sqrt{(1-\beta^2)}},$$

$$\mathbf{H} = \frac{e}{r'^3}\frac{(\mathbf{r}' \wedge \mathbf{v})/c}{\sqrt{(1-\beta^2)}}.$$

If $\mathbf{r} = (x, y, z)$ is the vector between the position of the charge, measured simultaneously in K, and the field point, then[158]

$$\mathbf{r}' = \left(\frac{x}{\sqrt{(1-\beta^2)}}, y, z\right), \qquad r' = \sqrt{\left(\frac{x^2}{1-\beta^2} + y^2 + z^2\right)},$$

$$\mathbf{E} = \frac{e}{r'^3}\frac{\mathbf{r}}{\sqrt{(1-\beta^2)}}, \qquad \mathbf{H} = \frac{e}{r'^3}\frac{(\mathbf{r} \wedge \mathbf{v})/c}{\sqrt{(1-\beta^2)}}. \qquad (246)$$

Thus, here too, the electric field is radial and the magnetic field is perpendicular to both the radius vector and the direction of motion. The equipotential surfaces in the moving system are not spheres but Heaviside ellipsoids, which were introduced into electrodynamics by Heaviside[159] as early as 1889. Such an ellipsoid is simply the result of carrying out a Lorentz transformation on a sphere.

The field (246) can also be obtained by taking a special case of the general formula (244). Let us introduce the vector X', starting from the world line of the charge (which is straight in our case) and normal to it, and whose end point is the field point. In the rest system K' its components are $(\mathbf{r}', 0)$. It is easily found that

$$X_i' = X_i + \frac{1}{c^2}u_i(X_r u^r), \qquad X_i' X'^i = |X'|^2 = \frac{1}{c^2}(X_r u^r)^2,$$

$$|X'| = -\frac{1}{c}(X_r u^r),$$

so that
$$4\pi F_{ik} = \frac{e}{c|X'|^3}(u_i X_k' - u_k X_i'). \qquad (246a)$$

(γ) *The field for hyperbolic motion.* The next-simplest case to that of uniform motion is that of "uniform" acceleration, i.e. of hyperbolic motion in relativity (see §26). The field of a point charge describing a

---

[158] This derivation is first found in Poincaré's paper (Cf. footnote 11, p. 2, *R.C. Circ. mat. Palermo*, *loc. cit.*, § 5).
[159] Cf. H. A. Lorentz, footnote 4, p. 1, *ibid.*, § 11 b, *q.v.* for references to the earlier literature.

hyperbolic motion was first determined by Born.[160] Sommerfeld[161] uses contour integration for calculating it. An elementary derivation is also given by v. Laue.[162] Let us take the origin of the coordinate system at the centre of the hyperbola and let the x^1x^4-plane coincide with the plane of the hyperbola. Then the point

$$\xi^1 = a \cos\frac{s}{a}, \qquad \xi^4 = a \sin\frac{s}{a}, \qquad \xi^2 = \xi^3 = 0$$

of the world line (196 a) of the charge, with which the field point $x^1, ..., x^4$ is associated through (241), is determined by

$$\cos(\psi - \varphi) = \frac{R^2 + a^2}{2\xi\rho}, \qquad \psi = \frac{s}{a}, \tag{247}$$

where $\qquad R = \sqrt{(x_i x^i)}, \qquad x^1 = \rho \cos\varphi, \qquad x^4 = \rho \sin\varphi. \tag{248}$

The components of the four-vector potential become

$$\phi_1 = \frac{e}{4\pi\rho}\frac{\sin\psi}{\sin(\psi - \varphi)}, \qquad \phi_2 = \phi_3 = 0, \qquad \phi_4 = -\frac{e}{4\pi\rho}\frac{\cos\psi}{\sin(\psi - \varphi)}. \tag{249}$$

In the coordinate system in which the charge is instantaneously at rest at time $t - (r/c)$, ϕ_1 vanishes at time t. In the system in which the field point and the centre of the hyperbola are simultaneous, we obtain for the field intensities

$$E_x = -\frac{ie}{4\pi\rho \sin^2(\psi - \varphi)} \cdot \frac{\partial\psi}{\partial\rho} = -\frac{ie[a\cos(\psi - \varphi) - \rho]}{4\pi a\rho^2 \sin^3(\psi - \varphi)}$$
$$= -\frac{e}{\pi}\frac{a^2[R^2 + a^2 - 2\rho^2]}{[(R^2 + a^2)^2 - 4a^2\rho^2]^{3/2}}, \tag{250}$$

$$E_y = -\frac{ie}{4\pi\rho \sin^2(\psi - \varphi)} \cdot \frac{\partial\psi}{\partial x^{(2)}} = \frac{iey}{4\pi a\rho^2 \sin^3(\psi - \varphi)}$$
$$= \frac{2e}{\pi}\frac{a^2\rho}{[(R^2 + a^2)^2 - 4a^2\rho^2]^{3/2}},$$

$$\mathbf{H} = 0$$

(where we have written y instead of $x^{(2)}$).

Hyperbolic motion thus constitutes a special case, for which there is no formation of a wave zone nor any corresponding radiation. (Radiation, on the other hand, does occur when two uniform, rectilinear, motions are connected by a "portion" of hyperbolic motion.)

For calculating the field due to hyperbolic motion, it would seem natural to introduce a (non-Galilean) reference system which moves with the

[160] Cf. footnote 114, p. 76, *ibid.*
[161] Cf. footnote 55, p. 22, *ibid.*, 33, p. 670.
[162] M. v. Laue, *Das Relativitätsprinzip* (1st edn., Braunschweig 1911), p. 108, § 18 (d).

charge. As x-coordinate we can introduce the quantity denoted by ρ above, as time coordinate the angle φ would be best, since it is a multiple of the proper time of the moving charge. The line element in this coordinate system becomes

$$ds^2 = (d\xi^1)^2 + (d\xi^2)^2 + (d\xi^3)^2 + (\xi^1)^2 (d\xi^4)^2$$

$$(\xi^1 = \rho, \qquad \xi^2 = x^2, \qquad \xi^3 = x^3, \qquad \xi^4 = \varphi). \tag{251}$$

The field equations in these coordinates can be written down immediately, if the methods of Part II are used. The problem then becomes a statical, but not one-dimensional, problem, and the calculations are not very much simplified. It is historically of interest that Born[163] already discussed this problem by introducing a moving system. The time parameter used by him is different from the one employed above, and he obtained the differential equations by reverting to the variational principle which he had earlier formulated in an invariant way (see § 31).

(δ) *Invariance of the light phase. Reflection at a moving mirror. Radiation pressure.* In § 6 we derived the relativistic formulae for the Doppler effect and aberration from the invariance of the light phase. The proof for the latter is obtained directly from the transformation formulae for the field intensities. Since, in addition, the phase of a *plane* wave is a linear function of the space–time coordinates, it can be written as a scalar product formed by the position vector and a wave four-vector l_i,

$$- \nu t + (\mathbf{k} \cdot \mathbf{r}) = l_i x^i. \tag{252}$$

Here, $\mathbf{k}$ is the three-dimensional propagation vector whose direction coincides with that of the wave normal and whose value is equal to that of the reciprocal wave-length (wave number.) If, in particular, the wave normal is parallel to the xy-plane, we have

$$l_i = \left(\frac{\nu}{c} \cos\alpha, \quad \frac{\nu}{c} \sin\alpha, \quad 0, \quad \frac{i\nu}{c} \right). \tag{252a}$$

In vacuo, l_i is a null vector. The transformation formulae (15) and (16) of § 6 then follow at once. Because of (204) it is easy to write down, in addition, the transformation formulae[164] for the amplitude A,

$$A' = A \frac{1 - \beta \cos\alpha}{\sqrt{(1 - \beta^2)}}. \tag{253}$$

If, furthermore, we consider the transformation of the volume V of a laterally bounded, finite, wave,

$$V' = V \frac{\sqrt{(1 - \beta^2)}}{1 - \beta \cos\alpha}, \tag{254}$$

we obtain for the total energy $E = \frac{1}{2} A^2 V$ of the wave the same formula $(228\,\text{a})$[164] as before. A comparison with (15) shows that energy and

[163] Cf. footnote 114, p. 76, *ibid.*
[164] Cf. footnote 12, p. 2, A. Einstein, *ibid.*, § 8.

amplitude transform in the same way as frequency; for the volume, on the other hand, the reverse is true,

$$\frac{E'}{\nu'} = \frac{E}{\nu}, \qquad \frac{A'}{\nu'} = \frac{A}{\nu}, \qquad V'\nu' = V\nu. \qquad (254\,\text{a})$$

The first of these relations was stressed by Einstein[165] as being of particular importance; Wien's law is connected with it.

Closely related to these transformation formulae for the frequency and direction of a plane wave are the laws of reflection at a moving mirror (assumed to be perfectly conducting and plane). These laws can evidently be reduced to those valid for a mirror at rest, by introducing a coordinate system K' moving with the mirror.[165] Here, too, the theory of relativity can only produce something new in the form of the derivation, but not as far as the result is concerned.[166] In fact, the formulae of the earlier theory are exact, since all quantities occurring in them are measured with the measuring rods and clocks of the *same* system, so that the Lorentz contraction and time dilatation cannot influence the result.

Let α_1 and α_2 be the angles, measured in K, which the normals of the incident and reflected waves make with the direction of motion of the mirror, α'_1 and α'_2 the corresponding angles in K', ν_1 and ν_2 the frequencies of the incident and reflected waves in K, $\nu'_1 = \nu'_2 = \nu'$ the corresponding frequencies in K'. If the mirror is moving parallel to its own plane, then $\alpha'_2 = 2\pi - \alpha'_1$ and it also follows from (15) and (16) that $\alpha_2 = 2\pi - \alpha_1$, $\nu_2 = \nu_1$. In other words, the law of reflection in this case is no different from that for a mirror at rest. Moreover, it is always only the velocity component in the direction of the normal to the mirror, which is of importance. It can therefore be assumed that the mirror moves normal to its own plane; let its velocity v be taken positive in the direction of the inner normal; α_1 and α_2 as well as α'_1 and α'_2 are now the angles of incidence and reflection, respectively. Then

$$\alpha'_2 = \pi - \alpha'_1.$$

From (15) and (16) it then follows that

$$\nu^2(1 - \beta \cos \alpha_2) = \nu_1(1 - \beta \cos \alpha_1), \qquad (255)$$

$$\nu_2 \sin \alpha_2 = \nu_1 \sin \alpha_1, \qquad (256)$$

$$\tan \frac{\pi - \alpha_2}{2} = \frac{1 + \beta}{1 - \beta} \tan \frac{\alpha_1}{2}. \qquad (257)$$

In addition,

$$\frac{\cos \alpha_1 - \beta}{1 - \beta \cos \alpha_1} = - \frac{\cos \alpha_2 - \beta}{1 - \beta \cos \alpha_2},$$

[165] Cf. footnote 12, p. 2, *ibid.*, § 7.
[166] Cf. detailed discussions of the reflection laws at a moving mirror (published before the advent of relativity) by W. Hicks, *Phil. Mag.*, **3** (1902) 9; M. Abraham, *Boltzmann-Festschrift* (1904) p. 85; *Ann. Phys., Lpz.*, **14** (1904) 236; *Theorie der Elektrizität*, Vol. 2 (1st edn., Leipzig 1905) p. 343, § 40; also E. Kohl, *Ann. Phys., Lpz.*, **28** (1909) 28.

from which we obtain

$$\cos \alpha_2 = -\frac{(1 + \beta^2)\cos \alpha_1 - 2\beta}{1 - 2\beta \cos \alpha_1 + \beta^2} \tag{257a}$$

and

$$\nu_2 = \nu_1 \frac{1 - 2\beta \cos \alpha_1 + \beta^2}{1 - \beta^2}. \tag{258}$$

A very elegant method for deriving these formulae is due to Bateman.[167] To obtain, in K', the phase of the reflected from that of the incident wave, one has simply to replace x' by $-x'$. This amounts to going over to the mirror image. To obtain the corresponding transformation in K, one has first to go over to K', by means of an imaginary rotation through an angle $+\varphi$ (determined by (187)), then reverse the x-axis, and finally get back to K by a rotation $-\varphi$. *These operations, however, are equivalent to a rotation through 2φ and a subsequent reversal of the x-axis.*

If we therefore put

$$\tan 2\varphi = \frac{iU}{c},$$

then, from (187),

$$U = \frac{2c^2 v}{c^2 + v^2},$$

$$\cos 2\varphi = \frac{1 + \beta^2}{1 - \beta^2}, \qquad \sin 2\varphi = i\frac{1 - \beta^2}{2\beta}. \tag{259}$$

The transformations

$$\bar{x} = -\frac{x - Ut}{\sqrt{[1 - (U^2/c^2)]}} = -\frac{c^2 + v^2}{c^2 - v^2}x + \frac{2c^2 v}{c^2 - v^2}t,$$

$$\bar{t} = \frac{t - (U/c^2)x}{\sqrt{[1 - (U^2/c^2)]}} = \frac{c^2 + v^2}{c^2 - v^2}t - \frac{2v}{c^2 - v^2}, \tag{260}$$

give the connection between object (x, t) and image $(\bar{x}, \bar{t})$ in the moving mirror. A point of the moving mirror for which $x = vt$, is transformed into itself $(\bar{x} = x, \bar{t} = t)$; if the object moves with the same velocity as the mirror $(x = vt+a)$ then the same is true of the image $(\bar{x} = v\bar{t}+\bar{a})$, as it should be. The image of a point at rest in K moves with velocity U, which can also be obtained from the addition theorem for velocities by combining v with v. The phase of the wave reflected at the moving mirror is obtained directly from that of the incident wave by means of the substitution (260). The relations (257), (257 a), (258) can be written, quite analogously to (16 a), (16) and (15),

$$\tan\frac{\pi - \alpha_2}{2} = \sqrt{\left[\frac{1 + (U/c)}{1 - (U/c)}\right]}\tan\frac{\alpha_1}{2}, \tag{257'}$$

[167] H. Bateman, *Phil. Mag.*, **18** (1909) 890.

$$\cos{(\pi - \alpha_2)} = \frac{\cos \alpha_1 - (U/c)}{1 - (U/c) \cos \alpha_1}, \qquad (257'\,a)$$

$$\nu_2 = \nu_1 \frac{1 - (U/c) \cos \alpha_1}{\sqrt{[1 - (U^2/c^2)]}}, \qquad (258')$$

by introducing everywhere $\pi - \alpha_2$ instead of α_2, because of the reversal of the x-axis. Varičak[168] interprets these formulae in terms of the Bolyai–Lobachevski geometry.

The relations (254a) allow us at once to write down the change in amplitude for reflection at a moving mirror.[168a]

$$\frac{A_2}{\nu_2} = \frac{A_1}{\nu_1}, \qquad A_2 = A_1 \frac{1 - 2\beta \cos \alpha_1 + \beta^2}{1 - \beta^2}. \qquad (261)$$

The difference, per unit time and unit area, between the outgoing energy

$$\tfrac{1}{2}A_2{}^2(c \cos \alpha_2 - v)$$

and the incident energy

$$\tfrac{1}{2}A_1{}^2(- c \cos \alpha_1 + v)$$

must be equal to the work done per unit time, pv, by the radiation pressure p.[168a] From this, p is obtained in agreement with the result given by pre-relativistic theory

$$p = A_1{}^2 \frac{(\cos \alpha_1 - \beta)^2}{1 - \beta^2} = A'_1{}^2 \cos^2 \alpha' = p'. \qquad (262)$$

The radiation pressure is an invariant. In §45 it will be shown that this is so for every kind of pressure.

(ϵ) *The radiation field of a moving dipole.* The field of the Hertz oscillator is contained in (243) as a special case. If, in addition, we are only concerned with the field in the wave zone (i.e. at large distances), X_i can be measured from the centre of the dipole and we can use for u_i the velocity vector of the dipole centre instead of that of the separate charges. Let $\mathbf{v}$, $\dot{\mathbf{v}}$ be the velocity of the dipole and the acceleration of the oscillating charge *at time* $t - r/c$; $\mathbf{r}$ the *retarded* and $\mathbf{R} = \mathbf{r} - r(\mathbf{v}/c)$ the *simultaneous* position vectors of the dipole, referred to the field point; $\mathbf{r}_1$ the unit vector $\mathbf{r}/r$, $\mathbf{R}_1$ the corresponding vector $\mathbf{R}_1 = \mathbf{R}/r = \mathbf{r}_1 - (\mathbf{v}/c)$; θ the angle between $\mathbf{v}$ and $\mathbf{r}$. Because of (241) we then obtain

$$\left.\begin{aligned}
\mathbf{E} &= \frac{e}{4\pi c^2 r (1 - \beta \cos \theta)^3}\{(\mathbf{r}_1 \cdot \dot{\mathbf{v}})\,\mathbf{R}_1 - \dot{\mathbf{v}}(\mathbf{R}_1 \cdot \mathbf{r}_1)\} \\[2mm]
&= \frac{e}{4\pi c^2 r (1 - \beta \cos \theta)^3}\mathbf{r}_1 \wedge (\mathbf{R}_1 \wedge \dot{\mathbf{v}}), \\[2mm]
\mathbf{H} &= \frac{e}{4\pi c^2 r (1 - \beta \cos \theta)^3}\left\{(\mathbf{r}_1 \cdot \dot{\mathbf{v}})\frac{(\mathbf{v} \wedge \mathbf{r}_1)}{c} + (\dot{\mathbf{v}} \wedge \mathbf{r}_1)\right\} = \mathbf{r}_1 \wedge \mathbf{E}.
\end{aligned}\right\} \qquad (263)$$

[168] Cf. footnote 111, p. 74, *ibid.*
[168a] Cf. footnote 12, p. 2, A. Einstein, *ibid.*, § 7.

The theory of relativity enables us to derive these formulae (first found by Heaviside[169] and later, more accurately, by Abraham[170]) from Hertz's formulae for the field of a dipole at rest. The simplest procedure is to prove, using Poincaré's[171] method, that **E** and **H** are normal to each other and to $\mathbf{r}_1$, and that they have the same values also in the moving system. This fact can be expressed by means of the invariant vector equations

$$F_{ik} X^k = 0, \qquad F^*_{ik} X^k = 0.$$

One then only needs to calculate the energy density from the transformation formula for the tensor S_{ik}.

M. v. Laue[172] investigated the momentum and energy radiated by a moving dipole, from the relativistic point of view. Equations (228 b) must hold here, since the *existence* of the wave field is independent of the presence of the electric charges. Taking the time dilatation into account, we have for the energy radiated per unit time

$$-\frac{dE}{dt} = -\frac{dE'}{dt'}.$$

But in the rest system

$$-\frac{dE'}{dt'} = \frac{e}{6\pi c^3} \dot{\mathbf{v}}'^2,$$

and the transformation formulae (193) for the acceleration then lead directly to

$$-\frac{dE}{dt} = \frac{e^2}{6\pi c^3} \left[\frac{\dot{v}_x^2}{(1-\beta^2)^3} + \frac{\dot{v}_y^2 + \dot{v}_z^2}{(1-\beta^2)^2} \right] = \frac{e^2}{6\pi c^3} \frac{1}{(1-\beta^2)^2} \left[\dot{\mathbf{v}}^2 + \frac{(\mathbf{v}\cdot\dot{\mathbf{v}})^2}{c^2(1-\beta^2)} \right],$$

$$\tag{264}$$

$$-\frac{dG}{dt} = -\frac{\mathbf{v}}{c^2} \frac{dE}{dt}.$$

These are in agreement with Abraham's calculations from the field (263). The radiated energy is the sum of the contributions which would be produced by the longitudinal and transverse components of **v** separately.

Looking at this process as seen from system K', it is of interest to note that the velocity of the dipole is not changed by the radiation (it remains zero in K'). Because of the inertia of the energy, however, the momentum conservation law is not violated, in spite of the momentum radiation (264) (cf. § 41).

[169] O. Heaviside, *Nature, Lond.*, **67** (1902) 6; cf. also H. A. Lorentz, footnote 4, p. 1, *ibid.*, § 14, p. 180 and the references quoted there.
[170] M. Abraham, *Ann. Phys., Lpz.*, **14** (1904) 236; *Theorie der Elektrizität*, Vol. 2. (1st edn., 1905), §§ 13–15.
[171] Cf. footnote 11, p. 2, *R.C. Circ. mat. Palermo, loc. cit.*
[172] M. v. Laue, *Verh. dtsch. phys. Ges.*, **10** (1908) 888; *Ann. Phys., Lpz.*, **28** (1909) 436.

(ζ) *Radiation reaction*. When $\mathbf{v} = 0$ at a given moment, the magnitude of the radiation reaction is given by[173]

$$\mathbf{K} = \frac{e^2}{6\pi c^3}\ddot{\mathbf{v}}.$$

From this, Laue[174] and Abraham[175] independently derived an expression for the radiation reaction on a moving charge, using a Lorentz transformation. For this, it is sufficient to find a vector K_i (see (219)) whose three space components coincide with the above expression for $\mathbf{K}$ for $\mathbf{v} = 0$, and whose time component vanishes. For this purpose, let us write

$$K_i = \frac{e^2}{6\pi c^3}\left\{\frac{d^2 u_i}{d\tau^2} + \alpha u_i\right\} \tag{265}$$

and determine α from the condition $K_i u^i = 0$. Because of (159) and (192), we find that

$$\alpha = \frac{1}{c^2}u^k\frac{d^2 u_k}{d\tau^2} = -\frac{1}{c^2}\frac{du_k}{d\tau}\frac{du^k}{d\tau}, \tag{266}$$

and the resulting expression for $\mathbf{K}$ is then

$$\mathbf{K} = \frac{e^2}{6\pi c^3}\frac{1}{(1-\beta^2)}\left\{\ddot{\mathbf{v}} + \dot{\mathbf{v}}\frac{3(\mathbf{v}\cdot\dot{\mathbf{v}})}{c^2(1-\beta^2)} + \frac{\mathbf{v}}{c^2(1-\beta^2)}\left[(\mathbf{v}\cdot\ddot{\mathbf{v}}) + \frac{3(\mathbf{v}\cdot\dot{\mathbf{v}})^2}{c^2(1-\beta^2)}\right]\right\}. \tag{265a}$$

Abraham[175] also proves that the time integral over $\mathbf{K}$, taken over the duration of the radiation process, is equal to the radiated momentum, and similarly that the time integral over $(\mathbf{v}\cdot\mathbf{K})$ is equal to the radiated energy. For hyperbolic motion, $\mathbf{K}$ vanishes, as it should, since no radiation takes place (see above, under (γ)).

33. Minkowski's phenomenological electrodynamics of moving bodies

In principle, all problems concerning the electrodynamics of moving bodies are solved by the field equations (203) and (208) of electron theory. Because of our incomplete knowledge of the structure of matter, however, we are entitled to ask ourselves what statements the relativity principle allows us to make concerning (macroscopic) processes in moving bodies, assuming processes in bodies at rest to be experimentally known. This question was answered by Minkowski.[176] He showed that the equations for

[173] Cf. H. A. Lorentz, footnote 4, p. 1, *ibid.*, § 20, p. 190, Eq. (74).
[174] Cf. footnote 172, p. 98, *ibid.*
[175] M. Abraham, *Theorie der Elektrizität*, Vol. 2 (2nd edn., 1908) p. 387.
[176] Cf. footnote 54, p. 21, Minkowski, II; see also the derivation by A. Einstein and J. Laub, *Ann. Phys., Lpz.*, **26** (1908) 532, without the use of tensor calculus.

moving bodies follow unambiguously from the relativity principle and from Maxwell's equations for bodies at rest,

$$\operatorname{curl} \mathbf{E} + \frac{1}{c} \frac{\partial \mathbf{B}}{\partial t} = 0, \qquad \operatorname{div} \mathbf{B} = 0, \qquad (F)$$

$$\operatorname{curl} \mathbf{H} - \frac{1}{c} \frac{\partial \mathbf{D}}{\partial t} = \mathbf{J}, \qquad \operatorname{div} \mathbf{D} = \rho, \qquad (G)$$

$$\mathbf{D} = \epsilon \mathbf{E}, \qquad \mathbf{B} = \mu \mathbf{H}, \qquad \mathbf{J} = \sigma \mathbf{E}. \qquad (H)$$

Just as in the four-dimensional formulation of the equations of electron theory, we can combine first those equations which do not contain the charge density and current, and then the remaining set. This leads to the introduction of the two "surface" tensors

$$(F_{41}, F_{42}, F_{43}) = i\mathbf{E}, \qquad (F_{23}, F_{31}, F_{12}) = \mathbf{B}, \qquad (267)$$

$$(H_{41}, H_{42}, H_{43}) = i\mathbf{D}, \qquad (H_{23}, H_{31}, H_{12}) = \mathbf{H}, \qquad (268)$$

and of the four-vector

$$(J^1, J^2, J^3) = \mathbf{J}, \qquad J^4 = i\rho, \qquad (269)$$

with the corresponding transformation formulae

$$\left. \begin{aligned} \mathbf{E}'_{\parallel} = \mathbf{E}_{\parallel}, \qquad \mathbf{E}'_{\perp} &= \frac{\{\mathbf{E} + (1/c)(\mathbf{v} \wedge \mathbf{B})\}_{\perp}}{\sqrt{(1 - \beta^2)}} \\ \mathbf{B}'_{\parallel} = \mathbf{B}_{\parallel}, \qquad \mathbf{B}'_{\perp} &= \frac{\{\mathbf{B} - (1/c)(\mathbf{v} \wedge \mathbf{E})\}_{\perp}}{\sqrt{(1 - \beta^2)}} \end{aligned} \right\} \qquad (267a)$$

$$\left. \begin{aligned} \mathbf{D}'_{\parallel} = \mathbf{D}_{\parallel}, \qquad \mathbf{D}'_{\perp} &= \frac{\{\mathbf{D} + (1/c)(\mathbf{v} \wedge \mathbf{H})\}_{\perp}}{\sqrt{(1 - \beta^2)}} \\ \mathbf{H}'_{\parallel} = \mathbf{H}_{\parallel}, \qquad \mathbf{H}'_{\perp} &= \frac{\{\mathbf{H} - (1/c)(\mathbf{v} \wedge \mathbf{D})\}_{\perp}}{\sqrt{(1 - \beta^2)}} \end{aligned} \right\} \qquad (268a)$$

$$\mathbf{J}'_{\parallel} = \frac{\mathbf{J}_{\parallel} - \beta\rho}{\sqrt{(1 - \beta^2)}}, \qquad \mathbf{J}'_{\perp} = \mathbf{J}_{\perp}, \qquad \rho' = \frac{\rho - (1/c)(\mathbf{v} \cdot \mathbf{J})}{\sqrt{(1 - \beta^2)}}. \qquad (269a)$$

When the substance is at rest in K', $\mathbf{v}$ is the velocity of the substance in K (in contrast to the velocity $\mathbf{u}$ of the electrons) and $\beta = v/c$. The equations (F) and (G) remain also valid for moving bodies, and can be written in the form

$$\frac{\partial F_{ik}}{\partial x^l} + \frac{\partial F_{kl}}{\partial x^i} + \frac{\partial F_{li}}{\partial x^k} = 0 \qquad \left(\text{or, respectively, } \frac{\partial F^{*ik}}{\partial x^k} = 0\right), \qquad (270)$$

$$\frac{\partial H^{ik}}{\partial x^k} = J^i. \qquad (271)$$

These formulae are rigorously valid only for uniformly moving bodies and, because of the additivity of the fields, also when several bodies are present which move uniformly with *different* velocities and are separated by vacuum regions. The approximation to which (270) and (271) are correct will generally be the better, the smaller the acceleration of the substance.

As for the physical meaning of the quantities occurring in these equations, we can say that $\mathbf{E}$, $\mathbf{D}$ ($\mathbf{B}$, $\mathbf{H}$) represent the force acting in vacuum on an electric (magnetic) unit pole at rest in K; in ponderable matter they have no immediately obvious meaning. Furthermore, $\mathbf{J}$, ρ can be denoted as the current and charge density also in system K. The justification for this in the case of a non-conductor, where $\mathbf{J}' = 0$, follows directly from (269 a). For here ρ becomes equal to $\rho'/\sqrt{(1-\beta^2)}$, i.e. $de = \rho\,dV$ is invariant, and $\mathbf{J} = \rho\mathbf{v}/c$ becomes equal to the convection current. Also, J^i satisfies quite generally the equation of continuity

$$\frac{\partial J^i}{\partial x^i} = 0. \tag{272}$$

Hence $\mathbf{J}$ is generally the sum of the conduction and convection currents and ρ is the charge density.

It is also of advantage to replace $\mathbf{E}(\mathbf{H})$ by the forces *measured in* K, $\mathbf{E}^*(\mathbf{H}^*)$, which act on an electric (magnetic) unit pole moving with the substance. From (213) and (267 a), (268 a) we have

$$\mathbf{E}^* = \mathbf{E} + \frac{1}{c}(\mathbf{v} \wedge \mathbf{B}), \qquad \mathbf{H}^* = \mathbf{H} - \frac{1}{c}(\mathbf{v} \wedge \mathbf{D}). \tag{273}$$

In contrast to $\mathbf{E}$ and $\mathbf{H}$, these vectors have a direct physical meaning also in the interior of ponderable bodies. In addition, the field equations (270), (271) take on a simple form when $\mathbf{E}^*$, $\mathbf{H}^*$ are introduced. If $\mathbf{A}$ is an arbitrary vector, let us define the operation $\underline{\dot{\mathbf{A}}}$ by[177]

$$\frac{d}{dt}\int \mathbf{A}_n\,d\sigma = \int \underline{\dot{\mathbf{A}}}_n\,d\sigma,$$

where the integration is to be carried out over a surface moving with the body. Then[177]

$$\underline{\dot{\mathbf{A}}} = \frac{\partial \mathbf{A}}{\partial t} + \mathbf{v}\,\operatorname{div}\mathbf{A} - \operatorname{curl}(\mathbf{v} \wedge \mathbf{A}),$$

and the field equations can be written

$$\left.\begin{aligned}
\operatorname{curl}\mathbf{E}^* &= -\frac{1}{c}\underline{\dot{\mathbf{B}}}, & \operatorname{div}\mathbf{B} &= 0 \\[2mm]
\operatorname{curl}\mathbf{H}^* &= \frac{1}{c}\underline{\dot{\mathbf{D}}} + \mathbf{J}_c, & \operatorname{div}\mathbf{D} &= \rho.
\end{aligned}\right\} \tag{274}$$

$\mathbf{J}_c$ is the conduction current,

$$\mathbf{J} = \rho\frac{\mathbf{v}}{c} + \mathbf{J}_c. \tag{275}$$

[177] Cf. footnote 48, p. 17, H. A. Lorentz, *ibid.*, § 4, p. 78, Eqs. (12) and (13).

Equations (274) also allow us to go over to the integral form.[178] From the transformation formulae (269 a) it follows that *the splitting of the current into a conduction and a convection current is not independent of the reference system. Even when there is no charge density and only a conduction current present in K′, there will appear a charge density, and hence also a convection current, in K*.[178a] The corresponding transformation formulae are obtained from (269 a) and (275),

$$\mathbf{J}'_{c\parallel} = \frac{\mathbf{J}_{c\parallel}}{\sqrt{(1-\beta^2)}}, \qquad \mathbf{J}'_{c\perp} = \mathbf{J}_{c\perp}, \tag{276}$$

$$\rho' = \rho\sqrt{(1-\beta^2)} - \frac{(1/c)(\mathbf{v}\cdot\mathbf{J}_c)}{\sqrt{(1-\beta^2)}}, \qquad \rho = \frac{\rho' + (1/c)(\mathbf{v}\cdot\mathbf{J}'_c)}{\sqrt{(1-\beta^2)}}. \tag{277}$$

(See § 34 on the electron-theoretical proof of these formulae.)

Equations (*F*), (*G*) (or (274)) constitute only a formal scheme without physical content, as long as we do not add the connecting equations which relate **E***, **H*** and **D**, **B**. These are found by means of equations (*H*), which have not so far been used. From (267 a), (268 a) and (273) we obtain immediately,

$$\left.\begin{aligned}
\mathbf{D} + \frac{1}{c}(\mathbf{v}\wedge\mathbf{H}) &= \epsilon\left\{\mathbf{E} + \frac{1}{c}(\mathbf{v}\wedge\mathbf{B})\right\} = \epsilon\mathbf{E}^*, \\
\mathbf{B} - \frac{1}{c}(\mathbf{v}\wedge\mathbf{E}) &= \mu\left\{\mathbf{H} - \frac{1}{c}(\mathbf{v}\wedge\mathbf{D})\right\} = \mu\mathbf{H}^*.
\end{aligned}\right\} \tag{278}$$

Solving for **D** and **B**, and eliminating **E***, **H***, these equations give

$$\left.\begin{aligned}
\mathbf{D} &= \frac{\epsilon(1-\beta^2)\mathbf{E} + (\epsilon\mu-1)\{[(\mathbf{v}\wedge\mathbf{H})/c] - \epsilon(\mathbf{v}/c)(\mathbf{v}\cdot\mathbf{E})/c\}}{1-\epsilon\mu\beta^2}, \\
\mathbf{B} &= \frac{\mu(1-\beta^2)\mathbf{H} - (\epsilon\mu-1)\{[(\mathbf{v}\wedge\mathbf{E})/c] - \mu(\mathbf{v}/c)(\mathbf{v}\cdot\mathbf{H})/c\}}{1-\epsilon\mu\beta^2},
\end{aligned}\right\} \tag{278a}$$

and eliminating **E**, **H** by means of (273),

$$\left.\begin{aligned}
\mathbf{D} &= \frac{\epsilon[\mathbf{E}^* - (\mathbf{v}/c)(\mathbf{v}\cdot\mathbf{E}^*)/c] - (1/c)(\mathbf{v}\wedge\mathbf{H}^*)}{1-\beta^2}, \\
\mathbf{B} &= \frac{\mu[\mathbf{H}^* - (\mathbf{v}/c)(\mathbf{v}\cdot\mathbf{H}^*)/c] + (1/c)(\mathbf{v}\wedge\mathbf{E}^*)}{1-\beta^2}.
\end{aligned}\right\} \tag{278b}$$

For *unmagnetizable bodies* ($\mu = 1$) these equations agree, *to first order*,

---

[178] Cf. H. A. Lorentz (see footnote 48, p. 17, *ibid.*, § 6, and footnote 4, p. 1, *ibid.*, § 33). The quantities E′, H′ used there are identical with our **E***, **H***, and the equations (III″a), (IV″a), *loc. cit.*, agree with (274). While equations (III″), (IV″) are of the same form as (*F*), (*G*), Lorentz's connection between H and H′ is different from that expressed by the second of the relations (273). On the other hand, his quantity **E** is identical with ours (cf. Eq. (106), *loc. cit.*, and the first of our equations (273)). See also Minkowski's own comparison of his formulae with those of Lorentz (cf. footnote 54, p. 21, Minkowski II, § 9).

[178a] H. A. Lorentz, 'Alte und neue Fragen der Physik', *Phys. Z.* **11** (1910), 1234; in particular, 1242. M. v. Laue, '*Das Relativitätsprinzip*', (1st edn., Braunschweig 1911) 119.

with the relation between **D**, **B** and **E'**, **H'** (corresponding to our **E***, **H***), given by Lorentz.[179]

Just as (278) is obtained from the first two relations (H), so the differential form of Ohm's law for moving bodies results from the last of Eqs. (H). From (276), (276 a) and (273) we have

$$\mathbf{J}_{c\,\|} = \sigma\sqrt{(1-\beta^2)}\mathbf{E}^*{}_\|, \qquad \mathbf{J}_{c\perp} = \frac{\sigma}{\sqrt{(1-\beta^2)}}\mathbf{E}^*{}_\perp, \qquad (279)$$

which can also be written

$$\mathbf{J}_c = \frac{\sigma}{\sqrt{(1-\beta^2)}}\left\{\mathbf{E}^* - \frac{\mathbf{v}}{c}\frac{(\mathbf{v}\cdot\mathbf{E}^*)}{c}\right\}. \qquad (279\text{a})$$

The transformation formula (277) for the charge density can now be put in the form

$$\rho = \frac{\rho' + \sigma(\mathbf{v}\cdot\mathbf{E}')/c}{\sqrt{(1-\beta^2)}} = \frac{\rho' + \sigma(\mathbf{v}\cdot\mathbf{E}^*)/c}{\sqrt{(1-\beta^2)}}. \qquad (277\text{a})$$

Minkowski[180] also gave Eqs. (278), (279) in a covariant form

$$\left.\begin{aligned}
H_{ik}v^k &= \epsilon F_{ik}v^k, \\
F_{ik}^* v^k &= \mu H_{ik}^* v^k
\end{aligned}\right\} \qquad (280)$$

or, $\qquad F_{ik}v_l + F_{kl}v_i + F_{li}v_k = \mu(H_{ik}v_l + H_{kl}v_i + H_{li}v_k),$

and $\qquad\qquad J_i + (v_k J^k)v_i = -\sigma F_{ik}v^k. \qquad (281)$

Here, v^k are the components of the four-velocity of the substance. To demonstrate the correctness of these equations, we need evidently only prove that, for a coordinate system K' moving with the substance, they go over into relations (H); this can be shown quite easily.

Each of the relations (280), (281) represents a system of four equations. The fourth equation, however, follows from the others. This can be seen by multiplying (280), (281) scalarly by v^i, in which case both sides vanish identically.

The *boundary conditions* are obtained from those of a body at rest by means of a Lorentz transformation. At the boundary surfaces of a moving body the tangential components of **E*** and **H*** and the normal component of **B** have to be continuous. It has to be assumed, though, that **v** is continuous. The same conditions hold for the case of a body bounded by vacuum, if in expressions (273) for **E***, **H*** we set **v** equal to the velocity of the body, on *both* sides of the equation. Also, for the surface charge density ω, $D_{n1} - D_{n2} = 4\pi\omega$. These conditions also follow directly from

[179] Cf. footnote 4, p. 1, *ibid.*, § 45, p. 227, Eq. (XXXIV''). For unmagnetizable bodies, according to Lorentz,

$$\mathbf{B} = \mathbf{H} = \mathbf{H}' + \frac{1}{c}(\mathbf{w}\wedge\mathbf{E}),$$

cf. *ibid.*, Eq. (XXX'), also § 42.

[180] H. Weyl, *Raum-Zeit-Materie* (1st edn., 1918), p. 153, Eq. (46).

(274) if we require that the time derivatives of the field quantities for a point moving with the substance (obtained by using the operator $(\partial/\partial t)+$ $(\mathbf{v} \cdot \text{grad}))$ must always remain finite.[181]

Just as Minkowski's field- and connecting-equations were derived from the corresponding laws for bodies at rest by means of a Lorentz transformation, so P. Frank[182] arrived at the equations of Hertz's theory[183] by means of a Galilean transformation.

34. Electron-theoretical derivations

The field equations of electron theory are covariant with respect to the Lorentz group and for stationary bodies they result in Maxwell's equations, when averaging processes are carried out. Therefore, for moving bodies, they must necessarily lead to the Minkowski field equations. Born[184] could in fact show this from Minkowski's posthumous notes, by regarding the motion of the electrons as a motion of the substance, subject to variations. From the first variation the electric polarization is obtained, from the second a further contribution to this, as well as the magnetization.

Another point to be clarified was, why Lorentz[185] arrived at equations, based on electron theory, which were different from those of Minkowski. For unmagnetizable bodies, P. Frank[186] was able to show that this was due to the fact that the Lorentz contraction and time dilatation had not been considered. Dällenbach[187] has given the natural extension of Lorentz's argument to the four-dimensional case, actually for arbitrarily constituted, moving, bodies. He defines the tensor F_{ik} as the average of the microscopic field tensor F_{ik}, the current vector J^i as the average of the contribution $(1/c)\rho_0 u^i_{\text{cond}}+(1/c)\rho_0 u^i_{\text{conv}}$ of the conducting electrons and the convective charges. The averaging is to be carried out over "physically infinitely small" world regions. From (208) we now see that it is essentially a question of finding the average of the current vector of the polarization electrons, $\rho_0 u^i_{\text{pol}}$. Using an argument which is perfectly analogous to that of Lorentz, only replacing space regions by world regions everywhere, we obtain

$$\overline{\rho_0 u^i_{\text{pol}}} = \frac{\partial M^{ik}}{\partial x^k}, \tag{282}$$

where, at first, M^{ik} is defined by

$$M^{ik} = \overline{\rho_0(x^i u^k)_{\text{pol}}}.$$

---

[181] The boundary conditions in Minkowski's electrodynamics are discussed by A. Einstein and J. Laub, *Ann. Phys.*, *Lpz.*, **28** (1909) 445, and M. v. Laue (cf. footnote 178 a. p. 102, *ibid.*, pp. 128 and 129).

[182] P. Frank, *Ann. Phys.*, *Lpz.*, **27** (1908) 897.

[183] E. Henschke, *Dissertation* (Berlin, 1912); *Ann. Phys.*, *Lpz.*, **40** (1913) 887, and I. Ishiwara, *Jb. Radioakt.* **9** (1912) 560; *Ann. Phys.*, *Lpz.*, **42** (1913) 986, derive the field equations from a generalization of the variational principle (232).

[184] Minkowski–Born, *Math. Ann.*, **68** (1910) 526; also published separately (Leipzig 1910); see also A. D. Fokker, *Phil. Mag.*, **39** (1920) 404.

[185] Cf. footnote 4, p. 1, *ibid.*, Part IV.

[186] P. Frank, *Ann. Phys.*, *Lpz.*, **27** (1908) 1059.

[187] W. Dällenbach, *Dissertation* (Zürich 1918); *Ann. Phys.*, *Lpz.*, **58** (1919) 523.

Since, however, the average of

$$\rho_0 x^i u^k + \rho_0 x^k u^i = \rho_0 \frac{d}{d\tau}(x^i x^k)$$

vanishes, we can put

$$M^{ik} = \overline{\tfrac{1}{2}\rho_0(x^i u^k - x^k u^i)_{\text{pol}}}. \tag{283}$$

Let us now *define* H^{ik} by

$$H^{ik} = F^{ik} - M^{ik}. \tag{284}$$

Then (271) follows from (208), by averaging. If the velocity of the polarization electrons relative to the centre of mass of the molecule is small compared with the light velocity, then the "surface" tensor M_{ik} is simply related to the electric polarization, defined in the usual way,

$$\mathbf{P} = N \sum \overline{e\mathbf{r}}$$

and the magnetization

$$\mathbf{M} = \tfrac{1}{2}N \sum \overline{e(\mathbf{r} \wedge \mathbf{u})}$$

(averages over time have been taken, N = number of molecules per unit volume, $\mathbf{u}$ = velocity of electrons, $\sum$ is taken over all electrons of a molecule). We have

$$\left. \begin{aligned} (M^{41}, M^{42}, M^{43}) &= \frac{-i\mathbf{P}}{\sqrt{[1 - (v^2/c^2)]}}, \\[2mm] (M^{23}, M^{31}, M^{12}) &= \frac{\mathbf{M}}{\sqrt{[1 - (v^2/c^2)]}} \end{aligned} \right\} \tag{285}$$

(v = velocity of the *substance*).

The definition (284) of H^{ik} then becomes, because of (267), (268),

$$\mathbf{D} = \mathbf{E} + \mathbf{P}, \qquad \mathbf{H} = \mathbf{B} - \mathbf{M}. \tag{284a}$$

From (285) follow the transformation formulae

$$\left. \begin{aligned} \mathbf{P'}_\| &= \mathbf{P}_\|, & \mathbf{P'}_\perp &= \frac{\{\mathbf{P} - (1/c)(\mathbf{v} \wedge \mathbf{M})\}_\perp}{\sqrt{[1 - (v^2/c^2)]}}, \\[2mm] \mathbf{M'}_\| &= \mathbf{M}_\|, & \mathbf{M'}_\perp &= \frac{\{\mathbf{M} + (1/c)(\mathbf{v} \wedge \mathbf{P})\}_\perp}{\sqrt{[(1 - (v^2/c^2)]}}. \end{aligned} \right\} \tag{285a}$$

If in system K' an electrically unpolarized particle is magnetized, then it will, in K, also be electrically polarized; if in K' an unmagnetized particle is electrically polarized, then it will, in K, also be magnetized[187a]. For this reason it is not correct to distinguish between magnetization electrons and (electrical) polarization electrons. We have therefore given them the same name, polarization electrons, and both electric and magnetic polarizations are to be understood. Formulae (285a) can naturally also be derived without the use of tensor calculus, by just using the definitions

[187a] Cf. footnote 178a, p. 102, H. A. Lorentz, *ibid.*

of $\mathbf{P}$ and $\mathbf{M}$. If, in particular, the substance is such that, in a system K' moving with it,

$$\mathbf{P}' = (\epsilon - 1)\mathbf{E}', \qquad \mathbf{M}' = (\mu - 1)\mathbf{H}',$$

then, because of the Lorentz covariance of these relations, the tensor equations

$$
\left.
\begin{aligned}
M_{ik}v^k &= -(\epsilon - 1)F_{ik}v^k, \\
M_{ik}v_l + M_{kl}v_i + M_{li}v_k &= (\mu - 1)\{H_{ik}v_l + H_{kl}v_i + H_{li}v_k\}
\end{aligned}
\right\} \tag{286}
$$

follow immediately and can be reduced to (280) with the help of (284). (286) can also be written in the form

$$
\left.
\begin{aligned}
\mathbf{P} - \frac{1}{c}(\mathbf{v}\wedge\mathbf{M}) &= (\epsilon - 1)\mathbf{E}^*, \\
\mathbf{M} + \frac{1}{c}(\mathbf{v}\wedge\mathbf{P}) &= (\mu - 1)\mathbf{H}^*.
\end{aligned}
\right\} \tag{286a}
$$

It still remains to derive theoretically the transformation formulae for the charge density and conduction current. If $N'_-(N'_+)$ and $\mathbf{u}'_+(\mathbf{u}'_-)$ are the number of positive (negative) particles per unit volume and their velocities, respectively, then by definition,

$$\rho' = e'_+ N'_+ - e'_- N'_-,$$

$$\mathbf{J}'_c = e'_+ N'_+ \mathbf{u}'_+ - e'_- N'_- \mathbf{u}'_-,$$

and in system K,

$$\rho = e_+ N_+ - e_- N_-,$$

$$\mathbf{J}_c = e_+ N_+(\mathbf{u}_+ - \mathbf{v}) - e_- N_-(\mathbf{u}_- - \mathbf{v}).$$

One can now introduce intermediate coordinate systems $K_+{}^0$ and $K_-{}^0$ in which the positive and negative particles, respectively, are at rest. From the addition theorem for velocities the following relations are then found,

$$N = N'\frac{1 + (\mathbf{v}\cdot\mathbf{u}')/c^2}{\sqrt{(1 - \beta^2)}},$$

$$(\mathbf{u} - \mathbf{v})_{\parallel} = \frac{(1 - \beta^2)\mathbf{u}'_{\parallel}}{1 + (\mathbf{v}\cdot\mathbf{u}')/c^2}, \qquad (\mathbf{u} - \mathbf{v})_{\perp} = \mathbf{u}_{\perp} = \frac{\sqrt{(1 - \beta^2)}\mathbf{u}'_{\perp}}{1 + (\mathbf{v}\cdot\mathbf{u}')/c^2},$$

(where the indices $+$ and $-$ have been omitted, since the formulae are identical for the two cases). From this, and from the invariance of the charge $(e = e')$, (276) and (277) are directly obtained. In this way we have arrived at an electron-theoretical explanation for the strange appearance of charge in moving current-carrying conductors.[187a]

35. Energy–momentum tensor and ponderomotive force in phenomenological electrodynamics. Joule heat

The principle of relativity enables us to deduce *unambiguously* the expressions for the energy–momentum tensor and the ponderomotive force for moving bodies from those for bodies at rest. Different expressions for the energy–momentum tensor have been suggested by different authors, though, and the question, which of these is to be preferred, cannot as yet be regarded as finally settled. Let us, first of all, discuss those results of relativity which are independent of the particular choice for the energy–momentum tensor.

We can combine the energy density W, energy current $\mathbf{S}$, momentum density $\mathbf{g}$ and stress tensor components T_{ik} ($i, k = 1, 2, 3$) (just as for the fields in a vacuum) into a single tensor S_{ik},

$$\left. \begin{aligned} S_{ik} &= -T_{ik} \quad \text{for} \quad i, k = 1, 2, 3, \\ (S_{14}, S_{24}, S_{34}) &= ic\mathbf{g}, \quad (S_{41}, S_{42}, S_{43}) = \frac{i}{c}\mathbf{S}, \\ S_{44} &= -W. \end{aligned} \right\} \tag{287}$$

For the present, nothing is said about the symmetry properties of this tensor. The equations

$$\mathbf{f} = \operatorname{div} \mathbf{T} - \dot{\mathbf{g}}, \tag{288}$$

$$\frac{\partial W}{\partial t} + \operatorname{div} \mathbf{S} + Q + A = 0 \tag{289}$$

give the ponderomotive force and the energy equation, similarly to (D) and (E) in § 30. In (289), Q is the Joule heat developed per unit time and unit volume, A the work done per unit time and unit volume,

$$A = \mathbf{f} \cdot \mathbf{v}. \tag{289a}$$

In a coordinate system K', in which the substance is momentarily at rest, A vanishes. It is natural to combine (288) and (289), to form the four-vector equation

$$f_i = -\frac{\partial S_i{}^k}{\partial x^k}. \tag{290}$$

The components f_i then have the following meaning,

$$(f_1, f_2, f_3) = \mathbf{f}, \qquad f_4 = \frac{i}{c}(Q + A). \tag{291}$$

It is seen from this that f_i is not now normal to v^i, but rather that

$$f_i v^i = -\frac{Q}{\sqrt{(1 - \beta^2)}}. \tag{292}$$

Since the right-hand side of (292), as well as the left-hand side, must be invariant, we obtain the transformation formula,

$$Q = Q'\sqrt{(1 - \beta^2)}. \tag{293}$$

Because of the invariance of the four-dimensional volume, this holds also for the *total* heat developed for a given process, in accordance with relativistic thermodynamics (cf. § 46), and it appears here as a consequence of the tensor character of S_{ik} and of the requirement that the force density can be derived from the stress tensor and momentum density as in (288).

The fact that $f_i v^i$ is different from zero leads to a peculiar dilemma. For the equations of motion can only be of the form

$$\mu_0 \frac{dv_i}{d\tau} = f_i \tag{221}$$

when $f_i v^i = 0$, since the left-hand side vanishes identically when multiplied scalarly by v^i. We thus have to choose between abandoning equations (290) for the four-force and abandoning the equations of motion (221). Minkowski[188] decided in favour of of the first alternative. This leads however to a transformation formula for the Joule heat different from (293), and thus to a contradiction with the requirements of relativistic thermodynamics. The correct formulation, due to Abraham[189], is as follows: It will be shown in general relativistic dynamics that an inertial mass must be ascribed to every kind of energy (see §§ 41 and 42). If, therefore, heat is developed, the rest-mass density does not remain constant and the equations of motion have to be written

$$\frac{d}{d\tau}(\mu_0 v_i) = f_i. \tag{294}$$

From this it follows by scalar multiplication with v^i, using (292), that

$$\frac{d\mu_0}{d\tau} = -\frac{1}{c^2}(f_i v^i) = +\frac{Q}{c^2}\frac{1}{\sqrt{(1-\beta^2)}}, \tag{295}$$

in other words,

$$\frac{d\mu_0}{d\tau} = \frac{Q}{c^2}, \tag{295a}$$

which is in agreement with the theorem of the inertia of energy (§ 41).

From (294) we obtain the remarkable result that the velocity of a body need not always have to undergo a change when it is acted upon by a force.[190] Consider, for instance, a current-carrying conductor which is at rest in K'. Since the stationary current (observed in system K') does not exert an overall force on it, it remains at rest. Nevertheless, according to (294), a force acts on it in system K. An analogous case was already met with in § 32 (ϵ).

We now come to the discussion of the various expressions for the energy–momentum tensor S_{ik} that have been put forward. As for bodies *at rest*, all authors agree to the extent that, for hysteresis-free media, the energy density W and the energy current S are given by

$$W = \tfrac{1}{2}\{\mathbf{E}\cdot\mathbf{D} + \mathbf{H}\cdot\mathbf{B}\} \quad \text{and} \quad \mathbf{S} = c(\mathbf{E}\wedge\mathbf{H}). \tag{296}$$

[188] Cf. footnote 54, p. 21, Minkowski I.
[189] M. Abraham, *R.C. Circ. mat. Palermo*, **28** (1909) 1; cf. also the discussion between Abraham and Nordström: G. Nordström, *Phys. Z.* **10** (1909) 681; M. Abraham, *Phys. Z.* **10** (1909) 737; G. Nordström, *Phys. Z.* **11** (1910) 440; M. Abraham, *Phys. Z.* **11** (1910) 527. Nordström's objections cannot be upheld.
[190] Cf. footnote 178a, p. 102, M. v. Laue, *ibid.*, p. 134.

But whereas Maxwell and Heaviside[191] suggest for the (three-dimensional) stress tensor, **T**,

$$T_{ik} = E_i D_k - \tfrac{1}{2}(\mathbf{E} \cdot \mathbf{D})\delta_i{}^k + H_i B_k - \tfrac{1}{2}(\mathbf{H} \cdot \mathbf{B})\delta_i{}^k, \quad (i, k = 1, 2, 3) \quad (297)$$

Hertz[191] makes use of the expression (symmetrical in i and k)

$$\begin{aligned} T_{ik} = \tfrac{1}{2}(E_i D_k + E_k D_i) - \tfrac{1}{2}(\mathbf{E} \cdot \mathbf{D})\delta_i{}^k + \\ + \tfrac{1}{2}(H_i B_k + H_k B_i) - \tfrac{1}{2}(\mathbf{H} \cdot \mathbf{B})\delta_i{}^k, \end{aligned} \quad (298)$$

which differs from (297) for the case of anisotropic media (crystals). Similarly, there are two alternatives for the momentum density **g**.

Either
$$\mathbf{g} = \frac{1}{c}(\mathbf{D} \wedge \mathbf{B}) \quad (299)$$

which, from (296), can also be written

$$\mathbf{g} = \frac{\epsilon\mu}{c^2}\mathbf{S} \quad (299a)$$

for homogeneous isotropic media.

Or
$$\mathbf{g} = \frac{1}{c}(\mathbf{E} \wedge \mathbf{H}) = \frac{1}{c^2}\mathbf{S}. \quad (300)$$

If the expressions for W, **S**, **T**, **g** are given for stationary bodies, then the corresponding expressions for moving bodies are uniquely determined, since the components of a tensor in *any* coordinate system can be derived from those in another. Corresponding to the above-mentioned ambiguities in the expressions for T_{ik} and **g**, the following forms for the tensor S_{ik} have mainly been discussed up to the present.

(i) *Minkowski's*[192] *expression for S_{ik}.* He makes use of (297) and (299), for stationary bodies. As is easily shown, this leads to

$$S_i{}^k = F_{ir}H^{kr} - \tfrac{1}{4}H_{rs}F^{rs}\delta_i{}^k, \quad (301)$$

and the expressions (296), (297) *and* (299) *also remain valid for moving bodies.* In addition, relation (223), which holds in vacuum, remains unaltered,

$$S_i{}^i = 0. \quad (223)$$

The four-force f_i is obtained from $S_i{}^k$ by means of (290). In the rest system K' its components have the values

$$(f'_1, f'_2, f'_3) = \rho'\mathbf{E}' + (\mathbf{J}_c' \wedge \mathbf{B}'), \qquad f'_4 = i(\mathbf{J}_c' \cdot \mathbf{E}'). \quad (302)$$

It should be mentioned that Dällenbach[193], too, has derived the Minkowski energy–momentum tensor, but his argument, based on electron theory, is not very cogent. He writes the tensor in a form which is

---

[191] For references cf. footnote 48, p. 17, H. A. Lorentz, *ibid.*, § 23.

[192] Cf. footnote 54, p. 21, Minkowski II. The same expressions for S_{ik} are also obtained by G. Nordström (*Dissertation* (Helsingfors 1908)) and, using a variational principle, by I. Ishiwara (cf. footnote 183, p. 104, *Ann. Phys., Lpz., loc. cit.*).

[193] Cf. footnote 187, p. 104, *ibid.*

also valid for arbitrary inhomogeneous and anisotropic media. Using another method, he obtains it from an action principle, which also gives him the field equations.[194]

(ii) *Abraham's*[195] *expression for* S_{ik}. The asymmetry of Minkowski's expression (301) for the energy–momentum tensor leads to results which are very peculiar, though not in contradiction with experiment. Torques, for instance, appear, which cannot be compensated for by a change in the electromagnetic angular momentum. For this reason, Abraham[195] constructed a symmetrical energy–momentum tensor, by assuming (298) and (300) for stationary bodies. For homogeneous isotropic media, this leads to

$$\left.\begin{aligned}
S_i{}^k &= \tfrac{1}{2}(F_{ir}H^{kr} + H_{ir}F^{kr}) - \tfrac{1}{4}F_{rs}H^{rs}\delta_i{}^k - \tfrac{1}{2}(\epsilon\mu - 1)(v_i\Omega^k + \Omega_i v^k) \\
&= F_{ir}H^{kr} - \tfrac{1}{4}F_{rs}H^{rs}\delta_i{}^k - (\epsilon\mu - 1)\Omega_i v^k \\
&= H_{ir}F^{kr} - \tfrac{1}{4}F_{rs}H^{rs}\delta_i{}^k - (\epsilon\mu - 1)v_i\Omega^k,
\end{aligned}\right\} \quad (303)$$

where the vector Ω^i (already to be found in Minkowski's work as "Ruhstrahlvektor") is defined by

$$F_i = F_{ik}v^k, \quad H_i = H_{ik}v^k, \quad \Omega^i = v_k F_l\{H^{ik}v^l + H^{kl}v^i + H^{li}v^k\}. \quad (304)$$

In a coordinate system K' moving with the substance, the components of these vectors have the values

$$\left.\begin{aligned}
(F'_1, F'_2, F'_3) = \mathbf{E}', \quad F'_4 = 0; \quad (H'_1, H'_2, H'_3) = \mathbf{D}', \quad H'_4 = 0; \\
(\Omega'_1, \Omega'_2, \Omega'_3) = c\mathbf{S}', \quad \Omega'_4 = 0.
\end{aligned}\right\} \quad (304\,a)$$

The identity of the three expressions (303) follows from their agreement in rest system K'. Relation (223) is also valid here. For moving bodies, W, $\mathbf{S}$, T_{ik}, $\mathfrak{g}$ are no longer given by (296), (298) and (300), and Abraham[195] has calculated the corresponding expressions explicitly, in addition to that for the ponderomotive force. The invariant formulation used here is due to Grammel.[196] In the expression for the ponderomotive force for stationary bodies an extra term

$$\frac{\epsilon\mu - 1}{c^2}\frac{\partial \mathbf{S}}{\partial t}$$

is added to (302) in Abraham's case. Because of the smallness of this term, it is hardly likely that an experiment could be devised for deciding in favour of one or other of the two approaches. It should also be mentioned that Laue[197] agrees with Abraham's assumptions.

The following electron-theoretical considerations, also due to Abraham[198], seem to us to constitute a very weighty argument in favour of the symmetry property of the phenomenological energy–momentum tensor. The

[191] W. Dällenbach, *Ann. Phys., Lpz.*, **59** (1919) 28.

[195] M. Abraham, *R. C. Circ. mat. Palermo*, **28** (1909) 1, and **30** (1910) 33; *Theorie der Elektrizität*, Vol. 2 (3rd edn., Leipzig 1914), p. 298 *et seq.*, §§ 38 and 39.

[196] R. Grammel, *Ann. Phys., Lpz.*, **41** (1913) 570.

[197] Cf. footnote 178 a, p. 102, *ibid.*, § 22, p. 135 *et seq.*

[198] M. Abraham, *Ann. Phys., Lpz.*, **44** (1914) 537.

four-force is to be regarded as the average of the microscopic four-force, i.e. from (290) the energy–momentum tensor, too, is to be regarded as the average of its microscopic counterpart.[199] On averaging, the symmetry property of a tensor is preserved (and so is relation (223)).†

(iii) *Einstein and Laub's*[200] *expression for* S_{ik}. Einstein and Laub arrived at a completely different result from that of Minkowski and Abraham for the ponderomotive force on stationary bodies (and thus also for the energy–momentum tensor). They found that the observed force density

$$\mathbf{f} = (\mathbf{J}_c \wedge \mathbf{B})$$

on a stationary current-carrying conductor is composed of a surface force $(1 - 1/\mu) \, (\mathbf{J}_c \wedge \mathbf{H}_{\text{ext}})$ ($\mathbf{H}_{\text{ext}}$ = external magnetic field) and the force

$$\mathbf{f} = (\mathbf{J}_c \wedge \mathbf{H}_{\text{int}})$$

which is to be regarded as the volume force proper, in contrast to (302), where the volume force is $(\mathbf{J}_c \wedge \mathbf{B})$. The energy–momentum tensor given by these authors will also have to be correspondingly modified. Gans[201] has disputed the correctness[202] of Einstein and Laub's arguments.

36. Applications of the theory

(α) *The experiments of Rowland, Röntgen, Eichenwald and Wilson.* The relativistic interpretation of these experiments can be taken over from electron theory[203], so long as one is dealing with unmagnetizable bodies, and so long as terms of higher order in v/c can be neglected.[204] This latter approximation we shall retain here for the moment, but at the same time admit arbitrary values of the permeability μ. The extension of the theory to magnetizable bodies can be regarded as a real step forward, and is due to Minkowski's electrodynamics.

Rowland's experiment demonstrates that the convection current generates the same magnetic field as a conduction current $\rho\mathbf{v}/c$. The explanation for this follows directly from the field equations (G) and the transformation formulae (269 a) for the current $\mathbf{J}$. Admittedly, this experiment was previously used as an argument for the existence of the aether. But from the relativistic point of view it must be said that, after all, it only proves the dependence on the reference system of the splitting of the electromagnetic field into an electrical and a magnetic part, as required by relativity.

Röntgen's experiment[203] proves that when a dielectric is moved in an

---

[199] Dällenbach's objections to this argument (cf. footnote 187, p. 104, *ibid.*) do not seem very sound.

† See suppl. note 11.

[200] A. Einstein and J. Laub, *Ann. Phys., Lpz.*, **26** (1908) 541.

[201] R. Gans, 'Über das Biot-Savartsche Gesetz', *Phys. Z.*, **12** (1911) 806.

[202] Grammel's statement (cf. footnote 196, p. 110, *ibid.*), that Einstein and Laub's expressions for the ponderomotive force contradicted the relativity principle is incorrect, since these expressions only claim to be valid for the rest system K'.

[203] Cf. H. A. Lorentz (footnote 48, p., 17 *ibid.*, § 17, and footnote 4, p. 1, *ibid.*, § 34), where references to earlier papers can be found.

[204] Cf. A. Weber, *Phys. Z.*, **11** (1910) 134.

electric field, a surface current is produced at its boundary, which generates a magnetic field. Later, Eichenwald showed that its magnitude is

$$|\mathbf{j}| = \beta|\mathbf{P}| = \beta(\epsilon - 1)|\mathbf{E}| \tag{305a}$$

where $\mathbf{P}$ is the polarization of the dielectric. In the actual experiment, the dielectric is made to rotate between the plates of a condenser. It will certainly be permissible, to a very good approximation, to apply the theory for uniformly moving bodies to this case. Let a dielectric move parallel to the condenser plates and let $E_n = \omega$ be the surface density of the (free) charge on them. Since $\mathbf{B}$ is source-free and equal to $\mathbf{H}$ in the external region, it is sufficient to investigate the curl of $\mathbf{B}$. Since, in addition, we are dealing with a stationary field, $\mathbf{E}$ and $\mathbf{H}$ are irrotational, from (F) and (G). We shall now neglect, throughout, quantities of higher order in v/c. Making use of the fact that $\mathbf{H}$ and $\mathbf{B}$ are themselves first-order quantities, we obtain from (278 a)

$$\mathbf{D} = \epsilon\mathbf{E},$$
$$\mathbf{B} = \mu\mathbf{H} - (\epsilon\mu - 1)\frac{(\mathbf{v} \wedge \mathbf{E})}{c}.$$

The curl of $\mathbf{B}$ is thus determined by the curl of $(\epsilon\mu - 1)$ $(\mathbf{v} \wedge \mathbf{E})/c$, which reduces here to the surface curl, $\mathbf{j}$, of magnitude

$$|\mathbf{j}| = \beta(\epsilon\mu - 1)|\mathbf{E}|. \tag{305b}$$

$|\mathbf{E}|$ is the value of $\mathbf{E}$ *inside* the dielectric, and $\mathbf{j}$ has the direction of $\mathbf{v}$. For $\mu = 1$ this reduces to the value (305 a) given by electron theory. For magnetizable bodies the effect was not investigated.

H. A. Wilson's experiment[205] is the counterpart to Eichenwald's. A dielectric cylinder was made to rotate between the plates of a short-circuited condenser in a magnetic field parallel to the condenser plates. It was observed that the plates were charged up. Let us again replace the rotation by a rectilinear motion, parallel to the plates but at right angles to the magnetic field. According to (F), we can first of all derive $\mathbf{E}$ from a potential φ. Since the plates are short-circuited, it follows that $\varphi_1 - \varphi_2 = 0$, thus also $\mathbf{E} = 0$, and $\mathbf{D}$ gives directly the charge density ω. The boundary conditions in this case require that $\mathbf{H}$ should be continuous. From (278 a) we therefore have

$$\omega = \frac{(\epsilon\mu - 1)\beta H}{1 - \epsilon\mu\beta^2} \simeq (\epsilon\mu - 1)\beta H. \tag{306a}$$

For unmagnetizable bodies, the result of electron theory follows

$$\omega = (\epsilon - 1)\beta H. \tag{306b}$$

[205] H. A. Wilson, *Philos. Trans.* A, **204** (1904) 121. See H. A. Lorentz, (cf. footnote 48, p. 17, *ibid.*, § 20; footnote 4, p. 1, *ibid.*, § 45) on a previous experiment (with negative result) by Blondlot, in which air was used as a dielectric, and on the viewpoint of the earlier electron theory. On the discussion of Wilson's experiment from a relativistic point of view, see A. Einstein and J. Laub, (cf. footnote 176, p. 99, *ibid.*); M. v. Laue, (cf. footnote 178 a, p. 102, *ibid.*, p. 129 *et seq.*); and H. Weyl, *Raum-Zeit-Materie* (1st edn., Berlin 1918), p. 155.

H. A. and M. Wilson[206] succeeded in measuring the effect also in a magnetizable insulator, which they constructed artificially by embedding steel
spheres in sealing wax. The result confirms the relativistic value (306 a)
for the charge density.

The earlier Hertz theory gives, instead of (305 a) and (306 a), the values

$$|\mathbf{j}| = \beta|\mathbf{E}|, \qquad \omega = \beta H,$$

in contradiction with experiment.

(β) *Resistance and induction in moving conductors.*[207] From (279) we find
that for a moving conductor of finite length

$$R_0 J = \int \mathbf{E}^* \cdot d\mathbf{s}, \tag{307}$$

where R_0 is the 'rest' resistance of the conductor and $J = |\mathbf{J}|A$. For the
change in the conductivity, given by (280), will just be compensated for
by the change in the length of the wire and in its cross-section 'A' due to
the Lorentz contraction, when the resistance is calculated. This is the
way in which the experiment by Trouton and Rankine[208] is seen from a
moving system K. The induction law for moving conductors follows from
the first of Eqs. (274),

$$\int \mathbf{E}^* \cdot d\mathbf{s} = \frac{d}{dt} \int B_n d\sigma; \tag{308}$$

on the other hand

$$\int \mathbf{E} \cdot d\mathbf{s} \neq \frac{d}{dt} \int B_n d\sigma.$$

But (308) is certainly in agreement with experiment, since it is $\mathbf{E}^*$,
and not $\mathbf{E}$, which determines the conduction current in (279).

(γ) *Propagation of light in moving media. The drag coefficient. Airy's
experiment.* It is not necessary to go back to the field equations in order
to find the laws governing the propagation of light in moving media.
They must be directly obtainable from the corresponding laws for stationary bodies, with the help of the Lorentz transformation. Consider first
of all a non-absorbing medium. The invariant light phase is again given
by (252), where now l_i has the components

$$l_i = \left(\frac{\nu}{w} \cos\alpha, \quad \frac{\nu}{w} \sin\alpha, \quad 0, \quad \frac{i\nu}{c} \right), \tag{309}$$

if the z-axis is taken to be perpendicular to the velocity of the body and to
the wave normal. In system K', moving with the body, we have in particular

$$w' = \frac{c}{n}. \tag{310}$$

[206] H. A. and M. Wilson, *Proc. Roy. Soc.* A, **89** (1913) 99.
[207] Cf. footnote 132, p. 82, M. Abraham, *ibid.*, p. 388; footnote 178 a, p. 102, M. v. Laue,
ibid., p. 126 *et seq.*; and H. Weyl, *Raum-Zeit-Materie* (1st edn., Berlin 1918), § 22.
[208] Cf. footnote 15, p. 4, F. T. Trouton and A. O. Rankine, *ibid.*

From this follow the transformation formulae

$$\nu = \nu' \frac{1 + (v/w')\cos\alpha'}{\sqrt{(1 - \beta^2)}}, \tag{311a}$$

$$\frac{\nu}{w}\cos\alpha = \nu'\frac{(1/w')\cos\alpha' + \beta/c}{\sqrt{(1 - \beta^2)}}, \qquad \frac{\nu}{w}\sin\alpha = \frac{\nu'}{w'}\sin\alpha'$$

$$\frac{1}{w}\cos\alpha = \frac{(1/w')\cos\alpha' + \beta/c}{1 + (v/w')\cos\alpha'}, \qquad \frac{1}{w}\sin\alpha = \frac{(1/w')\sin\alpha'\sqrt{(1 - \beta^2)}}{1 + (v/w')\cos\alpha'}$$

$$\tan\alpha = \frac{\sin\alpha'\sqrt{(1 - \beta^2)}}{\cos\alpha' + \beta w'/c} \tag{311b}$$

$$w = c\frac{1 + \beta n \cos\alpha'}{\sqrt{[(n\cos\alpha' + \beta)^2 + n^2\sin^2\alpha'(1 - \beta^2)]}}. \tag{311c}$$

Relation (311a) gives the Doppler effect, (311b) the aberration, (311c) the drag coefficient. They agree with the expressions of the earlier theory up to first-order terms. The latter would give

$$w = \frac{c}{n} + v\left(1 - \frac{1}{n^2}\right)\cos\alpha'. \tag{311d}$$

(See § 6 on the influence of the wave-length dependence of the refractive index.) The law of refraction at moving boundary surfaces can also be obtained from the stationary case, by means of a Lorentz transformation, but it leads to complicated formulae.

We next have to discuss the experimental result of Airy[208a], according to which the angle of aberration is not changed when the telescope is filled with water. The earlier theory[209] had to make use of rather involved arguments in order to explain this, because it had to describe the effect as seen from a reference system relative to which the observer (the earth) is moving. If, on the other hand, it is observed from the rest system, Airy's result is self-evident from the relativistic point of view. For if the telescope is pointed towards the apparent position of the fixed star, then the light waves sent out by it will have normal incidence on the telescope. If it is now filled with water, the light waves will be propagated normal to the boundary surface also in water. The Airy experiment, as seen from the rest system of the observer (earth), therefore only demonstrates the (relativistically) trivial fact that for a zero angle of incidence (normal incidence) the angle of refraction is zero, too.

It will be observed that the relations (311b, c) do *not* correspond to the addition theorem for velocities. Agreement with it is only obtained for $\alpha = 0$ (cf. § 6). Laue[210] traces this back to the difference in direction

[208a] G. B. Airy, *Proc. Roy. Soc.* **20** (1871) 35; **21** (1873) 121; *Phil. Mag.*, **43** (1872) 310.
[209] Cf. H. A. Lorentz, *Arch. néerl. Sci.*, **21** (1887) 103 (Collected Papers, XIV, p. 341) where earlier references are to be found.
[210] Cf. footnote 178a, p. 102, *ibid.*, p. 134.

between the light ray and the wave normal. If the ray velocity is defined, in direction and magnitude, by

$$\mathbf{w}_1 = \frac{\mathbf{S}}{W} \tag{312}$$

($\mathbf{S} = $ Poynting vector, $W = $ energy density),

then the transformation formulae for the components of $\mathbf{w}_1$ are to satisfy the addition theorem for velocities rigorously. Scheye's[211] calculation shows that this is in fact the case, if Minkowski's asymmetrical energy-momentum tensor is used. Moreover, it follows from the energy equation for this case that the phase velocity w is equal to the component of the ray velocity in the direction of the wave normal. If, on the other hand, we were to start from Abraham's tensor (304), the situation would become more complicated and the addition theorem would not hold for the ray velocity either.[†]

The generalization to absorptive (conducting) media does not offer anything new in principle. It may be mentioned that a light wave propagated in a moving conductor is connected with a periodically varying charge density, according to (277 a).

(δ) *Signal velocity and phase velocity in dispersive media.* In dispersive media, the case arises where the phase velocity of a light wave is $\geqslant c$. This appears to contradict the relativistic requirement that no disturbance must be propagated with a velocity greater than c (cf. § 6). The difficulty was removed by an investigation by Sommerfeld[212], in which he showed on the basis of electron theory that the *wave crest* is always propagated with the velocity of light in vacuum, c, so that in reality it is not possible to send out signals with velocity greater than c. This result was extended by Brillouin[213], who showed that, apart from the absorption region, the main portion of the signal is propagated with the group velocity.

(c) MECHANICS AND GENERAL DYNAMICS

37. Equations of motion. Momentum and kinetic energy

Relativistic mechanics[213a] starts from the assumption that in a coordinate system K' in which a particle is momentarily at rest the equations of motion of classical mechanics

$$m_0 \frac{d^2 \mathbf{r}'}{dt'^2} = \mathbf{K}' \tag{313}$$

[211] A. Scheye, 'Über die Fortpflanzung des Lichtes in einem bewegten Dielektrikum', *Ann. Phys., Lpz.*, **30** (1909) 805.

[†] See suppl. note 11.

[212] A. Sommerfeld, 'Heinrich-Weber-Festschrift'; *Phys. Z.* **8** (1907) 841; *Ann. Phys., Lpz.*, **44** (1914) 177.

[213] L. Brillouin, *Ann. Phys., Lpz.*, **44** (1914) 203.

[213a] In what follows, the term "relativistic mechanics" is always used to designate the mechanics of the special theory of relativity, i.e. the mechanics of the Lorentz group. It could be argued against this use of the term that classical mechanics, too, is relativistic, since it satisfies the postulate of relativity. But the word "relativistic" has already frequently been used with the specific meaning "relative with respect to the Lorentz group". A case in point is the term "special theory of relativity" itself.

are valid. The relativity principle then permits us to deduce unambiguously the equations of motion in any other coordinate system K, by
subjecting (313) to a Lorentz transformation. With this, however, we have
not yet given a definition for the force in system K. In the three equations
of motion a common factor still remains undetermined, which may depend
on the velocity in an arbitrary manner. There are two essentially different
ways in which this arbitrariness can be removed.

The first approach makes use of electrodynamical concepts. If we
assume the expression for the Lorentz force to hold for arbitrarily fast
moving charges, a transformation law for the force is then also implied
(cf. § 29). That all kinds of force transform in the same way follows from
the fact that two forces which compensate each other in system K'
must also do so in every other system K. Formulae (213), (214), (215) can
at once be generalized for the case of arbitrary forces. In place of (217)
we have the four-vector force–density—power–density,

$$(f_1, f_2, f_3) = \mathbf{f}, \qquad f_4 = i\frac{(\mathbf{f} \cdot \mathbf{u})}{c}, \tag{314}$$

which is normal to the four-velocity,

$$f_k u^k = 0. \tag{315}$$

We then have again the equations of motion

$$\mu_0 \frac{d^2 x^i}{d\tau^2} = f^i \quad \text{or} \quad \mu_0 \frac{du_i}{d\tau} = f_i, \tag{316}$$

in which μ_0 is the (invariant) rest-mass density. We can also introduce
the Minkowski force, K_i, defined by (219), and the equations of motion
(220).

From the equations

$$\frac{d}{dt}(m\mathbf{u}) = \mathbf{K}$$

and

$$\frac{d}{dt}(mc^2) = \mathbf{K} \cdot \mathbf{u}, \tag{317}$$

it follows that the momentum is given by[214]

$$\mathbf{G} = m\mathbf{u} = \frac{m_0}{\sqrt{(1 - \beta^2)}}\mathbf{u} \tag{318a}$$

and the kinetic energy by

$$E_{\text{kin}} = mc^2 + \text{const.} = \frac{m_0 c^2}{\sqrt{(1 - \beta^2)}} + \text{const.}$$

One could think of determining the constant in such a way that E_{kin}
is zero for a particle *at rest*. It is however more practicable to put the

214 Cf. footnote 129, p. 82, M. Planck, *ibid.*

constant itself equal to zero. The energy of a particle at rest then becomes $m_0 c^2$ and, in general,

$$E = mc^2 = \frac{m_0 c^2}{\sqrt{(1 - \beta^2)}}. \tag{318 b}$$

By a power series expansion we obtain, for small β,

$$E = m_0 c^2 (1 + \tfrac{1}{2}\beta^2) = E_0 + \tfrac{1}{2} m_0 v^2,$$

which is in agreement with classical mechanics. The convenience of choosing the constant in this way becomes evident if we observe that the quantities

$$(J_1, J_2, J_3) = c\mathbf{G}, \qquad J_4 = iE \tag{319}$$

are the components of a four-vector. For,

$$J_k = m_0 c u_k. \tag{320}$$

It further follows that exactly the same transformation formulae hold for the $\mathbf{G}$, E here as for the momentum and energy of a closed, force-free, electromagnetic system (light wave), cf. Eq. (228),

$$G'_x = \frac{G_x - (v/c^2)E}{\sqrt{(1 - \beta^2)}}, \qquad G'_y = G_y, \qquad G'_z = G_z,$$

$$E' = \frac{E - vG_x}{\sqrt{(1 - \beta^2)}}, \tag{321}$$

with the corresponding inverse formulae. They are also valid for the momentum and energy of a *system* of freely moving particles.

The equations of motion, and the expressions for momentum and energy, of relativistic mechanics go over into those of classical mechanics for small velocities, as was to be expected from the start. But we can go further than this: The deviations of relativistic from classical mechanics are of *second* order in v/c. This then must be the reason (as was pointed out by Laue[214a]) why the older electron theory, which was based on classical mechanics, could explain all first-order effects correctly.

Minkowski[215] also gave a further important interpretation of the equations of motion (316). Let us introduce the kinetic energy–momentum tensor Θ_{ik} by means of the relation

$$\Theta_{ik} = \mu_0 u_i u_k. \tag{322}$$

Its space components represent the tensor of the momentum current, the mixed components (apart from a factor ic) the momentum density, and the time component the energy density. Because of the continuity condition

$$\frac{\partial(\mu_0 u^k)}{\partial x^k} = 0, \tag{323}$$

[214a] Cf. footnote 178 a, p. 102, *ibid.*, p. 88.
[215] Cf. footnote 54, p. 21, Minkowski II.

the equations of motion can be written in the form

$$\frac{\partial \Theta_i{}^k}{\partial x^k} = f_i. \tag{324}$$

It should be pointed out here, that the equations of motion (317) result in a hyperbolic motion (discussed in § 26) for the case of a particle moving under a *constant* force.

38. Relativistic mechanics on a basis independent of electrodynamics

The derivation in the preceding section is unsatisfactory because it has to make use of electrodynamical concepts. An important contribution to the subject is therefore Lewis and Tolman's[215a] derivation, which does not do so. There, the primary concept is not the force, but the momentum. They postulate that with each particle a momentum vector, parallel to its velocity, and a scalar kinetic energy can be associated in such a way that *conservation laws* obtain. This means that the sum of the momenta and energies of the separate masses of a system is to remain constant, if neither momentum and energy nor heat are generated during interactions between such masses. In particular, this is to hold for the case of elastic collisions. Lewis and Tolman devised a "thought experiment" to show that the velocity dependence of momentum and kinetic energy is uniquely determined by the Lorentz-invariance requirement of these conservation laws.

Let the relative velocity v of two observers A and B be in the x-direction. They throw towards each other spheres of equal mass and with equal velocity u in the positive and negative y-directions, respectively, such that the line of centres is in the y-direction. The x-components of the velocities of the two spheres remain unchanged. Moreover, for symmetry reasons, A and B will observe the same motion of their respective spheres. The addition theorem for velocities (10) then gives the following values for the velocity components w_x, w_y and w'_x, w'_y of the colliding spheres in K and K':

Before collision:

Sphere A

$$w_x = 0, \quad w_y = u \qquad \bigg| \qquad w'_x = -v, \quad w'_y = u\bigg/\sqrt{\left(1 - \frac{v^2}{c^2}\right)}$$

Sphere B

$$w_x = v, \quad w_y = -u\bigg/\sqrt{\left(1 - \frac{v^2}{c^2}\right)} \; \bigg| \; w'_x = 0, \quad w'_y = -u.$$

After collision:

Sphere A

$$w_x = 0, \quad w_y = -u' \qquad \bigg| \qquad w'_x = -v, \quad w'_y = -u'\bigg/\sqrt{\left(1 - \frac{v^2}{c^2}\right)}$$

[215a] G. N. Lewis and R.C. Tolman, *Phil. Mag.*, **18** (1909) 510. Objections to their method by N. Campbell, *Phil. Mag.*, **21** (1911) 626, apply more to the form than to the main essence of the argument. This is seen from P. Epstein's paper (*Ann. Phys., Lpz.*, **36** (1911) 729) which shows that Lewis and Tolman's conclusions can be reached in a perfectly rigorous way.

Sphere B

$$w_x = v, \quad w_y = +u'\bigg/\sqrt{\bigg(1 - \frac{v^2}{c^2}\bigg)} \quad \bigg| \quad w'_x = 0, \quad w'_y = +u'.$$

If $w = |\mathbf{w}|$ is the absolute value of the velocity, the momentum can be written

$$\mathbf{G} = m(w)\,\mathbf{w}$$

where, by definition, m is called the mass and can only depend on the *absolute value* of the velocity. From the conservation of momentum in the x-direction,

$$u = u',$$

and from the conservation of momentum in the y-direction,

$$m\bigg(\bigg[v^2 + u^2\bigg(1 - \frac{v^2}{c^2}\bigg)\bigg]^{1/2}\bigg)u\bigg/\sqrt{\bigg(1 - \frac{v^2}{c^2}\bigg)} = m(u)u. \tag{α}$$

Dividing by u and going to the limit $u \to 0$, we obtain the result we had wanted to prove,

$$m(v)\bigg/\sqrt{\bigg(1 - \frac{v^2}{c^2}\bigg)} = m_0, \qquad m = \frac{m_0}{\sqrt{[1-(v^2/c^2)]}},$$

where $m(0) = m_0$. It is easy to see that with this expression for m, relation (α) is satisfied also for arbitrary u. If, next, the momentum is calculated, expression (318 b) for the kinetic energy is easily obtained by a Lorentz transformation. The force is now defined as the time derivative of the momentum, and the transformation laws for it follow directly. We have thus shown that it is possible to find a basis for relativistic mechanics, without taking recourse to electrodynamics.†

It should also be mentioned that the laws for elastic collisions in relativistic mechanics have been derived and discussed for the general case by Jüttner.[216]

39. Hamilton's Principle in relativistic mechanics

It was already shown by Planck[217] that the equations of motion (317) can be derived from a variational principle. If we introduce the Lagrangian

$$L = -m_0 c^2\bigg/\sqrt{\bigg(1 - \frac{u^2}{c^2}\bigg)}, \tag{325}$$

then

$$\int_{t_0}^{t_1} (\delta L + \mathbf{K}\cdot\delta\mathbf{r})\,dt = 0, \tag{326}$$

as can easily be checked. As in the case of Hamilton's principle in classical mechanics, the values t_0, t_1 and the end points of the path of integration

† See suppl. note 12.
[216] F. Jüttner, *Z. Math. Phys.*, **62** (1914) 410.
[217] Cf. footnote 129, p. 82, *ibid.*

are prescribed. The equations of motion can also be written in the form of Hamilton's equations. If we introduce, in place of the velocity components $\dot{x}$, $\dot{y}$, $\dot{z}$, the momenta

$$G_x = \frac{\partial L}{\partial \dot{x}}, \qquad G_y = \frac{\partial L}{\partial \dot{y}}, \qquad G_z = \frac{\partial L}{\partial \dot{z}}, \tag{327}$$

and form the Hamiltonian

$$H = \dot{x}\frac{\partial L}{\partial \dot{x}} + \dot{y}\frac{\partial L}{\partial \dot{y}} + \dot{z}\frac{\partial L}{\partial \dot{z}} - L = \frac{m_0 c^2}{\sqrt{[1 - (u^2/c^2)]}} = E_{\text{kin}} \tag{328}$$

$$= m_0 c^2 \sqrt{\left(1 + \frac{G_x^2 + G_y^2 + G_z^2}{m_0^2 c^2}\right)},$$

then

$$\dot{x} = \frac{\partial H}{\partial G_x}, \dots \text{etc.}$$

$$\left.\frac{dG_x}{dt} = K_x, \dots \text{etc.}\right\} \tag{329}$$

The action integral $\int L\, dt$ must be Lorentz-invariant. In fact, it is simply equal to

$$\int L\, dt = -m_0 c^2 \int d\tau \tag{330}$$

where τ is the proper time. The action principle (326) can then be written[218]

$$-m_0 c^2 \delta \int d\tau + \int K_i \delta x^i\, d\tau = 0 \tag{331}$$

or, simpler still,

$$\delta \int d\tau = 0, \tag{332}$$

if the *auxiliary condition*

$$K_i u^i = 0$$

is imposed on the variations δx^i. This formulation of the variational principle is due to Minkowski.[219]

The equations of motion (317) can also be changed into a form which corresponds to the virial theorem in classical mechanics. If we write

$$L + E_{\text{kin}} + \frac{d}{dt}(m\mathbf{u}\cdot\mathbf{r}) = \mathbf{K}\cdot\mathbf{r} \tag{333}$$

and if $\mathbf{r}$ remains within finite limits throughout the motion and the velocity $\mathbf{u}$ does not approach the velocity of light arbitrarily closely, then by averaging over time,

$$\bar{L} + \bar{E}_{\text{kin}} = \overline{\mathbf{K}\cdot\mathbf{r}}. \tag{333a}$$

[218] The δx^i are to vanish at the integration limits.
[219] Cf. footnote 54, p. 21, Minkowski II, Appendix.

40. Generalized coordinates. Canonical form of the equations of motion

In relativistic mechanics it is in general not possible to introduce a potential energy which depends only on the position coordinates, because interactions cannot be propagated with velocities greater than that of light, according to the basic postulates of the theory. There are however certain special cases in which it is nevertheless useful to introduce such a potential energy, for instance when a particle moves in a force field which is constant in time. It is just this case which plays an essential part in the fine-structure theory of the Balmer lines. We can write

$$\mathbf{K} = -\operatorname{grad} E_{\text{pot}}, \tag{334}$$

$$L = -mc^2 \sqrt{\left(1 - \frac{u^2}{c^2}\right)} - E_{\text{pot}}, \quad \delta \int L \, dt = 0 \tag{335}$$

$$H(G_x \ldots, x \ldots) = E_{\text{kin}} + E_{\text{pot}} = \dot{x}\frac{\partial L}{\partial \dot{x}} + \ldots - L \tag{336}$$

$$\left.\begin{aligned} \dot{x} &= \frac{\partial H}{\partial G_x}, \ldots \\ \frac{dG_x}{dt} &= -\frac{\partial H}{\partial x}, \ldots \end{aligned}\right\} \tag{337}$$

and thus bring the equations into a canonical form. We can also introduce generalized coordinates $q_1, \ldots, q_f$. The canonically conjugate momenta are then given by

$$p_k = \frac{\partial L}{\partial \dot{q}_k},$$

and we have

$$\left.\begin{aligned} H(p, q) &= \sum \dot{q}_k \frac{\partial L}{\partial \dot{q}_k} - L \\ \frac{dq_k}{dt} &= \frac{\partial H}{\partial p_k}, \quad \frac{dp_k}{dt} = -\frac{\partial H}{\partial q_k}. \end{aligned}\right\} \tag{338}$$

Moreover, the Hamilton–Jacobi equation holds here, just as in classical mechanics. It follows from their very derivation that the above formulae are only valid in a single coordinate system, distinguished by the problem in question.

41. The inertia of energy

The simple connection between kinetic energy and mass, (318 b), leads us to the assumption that to each energy E there corresponds[220] a mass $m = E/c^2$. From this it would follow that the mass of a body is increased

[220] A. Einstein, *Ann. Phys., Lpz.*, **18** (1905) 639 (also in the collection 'Relativitätsprinzip'). It is here that the theorem of the equivalence of mass and energy first occurs; cf. also *Ann. Phys., Lpz.*, **20** (1906) 627. G. N. Lewis, *Phil. Mag.*, **16** (1908) 705, starts, conversely, with the postulate $E = mc^2$ and, using the equation $\mathbf{u} \cdot d(m\mathbf{u})/dt = dE/dt$, derives the velocity dependence of the mass, $m = m_0/\sqrt{(1 - \beta^2)}$.

by heating it, and that a transfer of mass would take place by means of radiation between absorbing and emitting bodies. As for the second example, it can be verified in the following way. Let a body, at rest in K', emit radiation energy E_{rad} in such a way that the overall momentum radiated is zero, so that the body remains at rest in K'. In a coordinate system K, moving with relative velocity $\mathbf{v}$, a momentum

$$\mathbf{G}_{\text{rad}} = \frac{\mathbf{v}}{c^2} \frac{E'_{\text{rad}}}{\sqrt{(1 - \beta^2)}} = \frac{\mathbf{v}}{c^2} E_{\text{rad}}$$

will then be radiated, according to (228). Since the velocity $\mathbf{v}$ of the body is unchanged, this is only possible if its rest mass m_0 is decreased by an amount

$$\Delta m_0 = \frac{E'_{\text{rad}}}{c^2}.$$

By similar momentum considerations it can be shown that a mass has also to be ascribed to the thermal energy. That this is reasonable can be seen from the following. As mentioned before, the same transformation formulae (321) hold for the total momentum and energy of a system of particles as for a single particle. If the coordinate system K_0 is chosen in such a way that the total momentum vanishes there, then in system K we have again

$$\mathbf{G} = \frac{\mathbf{v}}{c^2} \frac{E_0}{\sqrt{(1 - \beta^2)}}, \qquad E = \frac{E_0}{\sqrt{(1 - \beta^2)}}.$$

The system thus behaves like a single particle with rest mass[221] $m_0 = E_0/c^2$. Evidently, an ideal gas is such a system of particles. E_0 here becomes $\sum m_0 c^2 + U$, where U is the thermal energy. Its inertia has thus been proved.

A still more general case was discussed by Lorentz.[222] Consider an arbitrary closed physical system, consisting of masses, springs in tension, light rays, etc. Let the system be at rest (i.e. have zero total momentum) in a coordinate system K_0. In any other coordinate system K it will then have a velocity $\mathbf{u}$, i.e. the same velocity which K_0 has relative to K. We can now make the extremely plausible assumption that the momentum $\mathbf{G}_1$ of the system in K is given by

$$\mathbf{G}_1 = m\mathbf{u} = \frac{m_0}{\sqrt{[1 - (u^2/c^2)]}} \mathbf{u},$$

as for a single particle.[223] Then the following transformation formula holds for $\mathbf{G}_1$,

$$G'_{1x} = \frac{G_{1x} - mv}{\sqrt{(1 - \beta^2)}}, \qquad G'_{1y} = G_{1y}, \qquad G'_{1z} = G_{1z}.$$

[221] A. Einstein, *Ann. Phys., Lpz.*, **23** (1907) 371.

[222] Cf. footnote 3, p. 1, *ibid.*; and 'Over de massa der energie' *Versl. gewone Vergad. Akad. Amst.*, **20** (1911) 87.

[223] If one assumes that the system is only acted upon by electromagnetic forces, this assumption need not be made. Cf. A. Einstein, *Jb. Radioakt.*, **4** (1907) 440. A bounded plane light wave forms an exception in so far as its momentum does not vanish in any coordinate system (cf. § 30). Since we have to put $u = c$ in the above formula for this case, we have to assign it a zero rest mass (cf. footnote 222).

We now let this system 1 interact with a system 2 which is only to consist of radiation. Let $\Delta \mathbf{G}_1$ and $\Delta \mathbf{G}_2$ be the changes in momentum, and ΔE_1, ΔE_2 the changes in energy, in the two systems. Then we must have

$$\Delta \mathbf{G}_1 + \Delta \mathbf{G}_2 = 0, \qquad \Delta \mathbf{G}'_1 + \Delta \mathbf{G}'_2 = 0, \qquad \Delta E_1 + \Delta E_2 = 0,$$

and since, because of (228),

$$\Delta G'_{2x} = \frac{\Delta G_{2x} - (v/c^2)\Delta E_2}{\sqrt{(1 - \beta^2)}},$$

it directly follows that

$$\Delta m = \frac{\Delta E_1}{c^2}. \tag{339}$$

This shows that it is quite immaterial what kind of energy we are dealing with.

We can thus consider it as proved that the relativity principle, in conjunction with the momentum and energy conservation laws, leads to the fundamental principle of the equivalence of mass and (any kind of) energy. We may consider this principle (as was done by Einstein) as the most important of the results of the theory of special relativity. Quantitative experimental evidence has not, so far, been produced for it. Already in his first published paper on the subject, Einstein[224] pointed to the possibility of using radioactive processes to check the theory. But the looked-for mass defects in the atomic weights of the radioactive elements[225] are too small to be measured experimentally. Langevin[226] first pointed out that it might be possible to explain the deviations from integral numbers of the atomic weights (referred to H = 1) of the elements—in so far as they are not due to isotopes—by means of the equivalent mass of the interaction energy between the nucleons. This suggestion has been much discussed recently.[227] Perhaps the theorem of the equivalence of mass and energy can be checked at some future date by observations on the stability of nuclei. There exist indications of a qualitative agreement.[227]†

42. General dynamics

The situation becomes simpler still if we go over to the energy density and momentum density, instead of considering the total energy and

[224] Cf. footnote 220, p. 121, A. Einstein, *ibid.*, **18**.

[225] M. Planck, *S. B. preuss. Akad. Wiss.* (1907) 542; *Ann. Phys., Lpz.*, **76** (1908) 1; A. Einstein (cf. footnote 223, p. 122, *ibid.*, p. 443).

[226] P. Langevin, *J. Phys. théor. appl.* (5) **3** (1913) 553. Langevin, at the time, aimed at deriving *all* deviations of atomic weights from integral values, from the equivalent mass of the internal energy of atomic nuclei. It was stressed by R. Swinne, *Phys. Z.*, **14** (1913) 145, that one would have to consider possible isotopes as well, such as have been shown to be actually present in most cases by Aston's experiments.

[227] W. D. Harkins and E. D. Wilson, *Z. anorg. Chem.*, **95** (1916) 1 and 20; W. Lenz, *S. B. bayer. Akad. Wiss.* (1918) 35; *Naturwissenschaften*, **8** (1920) 181; O. Stern and M. Vollmer, *Ann. Phys., Lpz.*, **59** (1919) 225; A. Smekal, *Naturwissenschaften*, **8** (1920) 206; *S. B. Akad. Wiss. Wien*, Abt. IIa, **129** (1920) 455.

† See suppl. note 13.

momentum. We saw in § 30, Eq. (225), that the electromagnetic four-force could be derived from the divergence of a stress tensor $S_i{}^k$. It leads to the obvious generalization that the same must hold for any kind of force. With the present state of our knowledge, this can be proved. For we know that all forces (elastic, chemical, etc.) can be reduced to electromagnetic forces—leaving gravitation aside here.[227a] An exception is formed by the forces which the electrons and hydrogen nuclei exert on themselves during their motion (cf. Part V). We shall therefore proceed in the following way: In expression (222) for the energy–momentum tensor, let us decompose the field tensor F_{ik} into the separate parts due to each charged particle. The tensor $S_i{}^k$ is then split into two parts, of which one consists of the products of the field-tensor components of different particles, the other of the products of the field-tensor components of one and the same particle. We shall only retain the former, which describes the interaction between the particles. If we now form the divergence, we obtain only the interaction forces. We can then write

$$\mu_0 \frac{du_i}{d\tau} = -\frac{\partial S_i{}^k}{\partial x^k}$$

and from (322), (324),

$$\frac{\partial(\Theta_i{}^k + S_i{}^k)}{\partial x^k} = 0.$$

The substance can thus be characterized by means of an energy–momentum tensor $T_i{}^k$ whose divergence vanishes,

$$T_{ik} = \Theta_{ik} + S_{ik}, \tag{340}$$

$$\frac{\partial T_i{}^k}{\partial x^k} = 0. \tag{341}$$

As in (224), the space components of T_{ik} represent the stresses, which can also be interpreted as the components of the momentum current, whereas the remaining components determine the momentum density $\mathbf{g}$, the energy current $\mathbf{S}$ and the energy density W:

$$T_{i4} = ic\mathbf{g}, \qquad T_{4i} = \frac{i}{c}\mathbf{S}, \qquad [\text{index } i = 1, 2, 3]$$

$$T_{44} = W. \tag{342}$$

The energy–momentum tensor has been represented here as the sum of a mechanical and an electromagnetical contribution. Cf. Part V on attempts to reduce the mechanical part, too, to an electromagnetic one. For the following, purely phenomenological, considerations it is not the *nature* of the energy–momentum tensor but solely the fact of its *existence* which is of importance. On the historical side it should be mentioned that the existence of such a tensor for the *mechanical* (elastic) energy was

[227a] We cannot, at present, say to what extent the dynamics which we are discussing here will be modified by quantum theory.

first asserted by Abraham[228] and conclusively formulated by Laue.[228] The *symmetry* of the energy–momentum tensor is ensured by going back to the mechanical and electromagnetic tensors; previously it had been introduced as a separate postulate. This symmetry property leads to a very important result. It follows from $T_{i4} = T_{4i}$, because of (342), that

$$\mathbf{g} = \frac{\mathbf{S}}{c^2}. \tag{343}$$

This is the theorem of the momentum of the energy current, first expressed by Planck[229], according to which a momentum is associated with each energy current. This theorem can be considered as an extended version of the principle of the equivalence of mass and energy. Whereas the principle only refers to the *total* energy, the theorem has also something to say on the *localization* of momentum and energy.

Just as in § 30 we can deduce from (341) that the total energy and momentum of a closed system form a four-vector

$$(J_1, J_2, J_3) = c\mathbf{G}, \qquad J_4 = iE. \tag{227}$$

Formulae (228), too, are valid here. The inertia of any form of energy, in particular that of *potential* energy, follows from them directly. Let us again note that the additive constant of the energy has been fixed in such a way that the energy of a stationary electron becomes equal to $m_0 c^2$. Only then does $E = mc^2$ hold in general.

The conservation of angular momentum

$$\mathbf{L} = \int (\mathbf{r} \wedge \mathbf{g}) dV \tag{344}$$

also follows from (341) in the usual way.[230] For the validity of this conservation law it is essential that the *space* components of T_{ik} should be symmetrical. If this symmetry is demanded for each reference system, it follows that the mixed components, too, must be symmetrical, which in turn implies the theorem of the momentum of the energy current.[230a]

43. Transformation of energy and momentum of a system in the presence of external forces

Formulae (228) are only valid when *all* the kinds of energy and momentum entering the problem are included in E and $\mathbf{G}$. If we are dealing with a gas under an external pressure, or with a system of stationary electric charges, we should also take into account the elastic energy of the container, or of the charged matter, respectively. This would be very inconvenient. We shall therefore solve the following general problem. The types

---

[228] M. Abraham, *Phys. Z.*, **10** (1909) 739; M. v. Laue (cf. footnote 178 a, p. 102, *ibid.*, and *Ann. Phys., Lpz.*, **35** (1911) 524; cf. W. Schottky, *Dissertation* (Berlin 1912).

[229] M. Planck, *Phys. Z.*, 9 (1908) 828.

[230] Cf. footnote 4, p. 1, H. A. Lorentz, *ibid.*, § 7.

[230a] In connection with the angular momentum theorem, it should be mentioned that P. Epstein (cf. footnote 215a, p. 118, *ibid.*) introduces the angular momentum into the theory as a "surface" tensor $N_{ik} = x_i K_k - x_k K_i$, where K_i is the Minkowski force.

of energy which alone will be considered here, are to produce a force f_i such that

$$\frac{\partial S_i{}^k}{\partial x^k} = -f_i, \tag{345}$$

where S_{ik} is the corresponding tensor. We want to obtain the transformation formulae for the total energy and total momentum. Let the system be at rest in coordinate system K' (i.e. the total momentum is to be zero in it ($\mathbf{G}' = 0$)) and let all state functions be time independent.

We now have two methods at our disposal. We can either transform first the energy and momentum density into the moving system. This is easily done by means of the transformation formulae for the components of a symmetrical tensor, and we can then integrate over the volume. Laue[231] proceeds in this manner. We obtain

$$\left. \begin{array}{l} G_x = \dfrac{1}{\sqrt{(1-\beta^2)}} \cdot \dfrac{v}{c^2} \left[E' + \int S'_{xx}\, dV' \right], \qquad G_y = \dfrac{v}{c^2} \int S'_{xy}\, dV', \ldots \\[3mm] \qquad\qquad E = \dfrac{1}{\sqrt{(1-\beta^2)}} \left[E' + \dfrac{v^2}{c^2} \int S'_{xx}\, dV' \right]. \end{array} \right\} \tag{346}$$

If, in particular, the stresses are a uniform scalar pressure, p,

$$\left. \begin{array}{l} G_x = \dfrac{1}{\sqrt{(1-\beta^2)}} \cdot \dfrac{v}{c^2} (E' + p'V'), \qquad G_y = G_z = 0 \\[3mm] \qquad\qquad E = \dfrac{1}{\sqrt{(1-\beta^2)}} \left(E' + \dfrac{v^2}{c^2} p'V' \right), \end{array} \right\} \tag{346a}$$

as was first discovered by Planck[232] in his basic paper on the dynamics of moving systems.

The second method is similar to that in § 21 for the proof of the vector character of J_k. But it is essential to realize that we cannot, in this case, simply replace the integral over the hyperplane $x'^4 = $ const. by the integral over the hyperplane $x^4 = $ const. In fact, the two integrals differ by

$$-\int f_i\, d\Sigma,$$

where the integration is to be carried out over the world region contained between the two hyperplanes. If the x-axis is taken in the direction of the velocity of K relative to K', it is easy to see that

$$\int f_i\, d\Sigma = \beta \int f'_i x'\, dV',$$

[231] M. v. Laue, *Ann. Phys., Lpz.*, **35** (1911) 524; cf. also, footnote 178a, p. 102 *ibid.*: p. 87, Eq. (102), and p. 153, Eq. (XXVII).
[232] Cf. footnote 225, p. 123, *ibid.*, cf. also A. Einstein, *Jb. Radioakt.*, **4** (1907) 411.

and after some further calculations we obtain

$$G_x = \frac{1}{\sqrt{(1 - \beta^2)}} \cdot \frac{v}{c^2} \left[E' + \int f'_x \, x' \, dV' \right]$$

$$G_y = \frac{v}{c^2} \int f'_y \, x' \, dV', \qquad G_z = \dots \qquad\qquad (347)$$

$$E = \frac{1}{\sqrt{(1 - \beta^2)}} \left[E' + \beta^2 \int f'_x \, x' \, dV' \right].$$

These formulae are due to Einstein[233] and Laue's formulae can be obtained from them by integrating by parts.

44. Applications to special cases. Trouton and Noble's experiment

It is easily seen that according to the transformation formulae of relativistic mechanics a moving rigid body is not in equilibrium when the resulting moment of the forces acting on it vanishes. Let us consider, for instance, a rod in a coordinate system K, which moves with velocity $\mathbf{u}$ in the direction of the x-axis.[234] In a system K', moving with the rod, let there be two equal and opposite forces acting along the direction of the rod at its two ends. Let α be the angle, measured in K', between the rod and the velocity $\mathbf{u}$ of K' relative to K (x'-axis). If x', y' are the differences of the coordinates of the two ends of the rod in K', and x, y the corresponding values in K, then

$$K'_x = |\mathbf{K}'| \cos\alpha, \qquad K'_y = |\mathbf{K}'| \sin\alpha,$$

and we have from (216) that

$$K_x = K'_x, \qquad K_y = K'_y \sqrt{(1 - \beta^2)},$$

in contrast to

$$x = x' \sqrt{(1 - \beta^2)}, \qquad y = y'.$$

In K, therefore, the force is not in the direction of the rod. The couple acting on the rod is

$$N_z = (1 - \beta^2) x' K'_y - y' K'_x = -\beta^2 x' K'_y = -\beta^2 l_0 |\mathbf{K}'| \sin\alpha \cos\alpha. \quad (348)$$

We have now to ask ourselves why, in spite of the existence of this couple, no rotation takes place? It will, first of all, be observed that the elastic forces which, in K', are in equilibrium with the external forces $\mathbf{K}$, transform in exactly the same way as the latter. There exists therefore in system K a moment of the elastic forces which cancels the external couple $\mathbf{N}$. There is a more sophisticated reason for the fact that the elastic forces are not in the direction of the rod, in this case. These forces cannot be represented exclusively as the divergence of a stress tensor. An extra term has to be added, which comes from the time derivative of the

[233] Cf. footnote 221, p. 122, *ibid.* A more general treatment can be found in A. Einstein, *Jb. Radioakt.*, **4** (1907) 446 and 447.
[234] Cf. footnote 215a, p. 118, P. Epstein, *ibid.*, p. 779.

momentum density (cf. § 42). The correct, quantitative, value of the turning couple is obtained in this way, as can be seen from the following argument. The couple $\mathbf{N}$ which is due to the elastic forces is equal to the negative time-derivative of the total elastic angular momentum $\mathbf{L}$, i.e. according to (344),

$$\mathbf{N} = -\frac{d\mathbf{L}}{dt} = -\frac{d}{dt} \int (\mathbf{r} \wedge \mathbf{g})\, dV. \tag{344a}$$

This derivation is quite analogous to that of Lorentz[234a] for the case of electromagnetic forces. Since in K' all functions of state are independent of the time, one easily deduces[234a] that

$$\mathbf{N} = -(\mathbf{u} \wedge \mathbf{G}). \tag{344b}$$

This reduces the determination of the turning couple to that of the total elastic momentum

$$\mathbf{G} = \frac{1}{c^2} \int \mathbf{S}\, dV.$$

For our case the energy current is always parallel to the direction of the rod and the integral over the cross-section of the rod, $\int S_n\, d\sigma = \int |\mathbf{S}|\, d\sigma$, is equal to the work done, $\mathbf{K} \cdot \mathbf{u}$, according to the energy conservation law. Thus

$$\mathbf{G} = \frac{1}{c^2}(\mathbf{K} \cdot \mathbf{u})\mathbf{r},$$

where $\mathbf{r}$ is the vector having components x, y. Substituting in (344b), this gives

$$|\mathbf{N}| = \frac{1}{c^2}(\mathbf{K} \cdot \mathbf{u})|\mathbf{u} \wedge \mathbf{r}| = \beta^2 K'_x\, y' = \beta^2 l_0 |\mathbf{K}'| \sin\alpha \cos\alpha,$$

which, indeed, just cancels the turning couple (348).

An analogous argument can be applied to the case of a right-angled lever, for which the existence of a turning couple was noticed by Lewis and Tolman[235] and explained by Laue[236] on the basis of the theorem of the momentum of the energy current.

If, next, the external forces which act on the rod are thought of as produced by two small spherical charges situated at its ends, only a small step is needed to arrive at Trouton and Noble's experimental arrangement.[237] These physicists investigated whether a charged condenser would tend to turn into a direction perpendicular to the earth's motion. In a coordinate system in which the condenser moves with velocity $\mathbf{u}$ in the x-direction, the electromagnetic field would, in general, exert a couple on the condenser.[238] Let α' be the angle which the normal to the plates makes with the velocity $\mathbf{u}$, W' the energy density, E' the electrostatic

[234a] Cf. footnote 4, p. 1, *ibid.*, §§ 7 and 21(a).

[235] G. N. Lewis and R. C. Tolman, *Phil. Mag.*, **18** (1909) 510.

[236] M. v. Laue, *Phys. Z.*, **12** (1911) 1008.

[237] Cf. footnote 6, p. 2, *ibid.*; see also footnote 4, p. 1, H. A. Lorentz, *ibid.*, § 56 (c).

[238] Cf. for this derivation, footnote 178 a, p. 102, M. v. Laue, *ibid.*, p. 99.

energy in a system K' moving with the condenser. The momentum in the moving system can be calculated with the help of (346). Since the field in K' only consists of a homogeneous electrostatic field between the condenser plates and is perpendicular to them, we have

$$E'_x = |\mathbf{E'}| \cos\alpha', \qquad E'_y = |\mathbf{E'}| \sin\alpha',$$

and we obtain for S'_{xx} and S'_{xy},

$$S'_{xx} = W' - E'_x{}^2 = W'(1 - 2\cos^2\alpha'),$$

$$S'_{xy} = -E'_x E'_y = -2W' \sin\alpha' \cos\alpha'.$$

On substitution in (346),

$$\left.\begin{aligned}
G_x &= \frac{u}{c^2}\frac{E'}{\sqrt{(1-\beta^2)}}(2 - 2\cos^2\alpha') = 2\frac{u}{c^2}\frac{E'}{\sqrt{(1-\beta^2)}}\sin^2\alpha' \\
G_y &= -2\frac{u}{c^2}E' \sin\alpha' \cos\alpha' = -\frac{u}{c^2}E' \sin 2\alpha'.
\end{aligned}\right\} \quad (349)$$

Apart from higher-order terms, therefore, the momentum is parallel to the condenser plates. From this, and from (344 b), a couple of magnitude

$$|\mathbf{N}| = uG_y = \beta^2 E' \sin 2\alpha' \tag{350}$$

is obtained. Nevertheless, no rotation was observed, as was to be expected from the relativity principle, anyhow. As early as 1904, H. A. Lorentz[239] gave the correct explanation that the elastic forces transform in precisely the same way as the electromotive forces. Laue's[240] conception goes deeper. According to this, the momentum of the elastic energy current produces a couple which exactly cancels the electromagnetic couple. Laue[241] also investigated how the couple (350) is brought about in

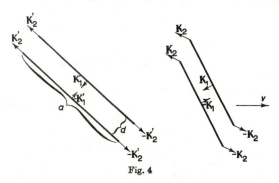

Fig. 4

detail. For this, it is important to note that in K', apart from the forces $|\mathbf{K'}_1| = E'/d$ perpendicular to the condenser plates, there act on each

[239] Cf. footnote 4, p. 1, *ibid.*, § 64; also cf. footnote 10, p. 2, *Versl. gewone Vergad. Akad. Amst., loc. cit.*
[240] M. v. Laue, *Ann. Phys., Lpz.*, **35** (1911) 524.
[241] M. v. Laue, *Ann. Phys., Lpz.*, **38** (1912) 370.

plate also forces which are perpendicular to each of its edges and in the plane of the plate. If the plates are rectangles with sides a, b, then the additional forces acting on edges b and a, and perpendicular to them, are $|\mathbf{K'_2}| = \frac{1}{2}E'/a$ and $|\mathbf{K'_3}| = \frac{1}{2}E'/b$, respectively. If the edges b are perpendicular to the velocity $\mathbf{u}$, then $\mathbf{K_3}'$ need not be considered. Fig. 4 illustrates the forces $\mathbf{K'_1}$, $\mathbf{K'_2}$ and $\mathbf{K_1}$, $\mathbf{K_2}$ acting in the systems K' and K, respectively.

The turning couple is obtained directly from a transformation of the forces into system K. The couple $\mathbf{K_1}$ contributes one half of the turning couple, the two couples $\mathbf{K_2}$ the other half. Expressions (347) for the momentum are also easily verified. We have

$$\int f'_x x' \, dV' = E'(\sin^2 \alpha' - \cos^2 \alpha')$$

$$\int f'_y x' \, dV' = 2E' \sin \alpha' \cos \alpha',$$

from which we regain (349).

Charge distributions other than those discussed above also lead to a turning couple, if they are in uniform motion. An example is the case of an ellipsoid.[242] A rotation, however, can never take place, because of the relativity principle. If the field is spherically symmetrical in the moving system K' then the momentum in K is parallel to $\mathbf{u}$, and the turning couple becomes zero, because of (344 b). In this case

$$\int S'_{xy} \, dV' = 0, \qquad \int S'_{xx} \, dV' = \int S'_{yy} \, dV' = \int S'_{zz} \, dV',$$

and it also follows from $S'_{xx} + S'_{yy} + S'_{xy} = W'$ that each of the last three integrals is equal to $\frac{1}{3}E'$. Therefore, from (346),

$$\mathbf{G} = \frac{\mathbf{u}}{c^2} \frac{\frac{4}{3}E'}{\sqrt{(1 - \beta^2)}}, \qquad E = \frac{E'(1 + \frac{1}{3}u^2/c^2)}{\sqrt{(1 - \beta^2)}}. \tag{351}$$

Cf. Part V on applications of these relations to the case of a single electron.

45. Hydrodynamics and theory of elasticity

The relativistic theory of elasticity took, historically, its origin from an endeavour to make use of the rigid-body concept also in the theory of relativity. One naturally had to look first for a definition of a rigid body which was Lorentz-invariant. Such a definition was first given by Born.[243] A body is to be considered rigid if a given volume element of it is undeformed in the coordinate system K_0 in which the volume element is momentarily at rest. The analytic formulation reads as follows: Let us characterize the flow of a deformable medium in Lagrange's way, by specifying the coordinates $x^1, ..., x^4$ as functions of the initial coordinates $\xi^1, ..., \xi^3$ and

[242] M. Abraham, *Ann. Phys., Lpz.*, **10** (1903) 174; and *Theorie der Elektrizität*, Vol. 2 (1st edn., 1905) p. 170 *et seq.*
[243] M. Born, *Ann. Phys., Lpz.*, **30** (1909) 1.

the proper time τ—or better still, of the coordinate $\xi^4 = ic\tau$, for symmetry reasons—

$$x^k = x^k(\xi^1, ..., \xi^4). \tag{352}$$

The world-line element

$$ds^2 = \sum (dx^k)^2$$

of two neighbouring space–time points then becomes a quadratic form of the differentials $d\xi^i$,

$$ds^2 = A_{ik} d\xi^i d\xi^k. \tag{353}$$

If we single out those world points which are simultaneous for an observer moving with the volume element at the given moment—they satisfy the equation

$$u_i dx^i = u_i \frac{\partial x^i}{\partial \xi^k} d\xi^k = 0, \tag{354}$$

where u_i = four-velocity—then $d\xi^4$ can be eliminated from (353) and the line element ds^2 can be written as a quadratic form of the three spatial differentials,

$$ds^2 = \sum_{i,k=1}^{3} p_{ik} d\xi^i d\xi^k. \tag{355}$$

The deviations of the p_{ik} from their initial values characterize the deformation of the volume element. For a rigid body they are to vanish, always, so that

$$\frac{\partial p_{ik}}{\partial \xi^4} = 0. \tag{356}$$

A simple argument by Ehrenfest[244] shows however that such a body cannot be set in rotation. If this were possible, the perimeters of the circles described by the points of the body would become smaller because of the Lorentz contraction, whereas their radii would remain unchanged since they are always perpendicular to the velocity. It was further proved, independently, by Herglotz[245] and Noether[246] that a rigid body in the Born sense has only three degrees of freedom, in contrast to the six degrees of freedom of a rigid body in classical mechanics. Apart from exceptional cases, the motion of the body is completely determined when the motion of a single of its points is prescribed. This in itself raised strong doubts as to the possibility of introducing the concept of a rigid body into relativistic mechanics.[247] The final clarification was brought about in a paper by Laue[248], who showed by quite elementary arguments that the number of kinematic degrees of freedom of a body cannot be limited, according to the theory of relativity. For, since no action can be propagated with a

[244] P. Ehrenfest, *Phys. Z.*, **10** (1909) 918.

[245] G. Herglotz, *Ann. Phys.*, *Lpz.*, **31** (1910) 393.

[246] F. Noether, *Ann. Phys.*, *Lpz.*, **31** (1910) 919.

[247] Cf. M. Born, *Phys. Z.*, **11** (1910) 233; M. Planck, *Phys. Z.*, **11** (1910) 294; W. v. Ignatowski, *Ann. Phys.*, *Lpz.*, **33** (1910) 607; P. Ehrenfest, *Phys. Z.*, **11** (1910) 1127; M. Born *Nachr. Ges. Wiss. Göttingen*, (1910) 161.

[248] M. v. Laue, *Phys. Z.*, **12** (1911) 85.

velocity greater than that of light, an impulse which is given to the body simultaneously at n different places, will, *to start off with*, produce a motion to which at least n degrees of freedom must be ascribed.

If thus the concept of a *rigid body* has no place in relativistic mechanics, it is nevertheless useful and natural to introduce the concept of a *rigid motion* of a body. We shall denote those motions as rigid for which Born's condition (356) is satisfied. Herglotz[249] afterwards developed a relativistic theory of elasticity which was based on the idea that stresses always occur when Born's condition (356) is violated. He derived the equations of motion from an action principle,

$$\delta \int \Phi \, d\xi_1 \dots d\xi_4 = 0, \tag{357}$$

where Φ is a function of the deformation quantities A_{ik}. It is chosen in such a way that, for the stationary case, Φ depends on the p_{ik} in exactly the same way as the Lagrangian in ordinary theory of elasticity. The resulting equations of motion can be fitted into Laue's scheme (340) and (341).

It should be mentioned as well that Laue[250] introduced also *relative*, in contrast to *absolute*, stresses. From (341) it follows that

$$\dot{g}_i = - \sum_{k=1}^{3} \frac{\partial T_i{}^k}{\partial x^k}. \qquad (i = 1, 2, 3) \tag{358a}$$

Since, on the left-hand side, we have the *local* and not the *total* rate of change of the momentum density, the space components of **T** do not represent the elastic stresses. The total rate of change of the momentum density, dg_i/dt, is determined by

$$\frac{dg_i}{dt} = \frac{\partial g_i}{\partial t} + \sum_{k=1}^{3} \frac{\partial}{\partial x^k}(g_i u_k),$$

so that

$$\frac{dg_i}{dt} = - \sum_{k=1}^{3} \frac{\partial T_i{}^k}{\partial x^k}, \tag{358b}$$

where

$$T_{ik} = T_{ik} - g_i u_k. \qquad (i, k = 1, 2, 3) \tag{359}$$

It is to be noted that the relative stresses T_{ik} are not symmetric. Their transformation laws are given by

$$\left. \begin{aligned} &T_{xx} = T^0{}_{xx}, & &T_{yy} = T^0{}_{yy}, & &T_{zz} = T^0{}_{zz} \\[2mm] &T_{xy} = \frac{T^0{}_{xy}}{\sqrt{(1 - \beta^2)}}, & &T_{xz} = \frac{T^0{}_{xz}}{\sqrt{(1 - \beta^2)}}, & &T_{yz} = T^0{}_{yz} \\[2mm] &T_{yx} = \sqrt{(1 - \beta^2)}T^0{}_{yx}, & &T_{zx} = \sqrt{(1 - \beta^2)}T^0{}_{zx}, & &T_{zy} = T^0{}_{zy}. \end{aligned} \right\} \tag{360}$$

[249] G. Herglotz, *Ann. Phys., Lpz.*, **36** (1911) 493.

[250] Cf. footnote 228, p. 102, *ibid.*; also footnote 178a, p. 102, *ibid.*, § 26. M. Abraham (*R. C. Circ. mat. Palermo*, **28** (1909) 1) had, already before that, introduced quite analogous relative stresses in the electrodynamics of moving bodies.

In contrast to the corresponding relations for the absolute stresses, the energy density W_0 does not appear here. If in the rest system the (three-dimensional) stress tensor is, in particular, a scalar

$$T^0{}_{ik} = p_0 \delta_i{}^k, \qquad (i, k = 1, 2, 3)$$

then we also have

$$T_{ik} = p_0 \delta_i{}^k.$$

The scalar pressure is an invariant:

$$p = p_0. \tag{361}$$

This follows already directly from the transformation formulae for the force and area, if one defines pressure as force per unit area[251] (cf. also § 32 (δ) on remarks about the invariance of radiation pressure).

The equations of motion take on a comparatively simple form for liquids, where the three-dimensional stress tensor degenerates into a scalar. This special case was treated, in addition to Herglotz,[252] by Ignatowski[253] and Lamla.[254] The results of these authors are all in agreement. If μ_0 is the rest-mass density, p the pressure, P the integral $\int dp/\mu_0$, as usually in hydrodynamics, and if we restrict ourselves to adiabatic processes, then the energy-momentum tensor is given by

$$T_i{}^k = \mu_0\left(1 + \frac{P}{c^2}\right)u_i u^k + p \,\delta_i{}^k. \tag{362}$$

From the equations

$$\frac{\partial T_i{}^k}{\partial x^k} = 0$$

follow, by scalar multiplication with u^i, the continuity equation

$$\frac{\partial(\mu_0 u^k)}{\partial x^k} = 0 \tag{363}$$

and the equation of motion

$$\mu_0\left(1 + \frac{P}{c^2}\right)\frac{du_i}{d\tau} = -\left[\frac{\partial p}{\partial x^i} + u_i \frac{d}{d\tau}\left(\frac{p}{c^2}\right)\right]. \tag{364}$$

For the stationary case, $T_0{}^0$ gives the usual expression for the energy density.

These arguments are only of value, in so far as they demonstrate the *possibility* of constructing a relativistic hydrodynamics and a theory of elasticity which are free from contradictions. Physically speaking they do not bring anything new, since for substances in which the velocity of the elastic waves is small compared with c, the equations of the relativistic

[251] Cf. footnote 232, p. 126, A. Einstein, *ibid.*, § 13; A. Sommerfeld, *Ann. Phys., Lpz.*, **32** (1910) 775. First expressed by M. Planck (cf. footnote 225, p. 123, *ibid.*).
[252] Cf. footnote 249, p. 132, *ibid.*
[253] W. v. Ignatowski, *Phys. Z.*, **12** (1911) 441.
[254] E. Lamla, *Dissertation* (Berlin 1911); *Ann. Phys., Lpz.*, **37** (1912) 772.

theory of elasticity do not differ in practice from those of the ordinary theory.

Herglotz and Lamla deduced from their equations that there must be a lower limit for the compressibility, since otherwise the elastic waves could be propagated with a velocity greater than that of light. It seems to us, however, that the relativity principle cannot make any statements on the magnitude of the cohesive forces. When the static compressibility approaches Herglotz's and Lamla's limit, the phenomenological equations will probably become incorrect. A *dispersion* of elastic waves will then occur, and the situation will be similar to that discussed in the case of light waves in § 36 (δ).

(d) THERMODYNAMICS AND STATISTICAL MECHANICS

46. Behaviour of the thermodynamical quantities under a Lorentz transformation

The manner in which the thermodynamical quantities transform for the transition to a moving coordinate system was derived by Planck[255] in his basic paper on the dynamics of moving systems. His starting point is a variational principle. It was shown by Einstein,[256] however, that the transformation formulae can also be derived directly; the variational principle then follows from them.

Let us start by collecting again the relations for volume, pressure, energy and momentum, where we assume that the elastic stresses only consist of a scalar pressure:

$$V = V_0 \sqrt{(1 - \beta^2)} \tag{7a}$$

$$p = p_0 \tag{361}$$

$$\left. \begin{aligned} \mathbf{G} &= \frac{\mathbf{u}}{c^2} \frac{1}{\sqrt{(1 - \beta^2)}} (E_0 + p_0 V_0) \\ E &= \frac{1}{\sqrt{(1 - \beta^2)}} \left(E_0 + \frac{u^2}{c^2} p_0 V_0 \right). \end{aligned} \right\} \tag{346a}$$

From this, it further follows that

$$E + pV = \frac{E_0 + p_0 V_0}{\sqrt{(1 - \beta^2)}}, \qquad \mathbf{G} = \frac{\mathbf{u}}{c^2} (E + pV). \tag{346b}$$

We now have to derive the corresponding relations for the quantity of heat, the temperature and the entropy. If dQ is the amount of heat transferred to the system, dA the work done by the external forces on the system, then

$$\left. \begin{aligned} dQ &= dE - dA \\ dA &= -p\,dV + \mathbf{u} \cdot d\mathbf{G}. \end{aligned} \right\} \tag{365}$$

[255] Cf. footnote 225, p. 123, *ibid*. Cf. also the paper by F. Hasenöhrl, *S.B. Akad. Wiss. Wien*, **116** (1907) 1391, in which he arrives at similar results in a different way, and independently, from Planck.
[256] Cf. footnote 232, p. 126, *ibid*., A. Einstein, §§ 15 and 16.

The second term is essential since, according to (346), it does not vanish even when the velocity of the system remains constant during a change of its state, as will be assumed from now on. We obtain

$$dQ = \frac{1}{\sqrt{(1 - \beta^2)}}dE_0 + \frac{u^2/c^2}{\sqrt{(1 - \beta^2)}}d(p_0 V_0) -$$

$$- \frac{u^2/c^2}{\sqrt{(1 - \beta^2)}}[dE_0 + d(p_0 V_0)] + \sqrt{(1 - \beta^2)}p_0 dV_0$$

$$= \sqrt{(1 - \beta^2)}(dE_0 + p_0 dV_0) = \sqrt{(1 - \beta^2)}dQ_0,$$

i.e.
$$Q = Q_0\sqrt{(1 - \beta^2)}. \tag{366}$$

This determination agrees with the transformation property of the Joule heat, which we had derived previously (cf. Eq. (293)).

If the system is given a velocity **u**, this can be regarded as an adiabatic process. The *entropy* therefore remains unchanged, and has the same value for a moving as for a stationary system. This means however that it is a Lorentz-invariant,

$$S = S_0. \tag{367}$$

If an amount of heat dQ is transferred infinitely slowly, then

$$dQ = T dS.$$

Using (366) and (367), we obtain

$$T = T_0\sqrt{(1 - \beta^2)}. \tag{368}$$

These relations enable us to write down, for each relation between the quantities p_0, V_0, E_0, G_0, T_0 in the stationary system, a corresponding relation for a moving system. In particular, the velocity dependence of the equation of state of a substance can be determined.

47. The principle of least action

In non-relativistic thermodynamics, the equation of state can be obtained from the action principle[257]

$$\int_{t_1}^{t_2} \{\delta(- F + E_\text{kin}) + \delta A\}dt = 0,$$

where F is the free energy,

$$F = E - TS.$$

Here, the independent variables are the position coordinates of the system, its volume, and its temperature. δA is the work done for a variation of these parameters. The function to be varied changes in the usual way for a change in the independent variables. The action function

$$L = - F + E_\text{kin}$$

[257] H. v. Helmholtz, *J. reine angew. Math.*, **100** (1886) 137 and 213 (*Collected Papers*, 3 (1895) 225).

consists here of two parts, of which the one depends only on the velocity, the other only the internal state (V, T) of the body. In relativistic mechanics, too, such an action function exists. But it cannot be split in the same way. Actually, for

$$L = -E + TS + \mathbf{u}\cdot\mathbf{G}, \tag{369}$$

$$\left. \begin{array}{ccc} \dfrac{d}{dt}\left(\dfrac{\partial L}{\partial \dot{x}}\right) = K_x, & \dfrac{d}{dt}\left(\dfrac{\partial L}{\partial \dot{y}}\right) = K_y, & \dfrac{d}{dt}\left(\dfrac{\partial L}{\partial \dot{z}}\right) = K_z, \\[3mm] \dfrac{\partial L}{\partial V} = p, & \dfrac{\partial L}{\partial T} = S. \end{array} \right\} \tag{370}$$

For it follows from

$$dE = \mathbf{K}\cdot d\mathbf{r} - p\,dV + T\,dS = \mathbf{u}\cdot d\mathbf{G} - p\,dV + T\,dS$$

that

$$dL = \mathbf{G}\cdot d\mathbf{u} + p\,dV + S\,dT.$$

But (370) are just the equations which follow from the action principle. We observe, in addition, that according to (318 a, b) and (325) we have to put

$$L = -E_{\text{kin}} + \mathbf{u}\cdot\mathbf{G}$$

for a particle. This can be regarded as a special case of (369). In the rest system K_0, L becomes identical with the (negative) free energy, $L_0 = -E_0 + T_0 S_0$. From (346), (367) and (368) we also obtain the transformation formula for L,

$$L = L_0 \sqrt{(1-\beta^2)}, \tag{371}$$

so that the action integral $\int L\,dt$ is an invariant, as required.

48.   The application of relativity to statistical mechanics

Liouville's theorem

$$dp_1 \ldots dq_N = dp_1{}^0 \ldots dq_N{}^0. \tag{372}$$

holds in the space of the canonical variables p_k, q_k (cf. § 40), since it is a direct consequence of Hamilton's equations. It is of course also valid in a space of other variables, $x_1 \ldots x_{2N}$, which are produced from the canonical variables with a functional determinant equal to unity,

$$dx_1 \ldots dx_{2N} = dx_1{}^0 \ldots dx_{2N}{}^0. \tag{372a}$$

As the general theorems of statistical mechanics are based on no other assumption than Liouville's theorem, they therefore remain unchanged in relativistic statistical mechanics.[258] They are formulated as follows:

(i) Let the energy be a function of the variables $x_1, \ldots, x_{2N}$—of which it will always be assumed that they satisfy condition (372 a)—given by

$$H(x_1, \ldots, x_{2N}) = E. \tag{373}$$

[258] We are not considering here modifications of the statistical theorems which are demanded by quantum theory.

Then the *entropy* is given by

$$S = k \log V, \tag{374}$$

where V is the volume enclosed by the energy surface $H = E$, or by the energy shell $E < H < E + dE$,

$$V = \int\limits_{H<E} dx_1 \dots dx_{2N}, \qquad \text{or} \qquad V = \int\limits_{E<H<E+dE} dx_1 \dots dx_{2N}. \tag{375}$$

(ii) The *free energy* $F = E - TS$ is given by

$$\left.\begin{aligned} F &= -kT \log Z \\[2mm] Z &= \int e^{-H/kT} dx_1 \dots dx_{2N}. \end{aligned}\right\} \tag{376}$$

(iii) *The equipartition law.* The time averages are

$$\left.\begin{aligned} \overline{x_i \frac{\partial H}{\partial x_i}} &= kT, \text{ for all } i \text{ from 1 to } 2N, \\[2mm] \overline{x_i \frac{\partial H}{\partial x_j}} &= 0, \text{ for } i \neq j. \end{aligned}\right\} \tag{377}$$

In particular, for canonical variables,

$$\overline{p_i \dot{q}_i} = kT, \qquad \overline{q_i \frac{\partial H}{\partial q_i}} = kT. \tag{377a}$$

Here we have disagreement with ordinary mechanics. For in the latter, $E_{\text{kin}} = \frac{1}{2} \sum p_i \dot{q}_i$, so that the first equation (377 a) simply states that the time averages of those contributions to the kinetic energy which correspond to the different degrees of freedom are each equal to $\frac{1}{2} kT$. *In relativistic mechanics the connection between the equipartition law and the average kinetic energy is lost.*

(iv) *The Maxwell–Boltzmann distribution law.* Let the energy function H of our system be separated into two parts

$$H = H_1(x_1, \dots, x_{2n}) + H_2(X_1, \dots, X_{2N}), \tag{378}$$

which depend on different variables. The number, $2n$, of variables in H_1 is to be much smaller than the corresponding number, $2N$, in H_2. In addition, the two groups of variables are supposed to be produced from different canonical variables, with a functional determinant equal to unity. Then the probability that the first group of variables should take on certain prescribed values $x_1 \dots x_{2n}$ within the range $dx_1 \dots dx_{2n}$, regardless of the values of the second group of variables, is given by

$$w(x_1, \dots, x_{2n}) \, dx_1 \dots dx_{2n} = A e^{-H_1(x)/kT} dx_1 \dots dx_{2n}. \tag{379}$$

The quantity A, which is independent of the x, is determined from the condition

$$\int w(x_1, \dots, x_{2n}) \, dx_1 \dots dx_{2n} = 1. \tag{379a}$$

The distribution law (379) is based on the assumption that the value of H_1 is small compared with the (constant) value of H.

49. Special cases

(α) *Black-body radiation in a moving cavity.* This case is of historical interest, since it can be treated entirely on the basis of electrodynamics, without relativity. When this is done, one comes to the inevitable conclusion that a momentum, and thus also an inertial mass, must be ascribed to the moving radiation energy. It is of interest that this result should have been found by Hasenöhrl[259] already before the theory of relativity had been formulated. Admittedly, his deductions were open to correction on some points. A complete solution of the problem was first given by Mosengeil.[260] Planck[261] derived many of his formulae for the dynamics of moving systems by generalizing Mosengeil's results.

The theory of relativity permits us to determine directly the temperature dependence of radiation pressure, momentum, energy and entropy, and the dependence of the spectral distribution on temperature and direction, by reducing the case of a moving to that of a stationary cavity.
For the latter we have

$$E_0 = aT_0^4 V_0, \qquad p_0 = \frac{1}{3}aT_0^4, \qquad S_0 = \frac{4}{3}aT_0^3 V_0 \qquad (380\text{a})$$

and from (369)

$$L = \frac{1}{3}aT_0^4 V_0.$$

Finally, the intensity of radiation in the frequency range $d\nu$ and solid angle $d\Omega$, is given by

$$K_0\nu_0 \, d\nu_0 \, d\Omega_0 = \frac{2h}{c^2} \frac{\nu_0^3 \, d\nu_0}{\exp(h\nu_0/kT_0) - 1} d\Omega_0. \qquad (381\text{a})$$

By means of the formulae of § 46 we then obtain

$$\left.\begin{aligned}
E &= E_0 \frac{1 + \frac{1}{3}\beta^2}{\sqrt{(1-\beta^2)}} = aT^4 V \frac{1 + \frac{1}{3}\beta^2}{(1-\beta^2)^3}, \\[2mm]
p &= p_0 = \frac{1}{3}aT^4 \frac{1}{(1-\beta^2)^2}, \qquad S = S_0 = \frac{4}{3}aT^3 V \frac{1}{(1-\beta^2)^2}, \\[2mm]
L &= L_0\sqrt{(1-\beta^2)} = \frac{1}{3}aT^4 V \frac{1}{(1-\beta^2)^2}, \\[2mm]
\mathbf{G} &= \frac{4}{3}\frac{\mathbf{u}}{c^2}\frac{1}{\sqrt{(1-\beta^2)}}E_0 = \frac{4}{3}aT^4 V \frac{1}{(1-\beta^2)^3}\frac{\mathbf{u}}{c^2}.
\end{aligned}\right\} \quad (380\text{b})$$

[259] F. Hasenöhrl, *S.B. Akad. Wiss. Wien*, **113** (1904) 1039; *Ann. Phys., Lpz.*, **15** (1904) 344 and **16** (1905) 589.

[260] K. v. Mosengeil, *Dissertation* (Berlin 1906); *Ann. Phys., Lpz.*, **22** (1907) 867; cf. also the account by M. Abraham, *Theorie der Elektrizität*, Vol. 2 (2nd edn.), p. 44.

[261] Cf. footnote 225, p. 123, M. Planck, *ibid.*

To obtain also the spectral distribution in the moving cavity, we make use of the relations

$$\nu' = \nu \frac{1 - \beta \cos\alpha}{\sqrt{(1 - \beta^2)}}, \qquad d\nu' = d\nu \frac{1 - \beta \cos\alpha}{\sqrt{(1 - \beta^2)}},$$

$$d\Omega' = \frac{1 - \beta^2}{(1 - \beta \cos\alpha)^2} d\Omega,$$

$$K'_{\nu'} d\nu' d\Omega' = K_\nu d\nu \, d\Omega \frac{(1 - \beta \cos\alpha)^2}{1 - \beta^2},$$

which are easily derivable from (15), (17) and (253). The last quantity must transform like the square of the amplitude A. It further follows that

$$K'_{\nu'} = K_\nu \frac{(1 - \beta \cos\alpha)^3}{(1 - \beta^2)^{3/2}},$$

i.e.
$$K_\nu d\nu \, d\Omega = \frac{2h}{c^2} \frac{\nu^3 \, d\nu}{\exp\{(h\nu/kT)(1 - \beta \cos\alpha)\} - 1} d\Omega. \qquad (381\,\text{b})$$

In addition, because of

$$K'_{\nu'} d\nu' = K_\nu d\nu \frac{(1 - \beta \cos\alpha)^4}{(1 - \beta^2)^2},$$

we have
$$K = \frac{ac}{4\pi} T^4 \frac{1}{(1 - \beta \cos\alpha)^4}. \qquad (382)$$

This formula gives the direction dependence of the total radiation intensity (integrated over all frequencies). It can, of course, also be obtained from (381 b) by integration over $d\nu$. The total energy is obtained from (382) by means of the relation

$$E = V \int \frac{1}{c} K \, d\Omega$$

and agrees with the first of equations (380 b).

Because of the extreme smallness of the expected effects it seems unlikely that the inertia of the radiation energy could be demonstrated experimentally.

(β) *The ideal gas.* We can naturally only expect a deviation due to relativistic effects (variability of mass), in the behaviour of an ideal gas from that calculated by classical mechanics, when the mean velocity of the molecules becomes comparable with the velocity of light. A criterion for this is the size of the quantity

$$\sigma = \frac{m_0 c^2}{kT}. \qquad (383)$$

For normal temperatures it is enormously large and reaches reasonable

proportions only for temperatures of about 10^{12} °K. To ask for any deviations from the classical behaviour of an ideal gas required by relativistic
mechanics is therefore not of practical, but only of theoretical, interest.
The answer was given by Jüttner.[262] The simplest way to proceed is to
calculate the free energy from Theorem (ii), § 48. Since the energy of a
particle is of the form

$$E = m_0 c^2 \left[1 + \frac{1}{m_0{}^2 c^2} (p_x{}^2 + p_y{}^2 + p_z{}^2) \right]^{1/2}$$

when expressed in terms of momenta, it follows that

$$F = - RT \log \bar{Z},$$

$$\bar{Z} = \bar{Z}^{1/L} = V \cdot \iiint \exp\left\{ - \frac{m_0 c^2}{kT} \left[1 + \frac{1}{m_0{}^2 c^2} (p_x{}^2 + p_y{}^2 + p_z{}^2) \right]^{1/2} \right\} \times$$

$$\times dp_x dp_y dp_z.$$

It is assumed here that the amount of gas present is equal to 1 gramme-
molecule; L is Avogadro's number, V the volume. The calculation then
gives the result

$$\left. \begin{array}{l} \bar{Z} = V m_0{}^3 c^3 \cdot 2\pi^2 (-i) \dfrac{H_2{}^{(1)}(i\sigma)}{\sigma}, \\[2ex] F = - RT \left\{ \log V + \log\left(- \dfrac{i H_2{}^{(1)}(i\sigma)}{\sigma} \right) + \text{const.} \right\}, \end{array} \right\} \quad (384)$$

where $H_n{}^{(i)}$ is the nth-order Hankel function of the ith kind, with $i = 1, 2$.
   All other thermodynamical quantities follow from the free energy in
the usual manner, e.g.

$$p = - \frac{\partial F}{\partial V}, \qquad E = F - T \frac{\partial F}{\partial T} = T^2 \frac{\partial}{\partial T}\left(\frac{F}{T} \right)$$

(independent variables V, T). From the first equation we have

$$p = \frac{RT}{V}. \qquad (385)$$

*The equation of state of an ideal gas remains unchanged in relativistic
mechanics.* This is connected with the fact that the volume dependence
of the free energy and of the partition function is not modified by relativistic mechanics; there is also an *a priori* reason for this. The situation is
different for the case of the temperature dependence of the energy. We
obtain

$$E = RT \left\{ 1 - \frac{i H_2{}'^{(1)}(i\sigma)}{H_2{}^{(1)}(i\sigma)} \sigma \right\}. \qquad (386)$$

[262] F. Jüttner, *Ann. Phys., Lpz.,* **34** (1911) 856.

For large σ we can replace the Hankel function by its asymptotic form

$$-iH_2^{(1)}(i\sigma) \simeq \frac{e^{-\sigma}}{\sqrt{(\tfrac{1}{2}\pi\sigma)}}.$$

By logarithmic differentiation,

$$-\frac{iH_2'^{(1)}(i\sigma)}{H_2^{(1)}(i\sigma)} = 1 + \frac{1}{2\sigma},$$

which, substituted in (386), gives

$$E = RT(\sigma + \tfrac{3}{2}) = Lmc^2 + \tfrac{3}{2}RT, \tag{386a}$$

in agreement with the earlier theory, as it should. Expression (386) for the energy could also have been obtained from Maxwell's distribution law. According to Theorem (iv), § 48, this differs from the distribution law of classical mechanics only in the manner in which the factor A depends on the temperature.

Jüttner[263] has also studied the influence of the *motion* of an ideal gas on its thermodynamical properties, from the point of view of relativistic dynamics. The corresponding relations can be written down at once on the basis of the transformation formulae of § 46. The case of an ideal gas is seen to be even more unfavourable than that of black-body radiation for the purpose of an experimental proof of the theorem of the equivalence of mass and energy.

[263] F. Jüttner, *Ann. Phys., Lpz.*, **35** (1911) 145.

PART IV. GENERAL THEORY OF RELATIVITY

50. Historical review, up to Einstein's paper of 1916[264]

Newton's law of gravitation, which requires *instantaneous* action at a distance, is incompatible with special relativity. The latter demands that the velocity of propagation should at most be equal to the light velocity[265] and that the gravitational laws should be Lorentz-covariant. Poincaré[266] already studied the problem of modifying Newton's law of gravitation in such a way that these requirements should be fulfilled. This can be done in several ways. All his "Ansätze" have this in common, that the force between two particles depends, not on their *simultaneous* positions, but on those differing by a time interval $t = r/c$, as well as on their velocities (and possibly also their accelerations). The deviations from Newton's law are always of second order in v/c, and thus remain always very small and are not in contradiction with experiment.[266a] Minkowski[267] and Sommerfeld[268] have put these attempts of Poincaré into a form corresponding to the four-vector calculus; a particular case is discussed by H. A. Lorentz.[268a]

The objection to all these considerations is that they take as their starting point a fundamental law of force, instead of Poisson's equation. Once the finite propagation of an effect has been demonstrated, one can only expect to arrive at *simple*, generally valid, laws if one describes it in terms of continuously varying functions of position and time (a *field*) and looks for the *differential equations* of this field. The problem thus consists in modifying the Poisson equation,

$$\Delta\Phi = 4\pi k\mu_0,$$

and the equation of motion of the particle,

$$\frac{d^2\mathbf{r}}{dt^2} = -\operatorname{grad}\Phi,$$

in such a way that they become Lorentz-invariant.

Before this problem was solved, however, the developments pointed in another direction. As soon as the physical deductions from the special theory of relativity had reached a certain stage, Einstein[269] at once attempted to extend the relativity principle to reference systems in non-

---

[264] Cf. the articles by J. Zenneck (*Encykl. math. Wiss.*, V2) and S. Oppenheim (*ibid.*, VI 2, 22, in particular Part V). We give here the historical development only in rough outline; for some of the details, see also the article by F. Kottler (*ibid.*, VI 2, 22).

[265] If one assumes that the velocity of propagation of the gravitational effects is independent of the state of motion of the bodies causing them, it even follows that it must be *exactly* equal to the light velocity.

[266] Cf. footnote 11, p. 2, *R.C. Circ. mat. Palermo, loc. cit.*

[266a] For a more detailed discussion, see W. de Sitter, *Mon. Not. R. Astr. Soc.*, **71** (1911) 388.

[267] Cf. footnote 54, p. 21, Minkowski III.

[268] Cf. footnote 55, p. 22, *ibid.*, **33**.

[268a] H. A. Lorentz, *Phys. Z.* **11** (1910) 1234; cf. also footnote 40, p. 13, *ibid.*, p. 19.

[269] Cf. footnote 232, p. 126, *ibid.*, Chap. V.

uniform motion. He postulated that the general physical laws should retain their form even in systems other than Galilean (§ 2). This was made possible by the so-called *principle of equivalence*. In the Newtonian theory a system in a homogeneous gravitational field is completely equivalent to a uniformly accelerated reference system, from a *mechanical* point of view.[269a] The postulate that, in addition, all other processes should take place in the same way in both systems, forms the content of Einstein's principle of equivalence. This principle is one of the bases of the general theory of relativity, which was developed by him at a later date (cf. § 51). Since one can compute the sequence of events in an accelerated system, the principle would make it possible to calculate the effect which a homogeneous gravitational field has on arbitrary processes. It is this feature which renders the principle of equivalence so powerful from a heuristic point of view. In this way Einstein derived the result that the rate of clocks at points of lower gravitational potential is slower than that for higher gravitational potentials; and he pointed out already then that this entails a shift towards the red of spectra emitted by the sun, compared with those on the earth (cf. § 53 (β)). A further result was that the velocity of light is not constant in a gravitational field, so that light rays become curved, and that not only an *inertial* but also a *gravitational* mass $m = E/c^2$ has to be ascribed to an energy E in all cases. In a subsequent paper, Einstein[270] showed that the bending of the light rays brings with it a displacement of the fixed stars seen at the edge of the sun, which can be checked by experiment. At the time, he calculated the size of this displacement to be 0·83″.

This theory of the homogeneous gravitational field implied breaking through the framework of the special theory of relativity. Because of the dependence of the light velocity and of the rate of a clock on the gravitational potential, the definition of simultaneity introduced in § 4 is no longer applicable, and the Lorentz transformation loses its meaning. *Seen from this point of view, the special theory of relativity can only be correct in the absence of gravitational fields.* Instead, once the gravitational potential is introduced as a physical quantity, the physical laws will have to be considered as relations between the other physical quantities and the gravitational potential. Their covariance will have to be demanded with respect to a wider transformation group, with suitable transformation properties for the gravitational potential. The problem next arose of how to set up such a theory which was to be based on the principle of equivalence and which would also apply to non-homogeneous gravitational fields. Einstein and Abraham[271] tried to characterize the

[269a] Strictly speaking, a uniformly accelerated motion has to be replaced by a hyperbolic motion (§ 26), which renders the transformation formulae for the coordinates more complicated. See H. A. Lorentz (footnote 40, p. 13, *ibid.*, p. 36) and P. Ehrenfest, *Proc. Acad. Sci. Amst.*, **15** (1913) 1187.

[270] A. Einstein, 'Über den Einfluss der Schwerkraft auf die Ausbreitung des Lichtes', *Ann. Phys., Lpz.*, **35** (1911) 898; also contained in the collection of papers, *Das Relativitätsprinzip* (3rd edn., 1920).

[271] A. Einstein, *Ann. Phys., Lpz.*, **38** (1912) 355 and 443; M. Abraham, *Phys. Z.*, **13** (1912) 1, 4 and 793; discussion between Einstein and Abraham, *Ann. Phys., Lpz.*, **38** (1912) 1056 and 1059; **39** (1912) 444 and 704.

general static gravitational field by the value of the light velocity c at each space-time point; c would thus play the rôle of a gravitational potential. They also attempted to find the differential equations which c had to satisfy. Apart from the fact that these theories only considered special gravitational fields, they led to difficulties in other ways, too.

For this reason Nordström[272] attempted to adhere consistently to the strict validity of the principle of special relativity. In his theory, the velocity of light is constant and a deflection of light in a gravitational field does not take place. The theory solves in a logically quite unexceptionable way the problem sketched out above, of how to bring the Poisson equation and the equation of motion of a particle into a Lorentz-covariant form. Also, the energy–momentum law and the theorem of the equality of inertial and gravitational mass are satisfied. If, in spite of this, Nordström's theory is not acceptable, this is due, in the first place, to the fact that it does not satisfy the principle of *general* relativity (or at least not in a simple and natural way, cf. § 56). Secondly, it is in contradiction with experiment: it does not predict the bending of light rays and gives the displacement of the perihelion of Mercury with the wrong sign. (It is in agreement with Einstein's theory with regard to the red shift.) Mie[273], too, set up a gravitational theory based on the principle of special relativity. But since the theorem of the equality of inertial and gravitational mass is not satisfied *rigorously* in this theory, there never seemed much likelihood of its being correct.

Einstein, however, was not deflected by the difficulties of the problem in his endeavour to put the physical laws into such a form that they would be covariant under the widest possible group of transformations. In a paper[274] in collaboration with Grossmann he succeeded in making a definite advance in this direction. If the square of the line element is transformed into an arbitrary curvilinear space-time coordinate system it becomes a quadratic form in the coordinate differentials, with ten coefficients g_{ik} (cf. § 51). The gravitational field is now determined by this ten-component tensor of the g_{ik}, and no longer by the scalar light velocity. At the same time the equation of motion of a particle, the energy–momentum law and the electromagnetic field equations for the vacuum were all given a definite, *generally* covariant, form by introducing[275] the g_{ik}. Only the differential equations for the g_{ik} themselves were not

[272] G. Nordström, *Phys. Z.*, **13** (1912) 1126; *Ann. Phys., Lpz.*, **40** (1013) 856; **42** (1913) 533; **43** (1914) 1101; *Ann. Acad Sci. fenn.*, **57** (1914 and 1915); also M. Behacker, *Phys. Z.*, **14** (1913) 989; A. Einstein and A. D. Fokker, *Ann. Phys., Lpz.*, **44** (1914) 321. A summarizing report was given by M. v. Laue, *Jb. Radioakt.*, **14** (1917) 263, and a review article on the earlier gravitational theories by M. Abraham, *Jb. Radioakt.*, **11** (1914) 470.

[273] G. Mie, *Ann. Phys., Lpz.*, **40** (1913) 1, Chap. V, Gravitation; *Elster-Geitel-Festschrift* (1915) p. 251.

[274] A. Einstein and M. Grossmann, *Z. Math. Phys.*, **63** (1914) 215. Summarizing report: A. Einstein, 'Zum gegenwärtigen Stand des Gravitationsproblems', *Phys. Z.*, **14** (1913) 1249; subsequent discussion by Mie, Einstein and Nordström in *Phys. Z.*, **15** (1914) 115, 169, 176 and 375.

[275] It is of interest that, without being connected with the theory of gravitation, the relevant formal developments as well as the electromagnetic field equations in a generally covariant form had already previously been given by F. Kottler, *S.B. Akad. Wiss. Wien*, **121** (1912) 1659.

generally covariant yet. In a subsequent paper[276], Einstein tried to establish these differential equations in a more rigorous manner and he even believed to have proved that the equations which determine the g_{ik} themselves could not be generally covariant. In the year 1915, however, he realized that his gravitational field equations were not uniquely determined by the invariant-theoretical conditions which he had formerly laid down for them. To restrict the number of alternatives, he reverted to the postulate of general covariance which he had previously "abandoned only with a heavy heart". Making use of Riemann's theory of curvature, he in fact succeeded in setting up generally covariant equations for the g_{ik} themselves which met all the physical requirements (cf. § 56).[277] In a further paper he was able to show[278] that his theory explained the perihelion displacement of Mercury quantitatively and that it predicted the bending of light rays in the gravitational field of the sun. The size of the deflection was double that derived previously on the basis of the principle of equivalence for *homogeneous* fields. Soon after, Einstein's concluding paper, 'Die Grundlagen der allgemeinen Relativitätstheorie',[279] appeared. The principles and further developments of this theory will now be presented.

51. General formulation of the principle of equivalence. Connection between gravitation and metric

Originally, the principle of equivalence had only been postulated for *homogeneous* gravitational fields. For the general case, it can be formulated in the following way: *For every infinitely small world region (i.e. a world region which is so small that the space- and time-variation of gravity can be neglected in it) there always exists a coordinate system K_0 (X_1, X_2, X_3, X_4) in which gravitation has no influence either on the motion of particles or any other physical processes.* In short, in an infinitely small world region every gravitational field can be transformed away. We can think of the physical realization of the local coordinate system K_0 in terms of a freely floating, sufficiently small, box which is not subjected to any external forces apart from gravity, and which is freely falling under the action of the latter.

It is clear that this "transforming away" is only possible because the gravitational field has the fundamental property that it imparts the same acceleration to all bodies; or, stated differently, because the gravitational

[276] A. Einstein, 'Die formale Grundlage der allgemeinen Relativitatstheorie', *S.B. preuss. Akad. Wiss.* (1914) 1030.

[277] A. Einstein, *S.B. preuss. Akad. Wiss.* (1915) 778, 799 and 844. At the same time as Einstein, and independently, Hilbert, formulated the generally covariant field equations (D. Hilbert, 'Grundlagen der Physik', 1. Mitt., *Nachr. Ges. Wiss. Göttingen*, (1915) 395). His presentation, though, would not seem to be acceptable to physicists, for two reasons. First, the existence of a variational principle is introduced as an axiom. Secondly, of more importance, the field equations are not derived for an arbitrary system of matter, but are specifically based on Mie's theory of matter (discussed in more detail in Part V). We shall be discussing the other results of Hilbert's paper in §§ 56 and 57.

[278] A. Einstein, *S.B. preuss. Akad. Wiss.* (1915) 831.

[279] A. Einstein, *Ann. Phys., Lpz.*, **49** (1916) 769. (Also published separately, and in the collection of papers, *Das Relativitätsprinzip*.)

mass is always equal to the inertial mass. This statement nowadays rests on a very secure experimental basis. Eötvös[280] showed that the inertial and gravitational masses are equal to an order of accuracy of 1 in 10^8, while investigating the question whether the direction of the resultant of the earth's attraction and the centrifugal force of the earth's rotation depended on the material. In view of the theorem of the inertia of energy an investigation by Southerns[281] is also of interest, in which he showed that the ratios between mass and weight for uranium oxide and lead oxide respectively differed at most by a factor of $1 \div 2 \times 10^5$. For it follows from the principle of equivalence, together with the theorem of the inertia of energy, that to *each form of energy* a *weight* has to be ascribed, too. If now the internal energy generated during the radioactive decay of uranium possessed inertia, but no weight, then the above ratios would differ by $1 \div 26,000$. Eötvös[280] found this result confirmed, while considerably improving on its accuracy.

It is evidently natural to assume that the special theory of relativity should be valid in K_0. All its theorems have thus to be retained, except that we have to put the system K_0, defined for an infinitely small region, in place of the Galilean coordinate system of § 2. All systems K_0 which are derived from each other by a Lorentz transformation are on the same footing. In this sense we can therefore say that the invariance of the physical laws under Lorentz transformations also persists in infinitely small regions. We can now associate with two infinitely close point events a certain measurable number, their distance ds. For this, we only need to transform away the gravitational field and then form, in K_0, the quantity[282]

$$ds^2 = dX_1{}^2 + dX_2{}^2 + dX_3{}^2 - dX_4{}^2. \qquad (387)$$

Here, the differentials of the coordinates, $dX_1, ..., dX_4$ are to be determined directly by means of a standard measuring rod and clock. Let us now consider some other coordinate system K in which the values of the coordinates $x^1, ..., x^4$ are assigned to the world points in a completely arbitrary way, apart from the conditions of uniqueness and continuity. At each space-time point, the corresponding differentials dX_i will then be linear homogeneous expressions in the dx^k, and the line element ds^2 will be transformed into the quadratic form

$$ds^2 = g_{ik} dx^i dx^k, \qquad (388)$$

where the coefficients g_{ik} are functions of the coordinates. It is moreover obvious that for a transition to new coordinates, the g_{ik} transform in such a way that ds^2 remains invariant. The situation is thus completely

[280] R. Eötvös, *Math. naturw. Ber. Ung.* 8 (1890) 65; R. Eötvös, D. Pekár, and E. Fekete, *Trans. XVI. Allgemeine Konferenz der internationalen Erdmessung* (1909); cf. also *Nachr. Ges. Wiss. Göttingen* (1909), geschäftliche Mitteilungen, p. 37; and D. Pekár, *Naturwissenschaften*, 7 (1919) 327.

[281] L. Southerns, *Proc. Roy. Soc.* A, 84 (1910) 325.

[282] As opposed to other authors we write the line element with three positive signs and *one* negative sign, in general relativity too, and not the reverse. This has to be borne in mind when comparing our formulae with the usual ones.

analogous to that obtaining in the geometry of non-Euclidean multi-dimensional manifolds (§ 15). The system K_0 in a freely falling box takes the place of the geodesic system of § 16; the g_{ik} in it are constant, so long as their *second* derivatives can be neglected, and the line element is of the form (387) up to terms of *second* order. The totality of the g_{ik}-values at all world points will be called the *G-field*.

The equation of motion of a particle which is subjected to no forces other than gravity can now be set up very easily. *The world line of such a particle is a geodesic line* (§ 17), and we have from (81) and (80),

$$\delta \int ds = 0, \tag{81}$$

$$\frac{d^2 x^i}{ds^2} + \Gamma^i{}_{rs} \frac{dx^r}{ds} \frac{dx^s}{ds} = 0, \tag{80}$$

where $\Gamma^i{}_{rs}$ is defined by (66) and (69). For in system K_0 the particle is in rectilinear uniform motion at a given moment, i.e. $d^2 X^i / ds^2 = 0$, which is at the same time the system of equations of the geodesic line in K_0. Now the statement, that the world line of a particle is a geodesic line, is invariant and therefore holds generally. (We have assumed here, however, that the *second* derivatives of the g_{ik} with respect to the coordinates do not appear in the equation of motion of the particle.) The validity of this simple theorem is not surprising. It is just due to the fact that the line element was defined in such a way that the world line of a particle becomes a geodesic line. We thus see that *the ten tensor components g_{ik} in Einstein's theory take the place of the scalar Newtonian potential Φ; the components $\Gamma^i{}_{rs}$, formed from their derivatives, determine the magnitude of the gravitational force.*

An exactly analogous argument can be carried out for the case of light rays. In system K_0 the light rays are straight lines[282a] and in addition satisfy the relation

$$dX_1^2 + dX_2^2 + dX_3^2 - dX_4^2 = 0.$$

Thus, the world lines of the light rays are geodesic null lines, quite generally (§ 22):

$$\frac{d^2 x^i}{d\lambda^2} + \Gamma^i{}_{rs} \frac{dx^r}{d\lambda} \frac{dx^s}{d\lambda} = 0 \tag{80a}$$

$$ds^2 = g_{ik} dx^i dx^k = 0. \tag{81a}$$

Kretschmann[283] and Weyl[284] showed in addition that an observation of the arrival of light signals is sufficient to determine the G-field in a particular coordinate system, without having to consider the motion of particles.

[282a] Naturally, the underlying assumption is that we are in the region of validity of geometrical optics. As soon as we are dealing with diffraction, this is no longer the case. Cf. also footnote 310 a, p. 157.

[283] E. Kretschmann, *Ann. Phys., Lpz.*, **53** (1917) 575.

[284] H. Weyl, *Raum-Zeit-Materie* (1st edn., 1918) p. 182; (3rd edn., 1919) p. 194.

There is, however, also a third way in which the G-field can be measured. With the help of measuring rods (or better, measuring threads) and clocks we could determine, for a given coordinate system, the dependence of the magnitude ds of the line element on the coordinate differentials dx^k along all the world lines originating from an arbitrary point. From this, the G-field follows immediately. *It thus characterizes not only the gravitational field but also the behaviour of measuring rods and clocks, i.e. the metric of the four-dimensional world which contains the geometry of ordinary three-dimensional space as a special case.* This fusion of two previously quite disconnected subjects—metric and gravitation—must be considered as the most beautiful achievement of the general theory of relativity. As was shown above and can be illustrated by means of simple examples, this fusion results conclusively from the principle of equivalence and from the validity of special relativity in the infinitely small. The motion of a particle under the sole influence of gravity can now be interpreted in the following way: The motion of the particle is *force-free*. It is not rectilinear and uniform because the four-dimensional space-time continuum is *non-Euclidean* and because in such a continuum a rectilinear uniform motion has no meaning and has to be replaced by motion along a geodesic line. Correspondingly the Galilean principle of inertia has to be replaced by

$$\delta \int ds = 0.$$

This has the great advantage over the former in that it is *generally* covariant. Gravitation in Einstein's theory is just as much of an *apparent force* as the Coriolis and centrifugal forces are in Newtonian theory. (We would be equally justified in taking the view that neither of the two forces should be called an apparent force in Einstein's theory.) It does not affect the argument that the gravitational force cannot, in general, be transformed away in finite regions, whereas the other forces can. The gravitational force can always be transformed away in infinitely small regions, and this fact alone is decisive. That the non-Euclidean character of the space-time world should show up so little in the behaviour of measuring rods and clocks, but very strongly in the deviation from rectilinear uniform motion of particles (i.e. for the case of gravity), is due to the magnitude of the velocity of light. This will be shown in § 53.

This fusion of gravity and metric leads to a satisfactory solution not only of the gravitational problem, but also of that of geometry. Questions concerning the truth of geometrical theorems and concerning the geometry actually applying in space, are meaningless so long as the geometry is dealing only with abstract ideas and not with experimental objects. If however we add to the theorems of geometry the definition that the length of an (infinitely short) line should be a number obtained in the usual way by means of a rigid rod or measuring thread, then geometry becomes a branch of physics and the above-mentioned questions acquire a certain meaning.[284a] The general theory of relativity now allows us

[284a] A. Einstein, *Über die spezielle und die allgemeine Relativitätstheorie*, (1st edn., Braunschweig 1917) p. 2.

immediately to make a general statement: Since gravitation is determined by the matter present, the same must then be postulated for geometry, too. *The geometry of space is not given a priori, but is only determined by matter.* (How this comes about in detail, will be shown in § 56.) A similar view had already been been put forward by Riemann.[285] But at the time it could at best remain a bold scheme, for the determination of the connection between geometry and gravitation only became possible after the metrical relationship between space and time had been recognized.

52. The postulate of the general covariance of the physical laws

It was this postulate which gave the real impetus to the general theory of relativity and to which the theory owes its name. It has various roots. In the first place, arbitrarily moving reference systems are *kinematically* completely equivalent. This makes it natural to assume that the equivalence should also apply in dynamical and physical respects. *A priori* this is of course not capable of proof, and only its success can show whether the assumption is correct.

It is however easy to understand that it will not suffice to introduce *arbitrarily moving* reference systems. As Einstein[286] showed for the example of a rotating reference system, the time intervals and spatial distances in non-Galilean systems cannot just be determined by means of a clock and rigid standard measuring rod; it is necessary to abandon Euclidean geometry. We have thus no choice but to admit all conceivable coordinate systems. The coordinates will have to be thought of as associated with the world points in a unique and continuous manner (Gaussian coordinate system). That such a description of the world is sufficient can be seen from the following argument due to Einstein[287]: All physical measurements amount to a determination of space-time coincidences; nothing apart from these coincidences is observable. If however two point events correspond to the same coordinates in *one* Gaussian coordinate system, this must also be the case in every other Gaussian coordinate system. We therefore have to extend the postulate of relativity: *The general physical laws have to be brought into such a form that they read the same in every Gaussian coordinate system, i.e. they must be covariant under arbitrary coordinate transformations.*[287a]

This covariance is made possible by incorporating the g_{ik} into the physical laws. Mathematically speaking, the general physical laws permit arbitrary point transformations after adjunction of the invariant quadratic form

$$ds^2 = g_{ik}\,dx^i\,dx^k.$$

[285] Cf. footnote 63, p. 34, *ibid.*, inaugural lecture.

[286] Cf. footnote 279, *ibid.*

[287] Cf. footnote 279, *ibid.* Cf. also E. Kretschmann, *Ann. Phys., Lpz.*, **48** (1915) 907 and 943.

[287a] P. Lenard, *Über das Relativitätsprinzip, Äther, Gravitation*, (Leipzig 1918; 2nd edn., 1920), see also 'Nauheimer Discussion', *Phys. Z.*, **21** (1920) 666. Lenard raises doubts on the use of coordinate systems of such generality and on the reality of the gravitational fields which would appear in them according to Einstein's theory. The author cannot agree with these objections.

In fact, every law of the special theory of relativity can be made generally covariant by a formal introduction of the g_{ik}, following the scheme given in Part II, as will be shown in § 54 by means of specific examples. For this reason Kretschmann[288] took the view that the postulate of general covariance does not make any assertions about the physical *content* of the physical laws, but only about their mathematical *formulation*; and Einstein[289] entirely concurred with this view. The generally covariant formulation of the physical laws acquires a physical content only through the principle of equivalence, in consequence of which gravitation is described *solely* by the g_{ik} and these latter are not given independently from matter, but are themselves determined by field equations. Only for this reason can the g_{ik} be described as *physical quantities*.[290] On the other hand, as was stressed by Einstein[289], the postulate of general covariance can also be given another meaning. The differential equations of the G-field itself have to be determined in such a way that they are as clear and simple as possible, from the point of view of the general theory of covariants. This heuristic aspect of the covariance principle has stood the test very well (§ 56).

Attempts have been made to normalize the coordinate system somehow, in spite of the general covariance requirement. In particular, the investigations of Kretschmann[288] and Mie[291] deal with this question. However, all suggested normalizations seem to be possible, or of practical importance, only in special cases. For the general case, and for questions of principle, the general covariance requirement is indispensable.

53. Simple deductions from the principle of equivalence

(α) *The equations of motion of a point-mass for small velocities*[292] *and weak gravitational fields*. The equation of motion (80) of a point-mass can be simplified considerably if the *velocity* of the point-mass is *small compared with the light velocity*, so that terms of order v^2/c^2 can be neglected. Let us also assume that the *gravitational field is weak*. This means that the g_{ik} should differ only very slightly from their normal values

$$g_{ik} = +1 \quad \text{for} \quad i = k = 1, 2, 3, \quad g_{44} = -1$$
$$g_{ik} = 0 \quad \text{for} \quad i \neq k,$$

so that the squares of such deviations can be neglected. Then we have

$$\frac{d^2 x^i}{dt^2} = -c^2 \Gamma^i{}_{44}, \quad \text{for} \quad i = 1, 2, 3; \quad x^4 = ct. \tag{389}$$

In addition, let us assume that the field is static or quasi-static, so that the time derivatives, too, of the g_{ik} can be neglected. Then $\Gamma^i{}_{44}$ can be

[288] Cf. footnote 283, p. 147, *ibid.*
[289] A. Einstein, *Ann. Phys., Lpz.*, **55** (1918) 241.
[290] H. Weyl, *Raum-Zeit-Materie* (1st edn., 1918), pp. 180–181; (3rd. edn., 1919), pp. 192, 193.
[291] G. Mie, *Ann. Phys., Lpz.*, **62** (1920) 46.
[292] Cf. footnote 279, p. 145, A. Einstein, *ibid.*, § 21.

replaced by $\Gamma_{i,44}$, or by $-\frac{1}{2} \partial g_{44}/\partial x^i$, and *the equations of motion* (389) *reduce to the Newtonian equations*

$$\frac{d^2 x^i}{dt^2} = -\frac{\partial \Phi}{\partial x^i}, \tag{390}$$

where we have put

$$\Phi = -\tfrac{1}{2}c^2(g_{44} + 1), \qquad g_{44} = -1 - \frac{2\Phi}{c^2}. \tag{391}$$

The additive constant, at first undetermined, in the expression for Φ is fixed in such a way that Φ vanishes when g_{44} has its normal value -1.

It is of interest to note that this particular approximation of the equations of motion contains only g_{44}, even though the deviations of the other g_{ik} from their normal values could be of the same order of magnitude as that of g_{44}. It is for this reason that it is possible to describe, approximately, the gravitational field in terms of a *scalar* potential.

(β) *The red shift of spectral lines.* For the same reason, we are also able to make a general statement regarding the influence of a gravitational field on clocks, even before the behaviour of the G-field is known, since such an effect is determined by g_{44}. A similar statement regarding the behaviour of measuring rods can only be made when the remainder of the g_{ik} are known.

Let us take a reference system K which rotates relative to the Galilean system K_0 with angular velocity ω. A clock at rest in K will then be slowed down the more, the farther away from the axis of rotation the clock is situated, because of the transverse Doppler effect. This can be seen immediately by considering the process as observed in system K_0. The time dilatation is given by

$$t = \frac{\tau}{\sqrt{[1 - (v^2/c^2)]}} = \frac{\tau}{\sqrt{[1 - (1/c^2)\omega^2 r^2]}}.$$

The observer rotating with K will not interpret this shortening of the time as a transverse Doppler effect, since after all the clock is at rest relative to him. But in K a gravitational field (field of the centrifugal force) exists with potential

$$\Phi = -\tfrac{1}{2}\omega^2 r^2.$$

Thus the observer in K will come to the conclusion that the clocks will be slowed down the more, the smaller the gravitational potential at the particular spot. In particular, the time dilatation is given, to a first approximation, by

$$t = \frac{\tau}{\sqrt{[1 + (2\Phi/c^2)]}} \simeq \tau\left(1 - \frac{\Phi}{c^2}\right), \qquad \frac{\Delta t}{\tau} = -\frac{\Phi}{c^2}.$$

Einstein[293] applied an analogous argument to the case of uniformly accelerated systems. We thus see that the transverse Doppler effect and

[293] Cf. footnote 270, p. 143, *ibid.*

the time dilatation produced by gravitation appear as two different modes of expressing the same fact, namely that a clock will always indicate the proper time

$$\tau = \frac{1}{ic} \int ds.$$

In general, the time $t = x^4/c$ will be different from the normal proper time τ of a clock at rest. For the world line element of a clock at rest is

$$(ds)^2 = g_{44} (dx^4)^2,$$

i.e. according to (157),

$$t = \frac{\tau}{\sqrt{(-g_{44})}} = \frac{\tau}{\sqrt{[1 + (2\Phi/c^2)]}} \simeq \tau\left(1 - \frac{\Phi}{c^2}\right), \qquad \frac{\Delta t}{\tau} = -\frac{\Phi}{c^2}. \quad (392)$$

Equation (392) has the following physical meaning: Consider two equal, originally synchronous, clocks at rest and let one of them be placed in a gravitational field for a certain length of time. Afterwards they will no longer be synchronous; the clock which had been placed in the gravitational field will have lost. As mentioned by Einstein[294], this is the basis of the explanation for the clock paradox described in § 5 [q.v.]. In the coordinate system K^* in which the clock C_2 is permanently at rest, a gravitational field exists during the time in which its motion is retarded, and the observer in K^* can regard this field as causing the clock C_2 to lose.

Relation (392) has an important consequence which can be checked by experiment: The transport of clocks can also be effected by means of a light ray, if one regards the vibration process of light as a clock. For if the gravitational field is static, the time coordinate can always be determined in such a way that the g_{ik} do not depend on it. Then the number of wave-lengths of a light ray contained between two points P_1 and P_2 will also be independent of the time. The frequency of the light ray, measured with the given time scale, will therefore be the same at P_1 and P_2, and thus independent of position.[295] On the other hand, the frequency measured in terms of the proper time does depend on position. If, therefore, a spectral line produced in the sun is observed on the earth, its frequency will, according to (392), be shifted towards the red compared with the corresponding terrestrial frequency. The amount of this shift will be

$$\frac{\Delta\nu}{\nu} = \frac{\Phi_E - \Phi_S}{c^2}, \qquad\qquad (393)$$

where Φ_E is the value of the gravitational potential on the earth, Φ_S that on the surface of the sun. The numerical calculation gives

$$\frac{\Delta\nu}{\nu} = 2 \cdot 12 \times 10^{-6}, \qquad\qquad (393\,\mathrm{a})$$

corresponding to a Doppler effect of $0 \cdot 63$ km/sec.

[294] A. Einstein, *Naturwissenschaften*, **6** (1918) 697.
[295] M. v. Laue, *Phys. Z.*, **21** (1920) 659, has confirmed this result by a direct calculation, based on the wave equation of light.

A great number of attempts have been made to investigate this relation experimentally. Jewell[296] already discovered the red shift of spectral lines of the sun, but he interpreted it as a pressure effect. When, at a later date, Evershed[297] established that the shift did not agree with the experimentally determined pressure shift, it was natural that the Einstein effect should be used for an explanation.[298] A more accurate examination showed however that different lines appeared to be shifted by different amounts, so that the Einstein effect alone, at any rate, was not capable of explaining the phenomenon in all its details. Much more suited for checking the Einstein theory are the more recent observations on the nitrogen band $\lambda = 3883$ Å (the so-called cyanogen band). For this is distinguished by the fact that it does not show any perceptible pressure effects. A comparison of the absorption lines of this band in the solar spectrum with the corresponding terrestrial emission lines was first carried out by Schwarzschild[299] and later, with greater accuracy, by St. John[300] at the Mount-Wilson Observatory and by Evershed and Royds[301]. These authors all found a considerably smaller shift of the lines than required by the theory; in fact, St. John found practically no shift at all. It thus seemed for a time as if the theory had been disproved by experiment.[302] In a series of more recent investigations, however, Grebe and Bachem[303] pointed out that the measured shifts have quite different values for different lines. They then showed by measuring the line [intensities] with a Koch recording microphotometer that it is the superposition of different lines in the solar spectrum which is the cause of this phenomenon, which at first appears so strange. *For the unperturbed lines, shifts were now obtained which were in agreement with the theoretical value* (393 a) *within the experimental errors.* It is true that only comparatively few lines are unperturbed. Recently, however, Grebe[304] found that also the average value of the shifts of 100 perturbed and unperturbed lines of the above-mentioned nitrogen band agree with the theory. Perot[304a], too, investigated this band for the red shift and found a positive result. All the same, his result can hardly be considered conclusive, since he did not take possible superpositions of the lines into account.

Freundlich[305] tried to demonstrate the gravitational shift of spectral lines also for the case of fixed stars. But for fixed stars it is only possible

[296] L. E. Jewell, *Astroph. J.*, **3** (1896) 89.

[297] J. Evershed, *Bull. Kodaikanal Obs.*, **36** (1914).

[298] Cf. footnote 270, p. 143, A. Einstein, *ibid.*; E. Freundlich, *Phys. Z.*, **15** (1914) 369.

[299] K. Schwarzschild, *S.B. preuss. Akad. Wiss.*, (1914) 120.

[300] C. E. St. John, *Astroph. J.*, **46** (1917) 249.

[301] J. Evershed and Royds, *Bull. Kodaikanal Obs.*, **39**.

[302] This discrepancy between Einstein's theory and the above-mentioned experimental results led Wiechert to construct a theory of gravitation which contained so many undetermined constants that it could be fitted to arbitrary empirical values for the red shift. the bending of light rays and the perihelion precession of Mercury: E. Wiechert, *Nachr., Ges. Wiss. Göttingen*, (1910) 101; *Astr. Nachr.*, No. 5054, p. 211, col. 275; *Ann. Phys., Lpz.*, **63** (1920) 301.

[303] L. Grebe and A. Bachem, *Verh. dtsch. phys. Ges.*, **21** (1919) 454; *Z. Phys.*, **1** (1920) 51, and **2** (1920) 415.

[304] L. Grebe, *Phys. Z.*, **21** (1920) 662 and *Z. Phys.*, **4** (1921) 105.

[304a] A. Perot, *C.R. Acad. Sci., Paris*, **171** (1920) 229.

[305] E. Freundlich, *Phys. Z.*, **16** (1915) 115; **20** (1919) 561.

to separate the gravitational from the Doppler effect if use is made of rather uncertain hypotheses. Freundlich's first results have moreover been rejected by Seeliger.[306]

Summarizing, then, we can say that the experimental results concerning the red shift seem now to be in favour of the theory, but that it has not, as yet, found final confirmation.†

(γ) *Fermat's principle of least time in static gravitational fields.* We assume that we are dealing with a *static* gravitational field, i.e. that the coordinate system can be chosen in such a way that all the g_{ik} are time independent and that the four-dimensional line element is of the form

$$ds^2 = d\sigma^2 - f^2 dt^2, \tag{394}$$

where $d\sigma^2$ is a positive definite quadratic form of the three space-coordinate differentials and f the position-dependent light velocity. We then have

$$g_{14} = g_{24} = g_{34} = 0, \qquad g_{44} = -\frac{f^2}{c^2}. \tag{394a}$$

The existence of the first three relations in all static G-fields is a separate hypothesis which can only be justified by means of the differential equations of the G-field itself. For the special case of *spherically symmetrical* static fields, admittedly, it can be seen *a priori* that the components g_{i4} $(i = 1, 2, 3)$ can always be made to vanish by suitably normalizing the time.[306a]

Let us investigate in particular the path of a light ray in such a field. According to § 51 it is determined by the condition that it should be a geodesic null line. For this special case, the condition can be written in the form of Fermat's principle, as was shown by Levi-Civita[307] and Weyl[308]. To demonstrate this, let us start from the variational principle (83) of § 15,

$$L = \tfrac{1}{2} g_{ik} \frac{dx^i}{d\lambda} \frac{dx^k}{d\lambda}, \qquad \delta \int L d\lambda = 0.$$

The coordinates are not to be varied here at the end points of the path of integration. If now we substitute for the g_{ik} the values obtained from (394), we have

$$L = \frac{1}{2}\left(\frac{d\sigma}{d\lambda}\right)^2 - f^2\left(\frac{dt}{d\lambda}\right)^2.$$

[306] H. v. Seeliger, *Astr. Nachr.*, **202** (1916) col. 83; cf. E. Freundlich, *ibid.*, col. 147.

† See suppl. note 14.

[306a] The Italian mathematicians make a distinction between the statical case, for which $g_{i4} = 0$ for $i = 1, 2, 3$, and the more general stationary case, for which the g_{ik} are only time independent, but $g_{i4} \neq 0$. Cf. in particular A. Palatini, *Atti. Ist. veneto*, **78** (2) (1919) 589, where the paths of point-masses and light rays are discussed in a general way for the stationary case; and A. De-Zuani, *Nuovo Cim.* (6) **18** (1919) 5.

[307] T. Levi-Civita, 'Statica Einsteiniana', *R.C. Accad. Lincei* (5) **26** (1917) 458; *Nuovo Cim.* (6) **16** (1918) 105.

[308] H. Weyl, *Ann. Phys., Lpz.*, **54** (1917) 117; *Raum-Zeit-Materie* (1st edn., 1918) pp. 195, 196; (3rd edn., 1920) pp. 209, 210.

In particular, for a variation of t, the variational principle produces the equation

$$\frac{d}{d\lambda}\left(f^2\frac{dt}{d\lambda}\right) = 0, \qquad f^2\frac{dt}{d\lambda} = \text{const.,}$$

and with a suitable normalization of the parameter λ we can put

$$f^2\frac{dt}{d\lambda} = 1. \tag{395}$$

Next, let us change the condition for the variation as follows:

(i) Only the *spatial* end points of the path are to remain fixed, and the time coordinate is to be varied at the initial and end points.

(ii) The path which is varied should also be a null line (but not necessarily geodesic). Because of this latter condition,

$$L \equiv 0, \qquad \delta L \equiv 0$$

at all points along the paths, of course. On the other hand, for a variation of the time coordinate,

$$\delta \int L \, d\lambda = -f^2\frac{dt}{d\lambda}\delta t\Big|_{t_1}^{t_2} + \int \frac{d}{d\lambda}\left(f^2\frac{dt}{d\lambda}\right)\delta t \, d\lambda.$$

This expression must therefore also vanish identically, if the path is a null line. The condition (395) for this null line to be geodesic can thus be replaced by

$$\delta t\Big|_{t_1}^{t_2} = \delta \int_{t_1}^{t_2} dt = 0$$

or, eliminating the time by means of the relation $L=0$, by

$$\delta \int \frac{d\sigma}{f} = 0. \tag{396}$$

This is just Fermat's principle of least time. It follows that, even when the gravitational field is static, *the light ray is not a geodesic line in three-dimensional space*. Such a geodesic line would, after all, be characterized by

$$\delta \int d\sigma = 0.$$

Only in the four-dimensional world is the world line of the light ray geodesic. A light ray will thus be curved in a gravitational field. The amount of curvature will however also depend on the form of $d\sigma$ and can, in contrast to the value of the red shift, only be determined when the field equations of the G-field are known (§ 58 (γ)).

In an analogous way, a variational principle can also be found for the path of a particle in a static gravitational field which no longer contains the time coordinate.[309] But it lacks the intuitive significance of the previous case.

54. Influence of the gravitational field on material phenomena[310]

It is convenient to follow Einstein in describing everything, apart from the G-field, as matter. The problem then consists in bringing the physical laws governing material processes into a generally covariant form. In principle it is solved by means of the following argument. Let a coordinate system K_0 be given, in which the g_{ik} have their normal values within a finite world region. The physical laws will then be of a form which is assumed to be valid in special relativity. We next introduce some other, arbitrarily moving, Gaussian coordinate system K and determine by straightforward calculation the form of the physical laws in K. By reason of the principle of equivalence it is clear that in this way we have simultaneously shown up the influence of gravitational fields on material processes. This result can then also be carried over to the case where no coordinate system can be found in which the gravitational field can be transformed away inside a finite region. Such a procedure is only possible if one makes use of the, admittedly somewhat arbitrary, hypothesis that the second derivatives of the g_{ik} do not occur in the relevant physical laws.

In a mathematical respect, the situation corresponds exactly to the transition from the tensor calculus of Euclidean to that of Riemannian geometry (§§ 13 and 20). Using the methods given in Part II we can therefore immediately write down every law of special relativity in its general covariant form, by replacing the tensor operations occurring there by the corresponding generalized operations of Riemannian geometry. One must of course be careful to distinguish between the contravariant and covariant components of a tensor and between tensors and tensor densities.

These general prescriptions will now be illustrated by taking Maxwell's equations for the vacuum as an example. Let us again define the field vector F_{ik} by (202). According to § 19 (140 b), the equations (203) remain valid in this case,

$$\frac{\partial F_{ik}}{\partial x^l} + \frac{\partial F_{li}}{\partial x^k} + \frac{\partial F_{kl}}{\partial x^i} = 0. \tag{203}$$

The second set of Maxwell's equations (208) will however have to be written in a slightly different form, because of (141 b). Let us introduce the contravariant components of the tensor density corresponding to F_{ik},

$$\mathfrak{F}^{ik} = \sqrt{(-g)} g^{\alpha i} g^{\beta k} F_{\alpha\beta}, \tag{397}$$

as well as the tensor density corresponding to the current vector,

$$\mathfrak{s}^i = \sqrt{(-g)} s^i. \tag{398}$$

[309] Cf. papers quoted in footnotes 307 and 308, p. 154.
[310] Cf. footnote 274, p. 144, A. Einstein and M. Grossmann, *ibid.*, Part I, § 6; footnote 276, p. 145, A. Einstein, *ibid.*, Part C; footnote 279, p. 145, A. Einstein, *ibid.*, Part D.

Then
$$\frac{\partial \mathfrak{F}^{ik}}{\partial x^k} = \mathfrak{s}^i,$$ (208 a)

from which also follows the generalization[310a] of the continuity equation (197),

$$\frac{\partial \mathfrak{s}^i}{\partial x^i} = 0.$$ (197 a)

Just as before, the ponderomotive force is, from (216),

$$f_i = F_{ik} s^k$$

and the corresponding tensor density,

$$\mathfrak{f}_i = \sqrt{(-g)} f_i = F_{ik} \mathfrak{s}^k.$$ (216 a)

The mixed components of the energy–momentum tensor density follow from (222) and are given by

$$\mathfrak{S}_i{}^k = F_{ir} \mathfrak{F}^{kr} - \tfrac{1}{4} F_{rs} \mathfrak{F}^{rs} \delta_i{}^k.$$ (222 a)

The generalization of (225) is of importance. From the rule (150a) of general tensor analysis we obtain

$$\left.\begin{array}{l} \dfrac{\partial \mathfrak{S}_i{}^k}{\partial x^k} - \mathfrak{S}_r{}^s \Gamma^r_{is} = -\mathfrak{f}_i \\[4mm] \dfrac{\partial \mathfrak{S}_i{}^k}{\partial x^k} - \tfrac{1}{2} \mathfrak{S}^{rs} \dfrac{\partial g_{rs}}{\partial x^i} = -\mathfrak{f}_i. \end{array}\right\}$$ (225 a)

or

The second term on the left-hand side is characteristic for the influence of the gravitational field. It can be seen from the calculation carried out in § 23 (α) that (225 a) really follows from (203), (208 a) and (216) in the general case, too.

Analogously, the equations of motion for liquids can be written down in a generally covariant form.[311] G. Nordström[312] considered Herglotz's general equations for elastic media. Just as (225 a) is derived from expression (225) for the ponderomotive force, so we obtain from the general energy–momentum law (341) *the energy–momentum law of matter in the presence of gravitational fields*,

$$\frac{\partial \mathfrak{T}_i{}^k}{\partial x^k} - \tfrac{1}{2} \mathfrak{T}^{rs} \frac{\partial g_{rs}}{\partial x^i} = 0.$$ (341 a)

In its physical aspect it differs very considerably from the previous form of the energy–momentum law. Whereas in the previous case a conservation law for the total energy and total momentum could be derived by integrating, this is no longer possible with the new formulation (341 a),

[310a] An application of these equations is given by M. v. Laue, *Phys. Z.*, **21** (1920) 659. He shows that for the world lines of light rays *in vacuo* (within the region of validity of geometrical optics) Eqs. (80) and (81) of the geodesic null line do in fact follow from them.

[311] Cf. footnotes 276 and 279, p. 145, A. Einstein, *ibid.*

[312] G. Nordström, *Versl. gewone Vergad. Akad. Amst.*, **25** (1916) 836.

because of the second term on the left-hand side. It simply means that energy and momentum can be transmitted from matter to the gravitational field and vice versa (for details cf. § 61). If there are no external forces acting, we can in particular introduce the energy-momentum tensor Θ_{ik} (given by (322)) for T_{ik}, and the expression

$$\mu_0 \sqrt{(-g)} g_{i\alpha} \frac{dx^\alpha}{d\tau} \frac{dx^k}{d\tau}$$

in place of $\mathfrak{T}_i{}^k$. Eqs. (341 a) then reduce to those of the geodesic line.†

55. The action principles for material processes in the presence of gravitational fields

It was first shown by Hilbert[313] that the energy–momentum tensor is related to the action function in a simple manner. This becomes clearly evident only in general relativity. It will be illustrated by taking as an example the mechanical-electrodynamical action principle of § 31, which we shall write down in Weyl's form (231 a)

$$W_1 = \int \{\tfrac{1}{2} F_{ik} F^{ik} - 2\phi_i s^i + 2\mu_0 c^2\} \, d\Sigma$$

$$= \int \tfrac{1}{2} F_{ik} F^{ik} \, d\Sigma - \int de \int 2\phi_i \, dx^i + 2\mu_0 c \int \sqrt{(-g_{ik} u^i u^k)} \, d\Sigma; \left.\begin{array}{c} \\ \\ \\ \end{array}\right\} \quad (231\,\text{b})$$

$$\delta W_1 = 0.$$

This action principle also remains valid in a gravitational field[313], provided the g_{ik} are not varied. (What has to be varied, independently, are the world lines of particles and the field potentials ϕ_i.)

Something new, however, is obtained when the g_{ik} are varied. The world lines of the substance and the potentials ϕ_i can now be left constant. Then the first integral will, according to § 23 (a), contribute an amount

$$- \int \mathfrak{S}^{ik} \delta g_{ik} \, dx = - \int S^{ik} \delta g_{ik} \, d\Sigma,$$

the second will contribute zero, and the third

$$- \int \mu_0 u^i u^k \delta g_{ik} \, d\Sigma = - \int \Theta^{ik} \delta g_{ik} \, d\Sigma.$$

Altogether, then,

$$\delta W = - \int \mathfrak{T}^{ik} \delta g_{ik} \, dx = + \int \mathfrak{T}_{ik} \delta g^{ik} \, dx. \quad (399)$$

We thus obtain the energy tensor of matter by varying the G-field in the action integral.[313] This rule holds generally, and not just in this particular case.

† See suppl. note 15.
[313] Cf. footnote 101, p. 68, *ibid.* Cf. also footnote 100, p. 68, H. A. Lorentz, *ibid.*; footnote 308, p. 154, H.Weyl, *Ann. Phys., Lpz., loc. cit.*; and H. Weyl, *Raum-Zeit-Materie* (1st edn., 1918) p. 215 *et seq.*, § 32; (3rd edn., 1920) p. 197.

It was proved for the case of the elastic energy tensor by Nordström.[314] See Part V, § 64, on Mie's theory.

This connection between the energy–momentum tensor of matter and the action function turns out to be of extreme importance for the applications of Hamilton's Principle in the general theory of relativity (see § 57). If moreover we set for δg_{ik} the variation $\delta^* g_{ik}$ generated by varying the coordinate system only, for which δW vanishes identically (§ 23), we can make the following statement on the basis of (169): *In every case in which the field laws of material processes can be derived from an action principle, and in which at the same time the energy tensor is obtained from the action integral by varying the G-field in the stated manner, the energy–momentum law* (341 a) *is a consequence of these field laws.* For making this deduction it is essential that the contributions of the *variation δ^* of the material quantities of state* should vanish in accordance with Hamilton's Principle.

56. The field equations of gravitation

The intrinsic and most important problem which the general theory of relativity has to solve, consists in setting up the laws of the G-field itself. It must naturally be required of these laws that they should be generally covariant. But in order to arrive at a unique determination of these laws, some further conditions have to be formulated. The guiding principles for this are as follows:

(i) According to the principle of equivalence the gravitational mass is equal to the inertial mass, i.e. it is proportional to the total energy. The same is therefore also true of the force which acts on a material system in a gravitational field. It is therefore natural to assume that, conversely, it is only the total energy which determines the gravitational field produced by a material system. According to special relativity, however, the energy density is not characterized by a scalar, but only by the 44-component of a tensor T_{ik}. At the same time, momentum and stresses appear on an equal footing with energy. We therefore formulate our assumption in the following way:

In the field equations for gravitation, no other material quantities than the total energy–momentum tensor should occur.

(ii) Going beyond this, Einstein started with the hypothesis, in analogy to the Poisson equation

$$\Delta\Phi = 4\pi k\mu_0,$$

that the energy tensor T_{ik} should be proportional to a differential expression of second order, formed from the g_{ik} alone. Because of the general covariance requirement, such an expression must evidently be a tensor, and it therefore follows from § 17, Eq. (113), that the differential equations of the G-field must be of the form

$$c_1 R_{ik} + c_2' R g_{ik} + c_3 g_{ik} = k T_{ik}. \tag{400}$$

Here, R_{ik} is the contracted curvature tensor defined by (94) and R the corresponding invariant (95). For their geometrical significance, cf. § 17.

[314] Cf. footnote 312, p. 157, *ibid*.

The essential features of these assumptions become evident if one makes a comparison with Nordström's theory, which can also be brought into a generally covariant form, according to Einstein and Fokker.[314a] In this theory, only the scalar $T = T_i{}^i$ occurs in the gravitational equations, and is proportional to the curvature invariant R. The remaining equations, not set up explicitly so far, have to contain the statement that it must always be possible to write the line element in the form

$$ds^2 = \Phi \sum (dx^i)^2$$

for a suitable choice of the coordinates, thus implying that the light velocity is constant. One can see that these field equations appear quite artificial and complicated from the point of view of the absolute differential calculus, when compared with the equations of Einstein's theory, in which all components of T_{ik} appear on the same footing.

(iii) To fix the (as yet undetermined) constants c_1, c_2, c_3 in (400) we shall have to consider the relationship between a generally relativistic theory and causality. Once we have found any kind of solutions of the general covariant field equations, we can derive from them an arbitrary number of other solutions by means of a different choice of coordinates. *The general solution of the field equations must therefore contain four arbitrary functions. Hence, there must exist 4 identities between the 10 field equations* (400) *for the 10 unknowns g_{ik}. In general, in a relativistic theory for m unknowns, there must exist no more than $m-4$ independent equations.* The contradiction with the causality principle is only apparent, since the many possible solutions of the field equations are only formally different. Physically they are completely equivalent. The situation described here was first recognized by Hilbert.[315]

We have thus arrived at the postulate that there must exist four identities between the ten equations (400). Now, we know that the tensor T_{ik} satisfies the energy–momentum law (341a) of § 54. This consists of just four equations. It is therefore extremely plausible to make the following assumptions about the content of the 4 stipulated identities: *The energy–momentum law* (341 a), p. 157, *has to be satisfied identically as a consequence of the field equations of gravitation. It is thus at the same time a result of the gravitational field equations and of the material field laws.*† This postulate evidently amounts to the requirement that the divergence of the left-hand side of (400), generalized in line with the tensor

---

[314a] Cf. footnote 272, p. 144, *ibid.*

[315] D. Hilbert, 'Grundlagen der Physik, I', *Nachr. Ges. Wiss. Göttingen*, (1915) 395. From a historical point of view it should be mentioned that E. Mach had already, on the basis of relativistic considerations, arrived at the result that the number of equations expressing the physical laws should in fact be less than the number of unknowns (*Die Geschichte und die Wurzel des Satzes von der Erhaltung der Arbeit*, (Prague 1877), pp. 36 and 37; *Mechanik*, (Leipzig 1883)).

Furthermore it deserves mentioning that Einstein had, for a time, held the erroneous view that one could deduce from the non-uniqueness of the solution that the gravitational equations could not be generally covariant (see footnote 276, p. 145, *ibid.*).

† See suppl. note 15.

calculus for Riemannian space (see Eq. (150)), should vanish identically.
If this operation is performed, one obtains from (182), (109) and (75),

$$(\tfrac{1}{2}c_1 + c_2)\sqrt{(-g)}\frac{\partial R}{\partial x^i}.$$

We must therefore have $c_2 = -\tfrac{1}{2}c_1$, so that, apart from the term $c_3 g_{ik}$, only the tensor defined by (124),

$$G_{ik} = R_{ik} - \tfrac{1}{2}g_{ik}R, \tag{124}$$

occurs in (400). We shall discuss the physical significance of the last term in (400) in § 62; for the moment we want to omit it. Its effect is extremely small for the cases we are going to discuss now, which provides a *post hoc* justification for its omission. With this proviso, the gravitational equations can therefore be written in the form

$$G_{ik} = -\kappa T_{ik}. \tag{401}$$

See § 58 (α) for the reason for the negative sign on the right-hand side. By contraction it further follows that

$$R = +\kappa T \tag{402}$$

and $$R_{ik} = -\kappa(T_{ik} - \tfrac{1}{2}g_{ik}T). \tag{401a}$$

This is the generally covariant form of the field equations of gravitation which were eventually found by Einstein in the year 1915, after many false starts.[316]

As was also mentioned before (in § 50, footnote (277)) the same equations were also derived by Hilbert in the same year. Whereas the variational principle forms the starting point there, it appears as a mathematical consequence in Einstein's paper and in our presentation, as will be shown in the next section.

57. Derivation of the gravitational equations from a variational principle[317]

The fact that the tensor G_{ik} satisfies the divergence equation (182) is, according to § 23, connected with its being derived from an invariant integral by a variation of the G-field,

$$\delta \int \Re\, dx = \int \mathfrak{G}_{ik}\,\delta g^{ik}\, dx, \tag{180}$$

if the variation of the field quantities vanishes at the boundary. We have also seen in § 55 that the invariant integral $\int \mathfrak{M}\, dx$, which produces the differential equations of the mechanical (elastic) and electromagnetic fields for the variation of the material field quantities, leads to the material energy–momentum tensor for a variation of the G-field,

$$\delta \int \mathfrak{M}\, dx = \int \mathfrak{T}_{ik}\,\delta g^{ik}\, dx. \tag{399a}$$

---

[316] A. Einstein, *S.B. preuss. Akad. Wiss.* (1915) 844. Before that, Einstein had also assumed $R_{ik} = \kappa T_{ik}$: *ibid.*, (1915) 778.

[317] Cf. papers quoted in § 23, footnotes 100 to 104, p. 68.

These two relations prompt us to combine all physical laws into a *single* action principle

$$\delta \int \mathfrak{W} \, dx = 0, \tag{403}$$

$$\mathfrak{W} = \mathfrak{R} + \kappa \mathfrak{M}. \tag{404}$$

Here the variations of the field quantities have to vanish at the boundary of the region of integration. This action function has the distinctive property that it can be split into two parts, of which the one is independent of the material quantities of state, the other independent of the derivatives of the g_{ik}. (See Part V on more general action functions, which do not have this property.)†

The action principle (403) also summarizes concisely, according to §§ 55, 56, the relations between the field equations of material processes and gravitation: From either follows the energy–momentum law (341 a) (§ 54). Eq. (184) of § 23, on the other hand, gives us the energy–momentum law in a different form. If we put

$$t_i{}^k = -\frac{1}{\kappa} U_i{}^k, \tag{405}$$

where the $U_i{}^k$ are defined by (183) and (185), we obtain, from (184) and (401),

$$\frac{\partial(\mathfrak{T}_i{}^k + t_i{}^k)}{\partial x^k} = 0. \tag{406}$$

By virtue of their derivation, these equations are generally covariant, although the quantities $t_i{}^k$ only transform like tensor components under linear transformations. In contrast to the form (341 a) of the energy–momentum law, *conservation laws* for the energy and momentum can be derived from (406) in *integral form*. Einstein[318] therefore calls the quantities $t_i{}^k$ the *energy–momentum components of the gravitational field* and considers them as equivalent in certain respects to the energy–momentum components $T_i{}^k$ of matter. See § 61 on the other physical consequences of this view.

Finally, the action principle (403) is also of *practical* value for the integration of the field equations in special cases. One can avoid, on occasion, having to refer back to the general differential equations. This permits a considerable shortening of the calculations. For details, see § 58 (β).

58. Comparison with experiment

(α) *Newtonian theory as a first approximation.*[319] We have seen in § 53(α) that for weak quasi-static gravitational fields the equations of motion

---

† See suppl. note 8.

[318] Cf. footnote 277, p. 145, *ibid.*, p. 778; footnote 279, p. 145, *ibid.*, Part C, § 15 *et seq.*; footnote 102, p.68, *ibid.*, § 3.

[319] See A. Einstein (cf. footnote 278, p. 145, *ibid.*, and footnote 279, p. 145, *ibid.*, Part E, § 21).

reduce to the Newtonian ones. To complete the proof that Newtonian is contained in relativistic theory as a limiting case, it still remains to be shown that the scalar potential (391) satisfies the Poisson equation

$$\Delta\Phi = 4\pi k\mu_0 \qquad (407\,\text{a})$$

for the above special case. For this purpose we form the 44-component of equation (401 a). For T_{ik} we can introduce the kinetic energy–momentum tensor $\mu_0\, u_i\, u_k$. If quantities of order u/c are neglected, all components of T_{ik} can evidently be put equal to zero, with the exception of T_{44}. The latter becomes

$$T_{44} = \mu_0 c^2$$

and from this we obtain

$$T = g^{ik} T_{ik} = g^{44} T_{44} = -\mu_0 c^2.$$

Equation (401 a) then becomes, first of all,

$$R_{44} = -\tfrac{1}{2}\kappa\mu_0 c^2. \qquad (408\,\text{a})$$

The value of R_{44} has to be obtained from (94). Since the time derivatives and products of the Γ^i_{rs} are neglected, this becomes simply

$$R_{44} = -\frac{\partial \Gamma^\alpha_{44}}{\partial x^\alpha},$$

and since

$$\Gamma^\alpha_{44} \simeq \Gamma_{\alpha,44} \simeq -\frac{1}{2}\frac{\partial g_{44}}{\partial x^\alpha},$$

$$R_{44} = +\frac{1}{2}\sum_\alpha \frac{\partial^2 g_{44}}{\partial x_\alpha{}^2} = \tfrac{1}{2}\Delta g_{44} = -\frac{\Delta\Phi}{c^2}, \qquad (408\,\text{b})$$

the last equation following from (391). If we now substitute this in (408 a), we obtain

$$\Delta\Phi = \tfrac{1}{2}\kappa c^4 \cdot \mu_0. \qquad (407\,\text{b})$$

Thus Poisson's equation is in fact valid. It constitutes a great achievement of the principle of general relativity that it leads to Newton's law of gravitation purely on the basis of the quite general postulates of § 56, without any further hypotheses. Moreover, we are now in a position to make some statement on the meaning and numerical value of the constant κ. For it follows from a comparison of (407 a) and (407 b) that

$$\kappa c^2 = \frac{8\pi k}{c^2} = \frac{8\pi}{c^2} 6{\cdot}7 \times 10^{-8} = 1{\cdot}87 \times 10^{-27}\ \text{cm g}^{-1}. \qquad (409)$$

At the same time it is seen that κ is positive, thus justifying the use of a negative sign on the right-hand side of (401). The general theory of relativity, therefore, does not provide a physical interpretation for the sign (gravitational attraction, and not repulsion) and numerical value of the gravitational constant, but takes these data from experiment.[320]

---

[320] The reason for writing κc^2 in (409) in place of κ (as is done by most other authors) is that T_{44} has, by definition, the dimensions of an energy density in our notation, whereas it has those of a mass density in theirs.

(β) *Rigorous solution for the gravitational field of a point-mass.* For the determination of the perihelion precession of Mercury and the bending of light rays one must calculate not only g_{44} for the field of a point-mass, but also the other g_{ik}, and g_{44} to the next-higher order of accuracy. As early as 1915, Einstein[321] solved this problem by a method of successive approximations. Schwarzschild[322] was the first, and after him independently Droste[323], to derive a *rigorous* solution for the G-field of a particle. The perihelion precession and bending of light follow in practically the same way as with Einstein. Considerable mathematical simplifications were achieved in a paper by Weyl[324], who introduced Cartesian instead of polar coordinates and reverted to the action principle instead of the general differential equations of the G-field.

Since the field of a particle is static and spherically symmetric, the square of the line element can be written in the form

$$ds^2 = \gamma[(dx^1)^2 + (dx^2)^2 + (dx^3)^2] + \\ + l(x^1 dx^1 + x^2 dx^2 + x^3 dx^3)^2 + g_{44}(dx^4)^2, \tag{410}$$

where γ, l and g_{44} are functions of $r = \sqrt{[(x^1)^2+(x^2)^2+(x^3)^2]}$ alone. This however does not yet specify the coordinate system uniquely. For, in a transformation

$$x'^i = \frac{f(r)}{r} x^i \tag{411}$$

[and thus $r' = \sqrt{[(x'^1)^2 + (x'^2)^2 + (x'^3)^2]} = f(r)$]

which contains the arbitrary function $f(r)$, the square of the line element retains the form (410). This enables us to normalize the coordinates still further. Two ways of normalizing, in particular, have often proved convenient:

(a) $\gamma = 1$:
$$ds^2 = (dx^1)^2 + (dx^2)^2 + (dx^3)^2 + l(x^1 dx^1 + x^2 dx^2 + x^3 dx^3)^2 + g_{44}(dx^4)^2;$$

$$\tag{410a}$$

(b) $l = 0$:
$$ds^2 = \gamma[(dx^1)^2 + (dx^2)^2 + (dx^3)^2] + g_{44}(dx^4)^2. \tag{410b}$$

Let us carry out the integration of the field equations in the coordinate system in which the square of the line element is of the form (410 a). In the space external to the mass, which alone need be considered here, the field equations are simply

$$R_{ik} = 0, \tag{412}$$

from (401 a). By calculation from (410a), and introducing the abbreviations

$$h^2 = 1 + lr^2, \qquad \Delta = \sqrt{(-g)} = h\sqrt{(-g_{44})}, \tag{413}$$

[321] Cf. footnote 278, p. 145, *ibid.*

[322] K. Schwarzschild, *S.B. preuss. Akad. Wiss.* (1916) 189.

[323] J. Droste, *Versl. gewone Vergad. Akad. Amst.*, **25** (1916) 163.

[324] H. Weyl, *Ann. Phys., Lpz.*, **54** (1917) 117; *Raum-Zeit-Materie* (1st edn., 1918), p. 199 *et seq.*; (3rd edn., 1920), p. 217 *et seq.*

the components R_{ik} of the curvature tensor can now be expressed in our case as follows,

$$R_{ik} = [R_{22}]\delta_i{}^k + ([R_{11}] - [R_{22}])\frac{x^i x^k}{r^2}, \qquad \text{for } i, k = 1, 2, 3, \quad (414)$$

$$
\left.
\begin{aligned}
[R_{11}] &= \frac{\Delta}{r^2 g_{44}} \cdot \frac{1}{2} \frac{d}{dr}\left(\frac{r^2 g'_{44}}{\Delta}\right) - \frac{2}{r}\frac{\Delta'}{\Delta}, \\
[R_{22}] &= -\frac{1}{r^2 \Delta}\frac{d}{dr}\left(\frac{r g_{44}}{\Delta}\right) - \frac{1}{r^2}, \\
R_{44} &= -\frac{g_{44}}{r^2 \Delta}\frac{1}{2}\frac{d}{dr}\left(\frac{r^2 g'_{44}}{\Delta}\right).
\end{aligned}
\right\}
\quad (415)
$$

It is seen that $[R_{11}]$ and $[R_{22}]$ are, respectively, the values of R_{11} and of R_{22} and R_{33} at the point $x^1 = r$, $x^2 = x^3 = 0$. These values of R_{ik} have now to be substituted in (412). We then obtain from the first and third equations of (415) that

$$\Delta' = 0, \qquad \Delta = \text{const.}$$

We next impose the condition that at infinity the g_{ik} have their normal values. It is this condition which in fact defines the problem (cf. § 62) in the first place. It then follows that

$$\Delta = 1 \tag{416}$$

and from the second equation (415),

$$g_{44} = -1 + \frac{2m}{r}, \tag{417}$$

where m is an integration constant. By a comparison with the Newtonian potential Φ, (391), it is seen that this constant m is connected with the mass M of the point-mass generating the field, by means of the formula

$$m = \frac{kM}{c^2}. \tag{418}$$

Since m has the dimensions of a length, we call this quantity the gravitational radius of the mass. It is easy to convince oneself that *all* the field equations are actually satisfied by (416) and (417).

According to Weyl one can avoid having to calculate the curvature components (415) by making use of the variational principle (403). Because of (177), this latter may also be written in the form

$$\delta \int \mathfrak{G}\, dx = 0 \tag{419}$$

for matter-free space. In our case we need introduce neither the time nor the coordinates x^1, x^2, x^3 separately, but can consider $\mathfrak{G}$ to be a function of r alone. After evaluation,

$$\mathfrak{G} = -\frac{2lr}{h^2}\Delta' = \left(\frac{1}{h^2} - 1\right)\frac{2\Delta'}{r},$$

so that (419) is of the form

$$\delta \int \left(\frac{1}{h^2} - 1 \right) r \Delta' \, dr = 0 \qquad (420)$$

because

$$dx = 4\pi r^2 \, dr.$$

Varying h, we obtain $\Delta' = 0$, $\Delta = \text{const.}$; varying Δ,

$$\frac{d}{dr} \left(\frac{1}{h^2} - 1 \right) r = 0, \qquad \left(\frac{1}{h^2} - 1 \right) r = \text{const.},$$

which again results in the field (416), (417), because of the definition (413) of Δ.

It follows from (413) that the square of the line element is of the form

$$ds^2 = (dx^1)^2 + (dx^2)^2 + (dx^3)^2 + \qquad (421\,\text{a})$$
$$+ \frac{2m}{r^2(r - 2m)} (x^1 \, dx^1 + x^2 \, dx^2 + x^3 \, dx^3)^2 - \left(1 - \frac{2m}{r} \right) (dx^4)^2.$$

The first part of this expression, which refers to three-dimensional space, can be visualized in the following way, according to Flamm.[325] On each plane passing through the origin (e.g. $x^3 = 0$) the geometry is the same as in the Euclidean space on the fourth-order surface

$$z = \sqrt{[8m(r - 2m)]}$$

which is generated by the rotation of the parabola

$$z^2 = 8m(x^1 - 2m), \qquad x^{(2)} = 0$$

about the z-axis. In fact, on this plane

$$ds^2 = (dx^1)^2 + (dx^2)^2 + \frac{2m}{r^2(r - 2m)} (x^1 \, dx^1 + x^2 \, dx^2)^2$$
$$= (dx^1)^2 + (dx^2)^2 + (dz)^2.$$

For $r = 2m$ the coordinate system becomes singular.

The second normalized form (410 b) is obtained from (411) by using the transformation

$$r = \left(1 + \frac{m}{2r'} \right)^2 r', \qquad x'^i = \frac{r'}{r} x^i. \qquad (i = 1, 2, 3) \qquad (422)$$

Then

$$ds^2 = \left(1 + \frac{m}{2r} \right)^4 [(dx^1)^2 + (dx^2)^2 + (dx^3)^2] - \left[\frac{1 - (m/2r)}{1 + (m/2r)} \right]^2 (dx^4)^2. \quad (421\,\text{b})$$

This coordinate system extends to $r = m/2$.

(γ) *Perihelion precession of Mercury and the bending of light rays.* We now come to the calculation of the paths of point-mass and light rays in the

[325] L. Flamm, *Phys. Z.*, **17** (1916) 448.

gravitational field (421). They are geodesic lines in the four-dimensional world, determined by the variational principle

$$\delta \int ds = 0 \tag{81}$$

or by the differential equations (80). From the latter we obtain by a simple calculation

$$\frac{d^2x^1}{d\tau^2} : \frac{d^2x^2}{d\tau^2} : \frac{d^2x^3}{d\tau^2} = x^1 : x^2 : x^3. \tag{423}$$

For the path of the point-mass, τ denotes the proper time, for that of the light ray an arbitrary parameter for which the differential equation (395) is satisfied. We can first of all deduce that the paths of point-mass and light ray lie in a plane; secondly, by taking x^3 perpendicular to this plane and introducing polar coordinates

$$x^1 = r \cos \varphi \qquad x^2 = r \sin \varphi, \tag{424}$$

the law of areas

$$r^2 \frac{d\varphi}{d\tau} = \text{const.} = B \tag{425}$$

is seen to be valid. On the other hand, it follows from (81) by varying the time, as in § 53 (γ), that

$$g_{44} \frac{dx^4}{d\tau} = \text{const.}$$

If we square this equation and eliminate $dx^4/d\tau$ by using the relations

$$g_{ik} \frac{dx^i}{d\tau} \frac{dx^k}{d\tau} = -c^2$$

for the case of a point-mass, and

$$g_{ik} \frac{dx^i}{d\tau} \frac{dx^k}{d\tau} = 0$$

for the case of the light ray, we obtain for the first case

$$\left(\frac{dr_*}{d\tau}\right)^2 + r^2 \left(\frac{d\varphi}{d\tau}\right)^2 - \frac{2mc^2}{r} - 2mr \left(\frac{d\varphi}{d\tau}\right)^2 = \text{const.} = 2E \tag{426a}$$

and for the second

$$\left(\frac{dr}{d\tau}\right)^2 + r^2 \left(\frac{d\varphi}{d\tau}\right)^2 - 2mr \left(\frac{d\varphi}{d\tau}\right)^2 = \text{const.} \tag{426b}$$

It is clear that (426 a) contains the energy conservation law. Both equations differ from the Newtonian equations only by their last term. If, next, we introduce φ as independent variable in place of τ, we obtain, with the help of (425),

$$B^2 \left[\frac{1}{r^4} \left(\frac{dr}{d\varphi}\right)^2 + \frac{1}{r^2} \right] - \frac{2mc^2}{r} - \frac{2mB^2}{r^3} = 2E, \tag{427a}$$

$$\frac{1}{r^4} \left(\frac{dr}{d\varphi}\right)^2 + \frac{1}{r^2} - \frac{2m}{r^3} = \text{const.} = \frac{1}{\Delta^2}. \tag{427b}$$

These equations completely determine the required paths. The last term on the left-hand side of (427 a) produces a displacement of the perihelion of planetary orbits in the direction of the planet's orbital motion which is of magnitude

$$\varDelta\pi = \frac{6\pi m}{a(1 - e^2)} \text{ per revolution} \qquad (428\,\text{a})$$

(a = semi-major axis, e = eccentricity).

Because of (418) and Kepler's third law,

$$\frac{4\pi^2 a^3}{T^2} = kM = mc^2, \qquad (T = \text{period})$$

this can also be written

$$\varDelta\pi = \frac{24\pi^3 a^2}{c^2 T^2(1 - e^2)}. \qquad (428\,\text{b})$$

There remains to be discussed equation (427 b) for the light ray. If the last term on the left-hand side were not present, the light ray would be a straight line, at a distance $\varDelta$ from the origin. The perturbation term produces a curvature of the light ray, concave towards the mass centre, which leads to a total deflection through an angle

$$\varepsilon = \frac{4m}{\varDelta}, \qquad (429)$$

where now $\varDelta$ denotes the distance of the origin from the asymptotic directions of the path. This method for calculating the bending of light rays is due to Flamm[326]. Einstein's[327] calculation was based on Huyghen's principle and leads to the same result, as it should, according to § 53 (γ).

The two consequences of Einstein's theory of gravitation developed here both admit of a check by experiment. As for the perihelion precession, given by (428), this is only of measurable size in the case of Mercury, where conditions are particularly favourable because of its small distance from the sun and the large eccentricity of its orbit. The theoretical value of the precession is

$$\varDelta\pi = 42\cdot89'', \qquad e\varDelta\pi = 8\cdot82'' \qquad \text{per century.[328]}$$

It has been known to astronomers since Leverrier's[329] time that a remainder is present in the perihelion precession of Mercury, which cannot be caused by perturbations due to the other planets. According to the renewed calculation by Newcomb[330] it is of magnitude

$$\varDelta\pi = 41\cdot24'' \pm 2\cdot09'' \qquad e\varDelta\pi = 8\cdot48'' \pm 0\cdot43''.$$

[326] Cf. footnote 325, p. 166, *ibid.*
[327] Cf. footnote 278, p. 145, *ibid.*, and footnote 279, p. 145, *ibid.*, § 22.
[328] Cf. numerical table in *Enzykl. math. Wiss.*, VI 2, 17 (J. Bauschinger), p. 887.
[329] U. J. Leverrier, *Ann. Obs. Paris*, **5** (1859).
[330] S. Newcomb, *Astr. pap.*, *Washington*, **6** (1898) 108.

The theoretical value thus falls within Newcomb's limits of accuracy. To what extent Newcomb's value itself can be regarded as reliable (or falsified by errors in the calculation, as claimed by some astronomers) will be discussed by F. Kottler in Article **VI** 2, 22 of the *Enzykl. math. Wiss.* That article will have to treat in detail effects of a non-relativistic nature acting on the perihelion of Mercury, such as spheroidal shape of the sun, rotation of the empirical with respect to the inertial system, non-planetary perturbing masses, in particular that of the zodiacal light (Seeliger[330a]). Compared with Seeliger's explanation, Einstein's has at least the advantage that no arbitrary parameters are needed. Thus, even if the degree of numerical agreement cannot as yet be estimated with certainty, the agreement of Einstein's and Newcomb's values constitutes, at any rate, a great success.

Recently, an earlier attempt by P. Gerber[331] has been discussed which tries to explain the perihelion advance of Mercury with the help of the finite velocity of propagation of gravitation, but which must be considered completely unsuccessful from a theoretical point of view. For while it leads admittedly to the correct formula (428)—though on the basis of false deductions—it must be stressed that, even so, only the numerical factor was new.

The theory of relativity has received a still more definitive confirmation than for the perihelion of Mercury for the case of the deflection of light rays. For, according to (429), a light ray passing the edge of the sun suffers a deviation of

$$\varepsilon = 1 \cdot 75''.$$

This can be checked by observing fixed stars near the sun during total eclipses of the sun. The expeditions dispatched to Brazil and the island of Principe on the occasion of the total eclipse of the sun on 29 May, 1919 found that the effect which had been predicted by Einstein did in fact exist.[332] Quantitatively, too, the agreement is a good one. The first-mentioned expedition found that the mean stellar deflection, extrapolated to the surface of the sun, was $1 \cdot 98'' \pm 0 \cdot 12''$, the second, $1 \cdot 61'' \pm 0 \cdot 30''$. Cf. Kottler's article in the *Enzykl. math. Wiss.*, **VI** 2, 22, for the extrapolation methods used for obtaining these numerical results.

The value originally calculated by Einstein (cf. § 50) on the basis of Newtonian theory for a particle moving with the velocity of light, was shown to be incompatible with the observations.†

[330a] H. v. Seeliger, *S.B. bayer. Akad. Wiss.*, **36** (1906) 595.

[331] P. Gerber, *Z. Math. Phys.*, **43** (1898) 93; *Jber. Real-Progymnasium Stargard* (1902). Reprinted in *Ann. Phys., Lpz.*, **52** (1917) 415; discussion: H. v. Seeliger, *Ann. Phys., Lpz.*, **53** (1917) 31; **54** (1917) 38; S. Oppenheim, *ibid.*, **53** (1917) 163; M. v. Laue, *ibid.*, **53** (1917) 214; *Naturwissenschaften*, **8** (1920) 735. Cf. Also J. Zenneck, *Enzykl. math. Wiss.*, **V** 2, § 24, and S. Oppenheim, *ibid.*, **VI** 2, 22, § 31 (b).

[332] F. W. Dyson, A. S. Eddington and C. Davidson, 'A determination of the deflection of light by the sun's gravitational field, from observations made at the total eclipse of 29 May, 1919.' *Phil. Trans.* A, **220** (1920) 291.

† See suppl. note 16.

59. Other special, rigorous, solutions for the statical case

The field (421) for a point-mass becomes singular for $r = 2m$ and $r = m/2$, respectively, and it is therefore of theoretical interest to investigate how the G-field is continued into the interior of the mass. For this, it will be necessary to make certain assumptions about the physical properties of the mass which generates the field, since otherwise the energy tensor T_{ik} is undetermined. The simplest assumption is that of an *incompressible liquid sphere*. For this case, Schwarzschild[333] performed the integration of the field equations, and the calculation was simplified by Weyl[334]. The energy tensor is, according to (362), given by

$$T_{ik} = \left(\mu_0 + \frac{p}{c^2} \right) u_i u_k + p g_{ik}$$

since $\mu_0 = $ const., $P = p/\mu_0$. The boundary conditions of the theory of elasticity demand the continuity of all the g_{ik} and the vanishing of the pressure p on the surface of the sphere. When this is taken into account, the field is uniquely determined. For the external region (where $r > r_0$, $r_0 = $ radius of the sphere) we obtain the same field as for a point-mass. The gravitational radius m is then

$$m = \frac{k\mu_0}{c^2} \frac{4\pi r_0{}^3}{3}. \tag{430}$$

For the interior of the sphere, however, we obtain the following relations, if we write the square of the line element in the normalized form (410 a) and assign the same meaning to h as in (413),

$$\frac{1}{h^2} = 1 - \frac{2m}{r_0{}^3} r^2, \quad \sqrt{(-g_{44})} = \frac{3h - h_0}{2h h_0}, \quad p = \mu_0 c^2 \frac{h_0 - h}{3h - h_0} \tag{431}$$

$$(h_0 = \text{values of } h \text{ on the surface}).$$

The square of the line element in the interior of the sphere thus becomes

$$\left. ds^2 = (dx^1)^2 + (dx^2)^2 + (dx^3)^2 + \frac{(x^1\,dx^1 + x^2\,dx^2 + x^3\,dx^3)^2}{a^2 - r^2} - \right. \\ \left. - \left(\frac{3h - h_0}{2h h_0} \right)^2 (dx^4)^2, \right\} \tag{432}$$

where

$$a = r_0 \Big/ \left(\frac{r_0}{2m} \right). \tag{433}$$

r_0 must be $> 2m$, in order that the line element should remain regular outside the sphere. As a comparison with (122) shows, the geometry of the three-dimensional space inside the liquid sphere has a positive constant curvature (spherical or elliptic); a has the meaning of a radius of curvature. Bauer[334a] has worked on the calculation of the G-field of *compressible* liquid spheres.

[333] K. Schwarzschild, *S.B. preuss. Akad. Wiss.* (1916) 424.
[334] H. Weyl, *Raum-Zeit-Materie* (1st edn., 1918), p. 208; (3rd edn., 1920), p. 225.
[334a] H. Bauer, *S.B. Akad. Wiss. Wien*, Abt. IIa, **127** (1918) 2141.

A further problem which admits of a rigorous solution is that of the field of an electrically charged sphere. In connection with the enquiry into the nature of the electron, it is of interest to investigate to what extent the electrostatic field of a charged sphere is influenced by its gravitational field and, conversely, to what extent a gravitational field is produced by the electrostatic energy. This problem was first solved by Reissner[335] and later by Weyl[336], who started from an action principle. It is seen that the electrostatic potential φ is exactly equal to the Coulomb potential

$$\varphi = \frac{e}{r}, \tag{434}$$

if we use ordinary c.g.s. and not Heaviside units. The G-field in the normalized form (410 a) is however no longer determined by (416), (417), but by

$$\Delta = 1, \qquad -g_{44} = \frac{1}{h^2} = 1 - \frac{2m}{r} + \kappa \frac{e^2}{r^2}. \tag{435}$$

The last term is the gravitational field produced by the electrostatic energy. Only at distances of the order of $a = \kappa e^2/m = e^2/Mc^2$ is this term comparable with the Newtonian term $2m/r$. For the electron, $a \sim 10^{-13}$ cm, which was called the "electron radius" in the earlier theories. The gravitational attraction exerted by an electron on another electron or on a charge element of its own is however always very much smaller than the electrostatic Coulomb repulsion—their ratio is

$$\frac{kM^2}{e^2} \sim 10^{-40}$$

—so that the electron can under no circumstances be held in equilibrium by the gravitational field (435) balancing its own repulsive forces.

Levi-Civita[337] also investigated the gravitational field produced by a *homogeneous* electric or magnetic field. If x^3 is taken in the direction of a magnetic field of intensity F, the square of the line element is of the form

$$ds^2 = (dx^1)^2 + (dx^2)^2 + (dx^3)^2 + \frac{(x^1\,dx^1 + x^2\,dx^2)^2}{a^2 - r^2} - $$
$$- \left[c_1 \exp\left(\frac{x^{(3)}}{a}\right) + c_2 \exp\left(-\frac{x^{(3)}}{a}\right) \right]^2 (dx^4)^2, \tag{436}$$

where $r = \sqrt{[(x^1)^2 + (x^2)^2]}$, c_1, c_2 are constants, $a = c^2/\sqrt{kF}$. The space is cylindrically symmetric about the direction of the field, and on each plane perpendicular to the field direction the same geometry holds as in Euclidean space on a sphere of radius a. The radius of curvature a is

[335] H. Reissner, *Ann. Phys., Lpz.*, **50** (1916) 106.

[336] Cf. footnote 324, p. 164, *Ann. Phys., Lpz., loc. cit.*; and *Raum-Zeit-Materie* (1st edn., 1918), p. 207; (3rd edn., 1920), p. 223.

[337] T. Levi-Civita, 'Realtà fisica di alcuni spazi normali del Bianchi', *R.C. Accad. Lincei* (5) **26** (1917), p. 519.

extremely large for fields of normal size, e.g. for $F = 25{,}000$ gauss, $a = 1{\cdot}5 \times 10^{15}$ km.

Weyl[338] and, in a series of papers, Levi-Civita[339] have also derived general solutions for arbitrary *cylindrically symmetric* charged and uncharged mass distributions. The G-field itself is then cylindrically symmetric and static. Corresponding to the non-linear character of the differential equations, g_{44} does not behave additively in the masses.

60. Einstein's general approximative solution and its applications

So far, it has only been possible to derive rigorous solutions of the gravitational field equations for the static case. It is therefore of great importance that Einstein[340] has indicated a method by which the G-field for masses moving with arbitrary velocities can be determined approximately, provided the masses are sufficiently small. For in this case, the g_{ik} differ only slightly from their normal values so that the squares of these deviations can be neglected. Also, only the *linear* part of the differential equations (401) of the gravitational field need be retained, so that they can be integrated very easily.

Let us again introduce the *imaginary* time coordinate $x^4 = ict$ so that we can put

$$g_{ik} = \delta_i{}^k + \gamma_{ik}. \tag{437}$$

Because of $g_{i\alpha}g^{k\alpha} = \delta_i{}^k$ it follows that

$$g^{ik} = \delta_i{}^k - \gamma_{ik}, \tag{437a}$$

apart from higher-order terms. It is to be noted here that the quantities γ_{ik} possess a tensor character only with regard to Lorentz transformations. Corresponding to expression (94) for the contracted curvature tensor, the field equations (401) are, to the desired approximation, of the form

$$\frac{\partial^2 \gamma}{\partial x^i\,\partial x^k} + \sum_\alpha \left[\frac{\partial^2 \gamma_{ik}}{(\partial x^\alpha)^2} - \frac{\partial^2 \gamma_{i\alpha}}{\partial x^k\,\partial x^\alpha} - \frac{\partial^2 \gamma_{k\alpha}}{\partial x^i\,\partial x^\alpha} - \left(\frac{\partial^2 \gamma}{(\partial x^\alpha)^2} - \sum_\beta \frac{\partial^2 \gamma_{\alpha\beta}}{\partial x^\alpha\,\partial x^\beta} \right)\delta_i{}^k \right]$$
$$= -2\kappa T_{ik}, \tag{438}$$

where the abbreviation

$$\gamma = \sum_\alpha \gamma_{\alpha\alpha} \tag{439}$$

is used. For simplification, let us next introduce the quantities

$$\gamma'_{ik} = \gamma_{ik} - \tfrac{1}{2}\delta_i{}^k \gamma \tag{440}$$

[338] Cf. footnote 324, p. 164, *Ann. Phys., Lpz., loc. cit.*; supplementary paper, *ibid.*, **59** (1919) 185.

[339] T. Levi-Civita, 'ds^2 einsteiniani in campi newtoniani I–IX', *R.C. Accad. Lincei* (5), **26** (1917); **27** (1918); **28** (1919). The general form of the differential equations of the G-field for the statical case is given by Levi-Civita in his "Statica Einsteiniana' (Cf. footnote 307, p. 154, *ibid.*).

[340] A. Einstein, *S.B. preuss. Akad. Wiss.* (1916) 688.

with the corresponding inverse equations for the unprimed quantities,

$$\gamma_{ik} = \gamma'_{ik} - \tfrac{1}{2}\delta_i{}^k \gamma', \tag{440a}$$

$$\gamma' = \sum_\alpha \gamma'_{\alpha\alpha} = -\gamma. \tag{439a}$$

We then obtain from (438),

$$\sum_\alpha \left[\frac{\partial^2 \gamma'_{ik}}{(\partial x^\alpha)^2} - \frac{\partial^2 \gamma'_{i\alpha}}{\partial x^k \, \partial x^\alpha} - \frac{\partial^2 \gamma'_{k\alpha}}{\partial x^i \, \partial x^\alpha} + \delta_i{}^k \sum_\beta \frac{\partial^2 \gamma'_{\alpha\beta}}{\partial x^\alpha \, \partial x^\beta} \right] = -2\kappa T_{ik}. \tag{438a}$$

These equations can be considerably simplified still further by normalizing the coordinate system in a suitable way. For the coordinate system itself is only determined to within quantities of the order of the γ_{ik}, by the condition that the g_{ik} should differ only slightly from their normal values. We can thus choose, in particular, the coordinates in such a way that the following equations hold in the normalized system,

$$\sum_\alpha \frac{\partial \gamma'_{i\alpha}}{\partial x^\alpha} = 0. \tag{441}$$

Hilbert[341] has given the mathematical proof that, for arbitrarily prescribed values of the γ'_{ik} in the original system, a coordinate system can always be found such that the new coordinate values differ from the old only by quantities of the order of the γ'_{ik}, and such that condition (441) is satisfied. There are exactly four functions available for satisfying the four equations (441).

Evidently the differential equations (438 a) will then simply reduce to

$$\Box \gamma'_{ik} = -2\kappa T_{ik}, \tag{442}$$

where we have written $\Box \gamma'_{ik}$ for $\sum_\alpha \partial^2 \gamma'_{ik}/(\partial x^\alpha)^2$, as in the special theory of relativity. The integration can be carried out in the well-known manner, by means of retarded potentials,

$$\gamma'_{ik}(x, y, z, t) = \frac{\kappa}{2\pi} \int \frac{T_{ik}[\bar{x}, \bar{y}, \bar{z}, t - (r/c)]}{r} \, d\bar{x} \, d\bar{y} \, d\bar{z}. \tag{443}$$

Because of the energy conservation law (341 a), p. 157, Eq. (441) is also satisfied to the required accuracy.

It can be seen from (443) that the gravitational effects are propagated with the velocity of light, just as electromagnetic disturbances. The shape of the gravitational waves in empty space follows from (441), (442), if we put $T_{ik} = 0$. In particular, for a plane wave

$$\gamma_{ik} = a_{ik} \exp \left[i\nu \left(t - \frac{x}{c} \right) \right] \tag{444}$$

[341] Cf. footnote 99, p. 63, *ibid.*

propagated along the x^1-axis, it follows from (441) that

$$a_{k4} = -ia_{k1}, \tag{445}$$

so that (442) is identically satisfied. Einstein[342] moreover showed that, for a suitable choice of the coordinates, also

$$a_{11} = a_{12} = a_{13} = 0, \qquad a_{22} = -a_{33}. \tag{446}$$

(See the following section on the emission and absorption of gravitational waves.)†

For the field of a point-mass at rest, (443) results in

$$\gamma'_{44} = -\frac{4m}{r}, \quad \text{all other} \quad \gamma'_{ik} = 0,$$

so that
$$\gamma_{44} = -\frac{2m}{r}, \qquad \gamma_{11} = \gamma_{22} = \gamma_{33} = +\frac{2m}{r}.$$

These are just the first-order quantities of the field (421 b). In addition, the calculation of the field of n moving particles can be carried out without any trouble.[343] The main result is that the deviations of their motions from the laws of Newtonian mechanics are only of second order in v/c, as required by experiment.

A deviation from Newtonian mechanics is also produced in the following circumstances. The relativistic theory of gravitation is in agreement with Newtonian theory in that the gravitational field of a sphere *at rest* is the same as the field of a point-mass. This, however, is not true for a *rotating* sphere. Thirring and Lense[344] applied Einstein's formulae to this case and calculated the corresponding perturbations of the planetary orbits and the orbit of the moon, caused by the spin of the central bodies. All such perturbations are probably too small to be observable. De Sitter[345] gives a general discussion of the deviations from classical mechanics, to be expected from Einstein's theory, in the perturbations of the planetary orbits and the orbit of the moon. Apart from the perihelion advance of Mercury, there exists no deviation which would be experimentally observable.

The most important application of Einstein's approximate solution (443), however, is Thirring's[346] investigation on the *relativity of the centrifugal force*. Since, in general relativity, processes can also be referred to a system which rotates relative to a Galilean reference system, one should be able to regard the centrifugal force also as a gravitational effect due to the relative rotation of the fixed-star masses. It might be thought that

[342] A. Einstein, 'Über die Gravitationswellen', *S.B. preuss. Akad. Wiss.* (1918) 154.

† See suppl. note 17.

[343] This is given by J. Droste, *Proc. Acad. Sci. Amst.*, **19** (1916) 447, using an integration method which differs slightly from Einstein's.

[344] H. Thirring and J. Lense, *Phys. Z.*, **19** (1918) 156.

[345] W. de Sitter, *Mon. Not. R. Astr. Soc.*, **76** (1916) 699 and **77** (1916) 155. Cf. also de Sitter's paper 'Planetary motion and the motion of the moon according to Einstein's theory', *Proc. Acad. Sci. Amst.*, **19** (1916) 367.

[346] H. Thirring, *Phys. Z.*, **19** (1918) 33. Cf. also the supplementary paper, *Phys. Z.*, **22** (1921) 29.

the possibility of such a viewpoint in general relativity was already guaranteed by the general covariance of the field equations. It will be discussed in detail in § 62 that this is not so, because the boundary conditions at infinity play an essential rôle in this case. Thirring, therefore, did not set himself the problem to prove the full equivalence of the relative rotation of all the fixed stars to a rotation of the reference system relative to a Galilean system. Instead, he modified the question in such a way that the difficulty of fixing the boundary conditions was eliminated.

Let us consider a rotating spherical shell which is present in an inertial system of Newton's gravitational theory, in addition to the very distant fixed stars which are at rest (or in rectilinear uniform motion, with very small velocities). From a relativistic point of view it is clear that centrifugal and Coriolis forces will appear inside the shell when the mass of the shell becomes comparable to that of the fixed-star system. According to the principle of continuity, we shall have to assume that such forces, however small, are also present when the mass of the shell is small. But in this latter case we are justified in applying formulae (443) directly, because the g_{ik} will evidently deviate from their normal values only very slightly. The calculation now shows that a point-mass inside the shell is indeed subjected to accelerations which are perfectly analogous to the Coriolis- and centrifugal accelerations of classical mechanics. If $\boldsymbol{\omega}$ is the angular velocity vector, $\mathbf{r}$ the perpendicular from the axis of rotation to the particle, $\mathbf{v}$ its velocity, then these accelerations are, of course, not exactly equal to

$$2\boldsymbol{\omega} \wedge \mathbf{v} + \omega^2 \mathbf{r},$$

as would be the case in classical mechanics for a reference system rotating with angular velocity $\boldsymbol{\omega}$ relative to the inertial system; but these two terms would have to be multiplied by coefficients of the order of magnitude of the ratio of the gravitational radius of the shell, $m = kM/c^2$, to its radius a. Since this ratio is quite minute for all masses obtainable in practice, there is no hope of verifying experimentally this decisive and important result; and we can understand the reason for the negative result of Newton's primitive experiment with the rotating pail, and of the improved experiment by B. and T. Friedländer[347], which attempted to demonstrate the presence of centrifugal forces inside a heavy rotating flywheel.

61. The energy of the gravitational field

We have already seen in § 54 that, in the presence of a gravitational field, the differential equations for the material energy tensor are not

$$\frac{\partial \mathfrak{T}_i{}^k}{\partial x^k} = 0, \tag{341}$$

as in special relativity, but are of the form

$$\frac{\partial \mathfrak{T}_i{}^k}{\partial x^k} - \tfrac{1}{2} \mathfrak{T}^{rs} \frac{\partial g_{rs}}{\partial x^i} = 0, \tag{341a}$$

[347] B. and T. Friedländer, *Absolute und relative Bewegung* (Berlin 1896).

so that we cannot derive from them the conservation laws

$$\int \mathfrak{T}_i{}^4 dx^1 \, dx^2 \, dx^3 = \text{const.}$$

for a closed system. In § 57 we showed however that, in view of the gravitational field equations (401), the energy–momentum law (341 a) leads to the system of equations

$$\frac{\partial(\mathfrak{T}_i{}^k + \mathfrak{t}_i{}^k)}{\partial x^k} = 0, \tag{406}$$

where $\mathfrak{t}_i{}^k$ are the quantities defined by (405), (183) and (185). From these equations we can obtain, once more, conservation laws for a closed system

$$J_i = \int (\mathfrak{T}_i{}^4 + \mathfrak{t}_i{}^4) \, dx^1 \, dx^2 \, dx^3 = \text{const.} \tag{447}$$

For this reason, Einstein calls the $\mathfrak{t}_i{}^k$ the energy components of the gravitational field and J_i the total momentum and energy of the closed system (cf. § 57).

On closer inspection, however, great difficulties become apparent, which oppose this point of view at first sight. In the final analysis, they are due to the fact that the t_{ik} do not form a tensor. Since these quantities do not depend on derivatives of the g_{ik} higher than the first, we can conclude immediately that they can be made to vanish at an arbitrarily prescribed world point for a suitable choice of the coordinate system (geodesic reference system).

But we can go still further: Schrödinger[348] found that all the energy components vanish identically for the field (421 a) of a point-mass which represents, at the same time, the field outside a liquid sphere. This result can also be extended to the case of the field (435) of a charged sphere. On the other hand, Bauer[349] showed that by simply introducing polar coordinates in the Euclidean line element of special relativity the energy components are found to have values different from zero, in fact the total energy becomes infinite! Also, the t_{ik} are certainly not symmetrical, and the energy density $-t_4{}^4$ is not everywhere positive. In the earlier field theories of gravitation already, it was the sign of the energy density of the gravitational field which had always led to difficulties.[349a]

In spite of these difficulties it would, on physical grounds, be hard to abandon the requirement that an analogue to the energy- and momentum-integrals of Newtonian theory should exist. Lorentz[350] and Levi-Civita[351] made the suggestion to denote as energy components of the gravitational

[348] E. Schrödinger, *Phys. Z.*, **19** (1918) 4.
[349] H. Bauer, *Phys. Z.*, **19** (1918) 163.
[349] Cf. e.g. footnote 272, p. 144, M. Abraham, *ibid.*, p. 570.
[350] Cf. footnote 100, p. 68, *ibid.*, **25**.
[351] T. Levi-Civita, *R.C. Accad. Lincei* (5) **26** (1917) 1st part, p. 381.

field, not the quantities t_{ik}, but $(1/\kappa)\, G_{ik}$. For, from (401),

$$T_{ik} + \frac{1}{\kappa} G_{ik} = 0$$

so that

$$\frac{\partial}{\partial x^k}\left(\mathfrak{T}_i{}^k + \frac{1}{\kappa}\mathfrak{G}_i{}^k\right) = 0.$$

But Einstein[352] quite rightly raised the objection that, with this definition of the gravitational energy, the total energy of a closed system would always be zero, and that the maintenance of this value of the energy does not require the continued existence of the system in some form or other. The usual kind of conclusions could not then be drawn from the conservation laws. In a rejoinder to Schrödinger's paper, Einstein[353] was further able to show that the t_{ik} certainly do not vanish everywhere for interactions between several masses.

A final clarification was eventually brought about in Einstein's paper 'Der Energiesatz in der allgemeinen Relativitätstheorie'[354] ('The energy conservation law in the general theory of relativity'). In this he proved that the expressions (447) for the total energy and momentum of a closed system are, to a large extent, independent of the coordinate system, although the localization of the energy will in general be completely different for different coordinate systems. The proof was later completed by Klein[354] (cf. § 21). According to this, one cannot assign any physical meaning to the values of the $t_i{}^k$ themselves, i.e. it is impossible to carry out a localization of energy and momentum in a gravitational field in a generally covariant and physically satisfactory way. But the integral expressions (447) have a definite physical meaning. The physical significance of equation (406) merely rests in the fact that it allows us to calculate the change in the material energy of a closed system in a simple fashion.

It is very simple to prove the invariance of the quantities J_k, defined by (447), for certain coordinate transformations, which will be discussed below. Let a bounded closed system be given. Outside a certain region B of it, the line element is to be that of special relativity (Galilean coordinate system). Let us first consider only such coordinates K which coincide with a Galilean coordinate system outside B. Thus polar coordinates, e.g., are excluded. Then the integrand of (447) will vanish outside the world canal B and all the assumptions of § 21 are satisfied. From what was said there, it follows that, *first*, the values of the energy and momentum integrals are independent of the choice of coordinates *inside B*, provided only the coordinates go over into a Galilean system outside B in a continuous manner. And *secondly*: The quantities J_i behave under *linear* coordinate transformations—thus implying changes in the coordinate values also outside B—like the covariant components of a vector. (See the next section on an analogous invariance theorem for the spatially bounded world.)

[352] Cf. footnote 342, p. 174, *ibid.*, § 6.
[353] A. Einstein, *Phys. Z.*, **19** (1918) 115.
[354] A. Einstein, *S.B. preuss. Akad. Wiss.* (1918) 448; see also footnote 94, p. 54, F. Klein, *ibid.*

We still have to discuss the question, whether the quantities $t_i{}^k$ are uniquely determined by the equations (406), i.e. whether we could not find quantities $w_i{}^k$, other than the $t_i{}^k$ defined by (405), (183) and (185), which would identically satisfy the system of equations (406) as a result of the field equations (401). As was first shown by Lorentz[350], and also stressed by Klein[355], this is indeed the case, as soon as the $w_i{}^k$ are allowed to contain also second derivatives of the g_{ik}. No physical arguments can be used against such a possibility. Different values for the total energy will naturally be obtained, depending on whether they are based on Einstein's $t_i{}^k$ or Lorentz's $w_i{}^k$.

Einstein[356] uses Eq. (406) for an important application to the case of the emission and absorption of gravitational waves, of which the field was discussed in § 60. If in a material system oscillations or other motions occur, it follows from the theory that the system will radiate waves: such a radiation is determined by the third time-derivatives of its moments of inertia,

$$D_{ik} = \int \mu_0 \, x^i \, x^k \, dx^1 \, dx^2 \, dx^3 \qquad (i, k = 1, 2, 3). \qquad (448)$$

The energy current of the radiated wave along the x^1-axis is given by

$$S_1 = \frac{k}{8\pi c^5 r^2} \left[\left(\frac{\dddot{D}_{22} - \dddot{D}_{33}}{2} \right)^2 + (\dddot{D}_{23})^2 \right] \qquad (449)$$

(k = usual gravitational constant)

and the total energy radiated in all directions per unit time is

$$-\frac{dE}{dt} = \frac{k}{10c^5} \left[\sum_{i,k} \dddot{D}_{ik}{}^2 - \tfrac{1}{3} \left(\sum_i \dddot{D}_{ii} \right)^2 \right]. \qquad (450)$$

The latter is always positive, because of (449). The radiated energy is so small that it does not give rise to any astronomical effects which would be noticeable during the relevant time intervals. The principle of it, however, is of importance in atomic physics. Einstein is of the opinion that Quantum Theory will also have to bring about a modification of the theory of gravitation.

The absorbed energy can similarly be calculated. Let a gravitational wave of the type (444), (445) and (446) in the x^1-direction be incident on a material system, whose dimensions are small compared with the wave-length of the incident wave. Then the energy absorbed per unit time is given by

$$\frac{dE}{dt} = \frac{1}{2} \left[\frac{\partial \gamma_{23}}{\partial t} \ddot{D}_{23} + \frac{1}{2} \frac{\partial (\gamma_{22} - \gamma_{33})}{\partial t} \frac{1}{2} (\ddot{D}_{22} - \ddot{D}_{33}) \right], \qquad (451)$$

where the quantities γ_{23}, γ_{22}, γ_{33} refer to the wave field.†

---

[355] Cf. footnote 104, p. 68, ibid., (1918) 235.
[356] Cf. footnote 342, p. 174, ibid.
† See suppl. note 17.

62. Modification of the field equations. Relativity of inertia and the space-bounded universe [357]†

(α) *The Mach principle.* In § 58 we discussed the perihelion precession of Mercury simply as such, without specifying how to determine physically the coordinate system K_0, relative to which this perihelion precession is to be measured. Such a coordinate system is distinguished from all other systems K, rotating uniformly relative to K_0, by the spherical symmetry of the G-field and above all by the behaviour of the g_{ik} in the spatial infinite; for the g_{ik} take on their normal values there. Certain coordinate systems K_0 are singled out from others by the boundary conditions for the spatial infinite, which have to be added to the differential equations of the G-field, in order to determine the g_{ik} completely from the positions and velocities of the masses, or, more generally, from the material energy tensor T_{ik}. This difficulty made itself felt particularly strongly in connection with the problem of the relativity of centrifugal forces (§ 60). While such singling out of certain coordinate systems by means of the boundary conditions is not logically incompatible with the general covariance postulate, it is nevertheless in contradiction to the spirit of a relativistic theory and must be considered as a severe epistemological defect. Einstein[358] throws it into sharp relief by means of a 'thought experiment', dealing with two liquid masses rotating relative to each other about their common axis. This defect is a feature, not only of classical mechanics and special relativity, but also of the gravitational theory based on Eqs. (401), which we developed above. It is only removed when the boundary conditions can be specified in a generally covariant way.

We therefore put forward the postulate: *The G-field is to be determined in a unique and generally covariant manner, solely by the values of the energy tensor* (T_{ik}). Since Mach[359] had already clearly recognized this defect in Newtonian mechanics, and had replaced absolute acceleration by acceleration relative to all the other masses in the universe, Einstein[360] called this postulate the 'Mach principle'. It has to be postulated, in particular, that the inertia of matter is solely determined by the surrounding masses. It must therefore vanish when all the other masses are removed, because it is meaningless, from a relativistic point of view, to talk of a resistance against *absolute* accelerations (relativity of inertia).‡

(β) *Remarks on the statistical equilibrium of the system of fixed stars. The λ-term.* Quite apart from the boundary conditions at the spatial infinite, one comes across a further difficulty if one wants to apply the previously used field equations to the system of fixed stars as a whole. Once this difficulty is overcome, the postulates set up in (α) will then also be fulfilled.

[357] The ideas presented in this section are developed by Einstein in his paper 'Kosmologische Betrachtungen zur allgemeinen Relativitätstheorie' (*S.B. preuss. Akad. Wiss.*, (1917) 142). (Also reprinted in the collection of papers *Das Relativitätsprinzip*.)

† See suppl. note 19.

[358] Cf. footnote 279, p. 145, *ibid.*, § 2.

[359] E. Mach, *Mechanik*, Chap. II, No. 6; 'Die Geschichte und die Wurzel des Satzes von der Erhaltung der Arbeit', Appendix note 1.

[360] A. Einstein, *Ann. Phys., Lpz.*, 55 (1918) 241.

‡ See suppl. note 17.

C. Neumann[361] and Seeliger[362] had already pointed out that Newton's law of gravitation can only then be strictly valid when the mass density of the universe converges to zero more rapidly than $1/r^2$ for $r \to \infty$. Otherwise, the force exerted by all the masses of the universe on a material particle would be undetermined. In a subsequent paper, Seeliger[363] further discusses the possibility that the mass density should remain different from zero at arbitrary distances, but that the Newtonian potential should be replaced by the potential

$$\Phi = A \frac{\exp(-r\sqrt{\lambda})}{r},$$

which decreases more rapidly with the distance. This potential had already been investigated mathematically by C. Neumann[364] in a different context. It amounts to replacing Poisson's equation

$$\Delta\Phi = 4\pi k\mu_0 \qquad\qquad (A)$$

by

$$\Delta\Phi - \lambda\Phi = 4\pi k\mu_0. \qquad\qquad (B)$$

The difficulty which arises in Newtonian theory then disappears.

According to Einstein, weighty arguments can be brought forward against the first possibility (strict validity of Newton's law and sufficiently rapid decrease of the mass density at infinity), by taking the view that the whole system of fixed stars must be in statistical equilibrium. If the potential were finite at great distances (and thus the mass density decreasing sufficiently strongly) it could happen that entire stars would leave the system. The system would thus become "depopulated" and this process would go on, according to the laws of statistical mechanics, for as long as the total energy of the stellar system is greater than the work required to remove a single star to infinity. Another alternative which had already been excluded by Neumann's and Seeliger's investigations was the case of an infinitely high potential value at very large distances (i.e. the mass density reaching to infinity or not decreasing sufficiently rapidly); according to Einstein this alternative is inadmissible because it would be in contradiction with the fact that the observed stellar velocities are comparatively small. If (B) is assumed valid, however, all these difficulties are removed. For in that case a uniform distribution of matter of density μ_0 and with the potential (constant in space)

$$\Phi = -\frac{4\pi k}{\lambda}\mu_0$$

is dynamically possible.

The situation in relativity is quite analogous to that in Newtonian theory. If Eqs. (401) are retained, it proves impossible, according to

[361] C. Neumann, *Abh. sächs. Ges. Wiss.*, **26** (1874) 97.

[362] H. v. Seeliger, *Astr. Nachr.*, **137** (1895) 129.

[363] H. v. Seeliger, *S.B. bayer. Akad. Wiss.*, **26** (1896) 373.

[364] C. Neumann, *Allgemeine Untersuchungen über das Newtonsche Prinzip der Fernwirkungen*, (Leipzig 1896), cf. in particular p. 1.

Einstein, to set up the boundary conditions in such a way as to counter the "depopulation argument" and to take the smallness of the stellar velocities into account as well. But the field equations can be modified in a way which is quite analogous to the transition from (A) to (B). In fact, when the field equations were set up in § 56, we had simply omitted the term $c_3 g_{ik}$, proportional to g_{ik}, in (400). This omission was neither imposed by the general covariance requirement, nor by the energy-momentum law of matter. The term can therefore now be included in the left-hand side of (401), where we write $-\lambda$ instead of c_3, following Einstein's notation, so that

$$G_{ik} - \lambda g_{ik} = -\kappa T_{ik}. \tag{452}$$

By contraction, this results in

$$R + 4\lambda = +\kappa T \tag{453}$$

and

$$R_{ik} + \lambda g_{ik} = -\kappa (T_{ik} - \tfrac{1}{2} g_{ik} T). \tag{452a}$$

It follows immediately on the basis of these modified field equations that a universe filled with constant mass density is in equilibrium. In fact, it is seen to be elliptic or spherical, and therefore bounded in space. In particular, if we start with

$$g_{ik} = \delta_i{}^k + \frac{x^i x^k}{a^2 - [(x^1)^2 + (x^2)^2 + (x^3)^2]}, \qquad g_{i4} = 0, \qquad g_{44} = -1, \tag{454}$$
$$(i, k = 1, 2, 3)$$

then, from § 18 (117), (118), (119) and (130),

$$R_{ik} = -\frac{2}{a^2} g_{ik}, \qquad R = -\frac{6}{a^2}, \qquad G_{ik} = \frac{1}{a^2} g_{ik}, \qquad \text{(for } i, k = 1, 2, 3)$$

$$R_{i4} = R_{44} = 0.$$

Also,

$$T_{44} = \mu_0 c^2, \text{ the remaining } T_{ik} = 0, \quad T = -\mu_0 c^2,$$

so that the Eqs. (452) are satisfied if

$$\lambda = \frac{1}{a^2} = \tfrac{1}{2} \kappa \mu_0 c^2 = \frac{4\pi k \mu_0}{c^2}. \tag{455}$$

On the other hand, the field equations (401) would result in $1/a^2 = \mu_0 = 0$, because $\lambda = 0$.

Since in this example, and presumably also for more general mass distributions, the world is found to be bounded in space, the boundary conditions at infinity are no longer required. *The field equations* (452) *thus not only resolve the disagreement of the small stellar velocities with statistical mechanics, but also remove the above-mentioned epistemological defect which is a feature of the previous version of the theory.* One will have to imagine that the solution (454) of the field equations represents the average behaviour of the world metric. It will only be in the neighbourhood of individual masses that the g_{ik} will deviate markedly from the values (454). For matter systems (such as the planetary system) whose dimensions are small compared with this certainly extremely large radius of curvature, the λ-term can be neglected and the solutions of the field equations (401) will remain valid. The Mach principle, too, seems to be upheld by the field equations (452), although a general proof for this has not yet been put forward. For, whereas the previous equations (401) have a general solution $g_{ik} =$ const.[365]† for matter-free space, this is not so for Eqs. (452), where $g_{ik} = 0$ for this case. This amounts to saying that in *perfectly empty space, no G-field exists at all*; neither the propagation of light nor the existence of measuring rods and clocks would then be possible. Also related to this is the fact that the postulate of the relativity of inertia is satisfied. Admittedly, de Sitter[366] has given a solution of the field equations (452) which is different from $g_{ik} = 0$ even for perfectly *empty* space a four-dimensional pseudo-spherical world

$$g_{ik} = \delta_i{}^k + \frac{x^i x^k}{a^2 - \sum\limits_{i=1}^{4} (x^i)^2}, \qquad (i, k = 1, 2, 3, 4; x^4 = ict) \qquad (456)$$

$$T_{ik} = \mu_0 = 0, \qquad \lambda = \frac{3}{a^2}, \qquad (457)$$

in contrast to Einstein's "cylindrical world", given by (454). But Einstein[367] was of the opinion that de Sitter's solution is not regular everywhere, so that it does not really represent the G-field of an empty world, but rather that of a world with a surface distribution of matter. Weyl[368] came to the same conclusion. However, this point has not yet been finally settled.

The astronomical implications of the field equations (452) are discussed by de Sitter[369] and Lense[370]. (Cf. also Kottler, *Enzykl. math. Wiss.* VI 2, 22.)

(γ) *The energy of the finite universe.* Eq. (452) can be derived from a variational principle, just as equations (401). One need only add the term $2\lambda\sqrt{(-g)}$ to the action function (404),

$$\delta \int \mathfrak{H} \, dx = 0 \qquad (458)$$

$$\mathfrak{H} = \mathfrak{R} + 2\lambda\sqrt{(-g)} + \kappa\mathfrak{M}. \qquad (459)$$

[365] That this is the only solution of the previous equations (401) for a space completely free of matter, has not yet been proved for the general case. For the *statical* case, however, a proof of this theorem was given by R. Serini, *R.C. Accad. Lincei* (5) **27** (1918) 1st part, p. 235.

† See suppl. note 18.

[366] W. de Sitter, *Proc. Acad. Sci. Amst.*, **19** (1917) 1217, and **20** (1917) 229.

[367] A. Einstein, *S.B. preuss. Akad. Wiss.*, (1918) 270; de Sitter's reply in *Proc. Acad. Sci. Amst.*, **20** (1918) 1309.

[368] H. Weyl, *Phys. Z.*, **20** (1919) 31; cf. also footnote 354, p. 177, F. Klein, *ibid.*, in particular § 9. The geometrical aspects of the question are discussed in detail in this paper.

[369] W. de Sitter, *Mon. Not. R. Astr. Soc.*, **78** (1917) 3.

[370] J. Lense, *S.B. Akad. Wiss. Wien.*, **126** (1917) 1037.

Moreover, the energy conservation law holds again in the form (406),

$$\frac{\partial(\mathfrak{T}_i{}^k + \tilde{\mathfrak{t}}_i{}^k)}{\partial x^k} = 0, \tag{460}$$

if one subtracts $(\lambda/\kappa)\delta_i{}^k$ from the previous quantities $t_i{}^k$,

$$\tilde{t}_i{}^k = t_i{}^k - \frac{\lambda}{\kappa}\delta_i{}^k. \tag{461}$$

These ideas are alluded to in Einstein's paper (quoted in footnote 357) and developed in greater detail by Klein.[371]

The question now arises, whether the theorem of the independence of the *total* value of the energy of the coordinate system (proved in § 61) also holds for the total energy of the finite universe. This has to be considered afresh, because the theorem had previously been proved on the assumption that the g_{ik} became equal to $\pm\delta_i{}^k$ outside the closed system under consideration. Evidently this is not now the case. Einstein[372] and Klein[372] concerned themselves with this question. It has to be proved that certain surface integrals vanish. It was already shown by Einstein that this is indeed the case for special coordinate systems, and Grommer[373] gives a general proof for it. It is further seen that both the total momentum and total energy of the finite universe vanish, in as far as they originate from the gravitational field, i.e.

$$\int \tilde{t}_i{}^4 \, dx^1 \, dx^2 \, dx^3 = 0. \tag{462}$$

If, however, in place of the Einstein energy components the Lorentz components (referred to in § 61) are used, which also contain second derivatives of the g_{ik}, the total energy of the gravitational field would no longer be equal to zero. This was shown by Klein.

[371] Cf. footnote 104, p. 68, *ibid.*, (1918) 235, appendix.
[372] Cf. footnote 354. p. 177, A. Einstein, *ibid.*, and F. Klein, *ibid.*
[373] J. Grommer, *S.B. preuss. Akad. Wiss.* (1919) 860.

PART V. THEORIES ON THE NATURE OF CHARGED ELEMENTARY PARTICLES

63. The electron and the special theory of relativity

For a long time endeavours have been made to derive all the mechanical properties of the electron from electromagnetic principles. In such a case the equation of motion

$$\frac{d\mathbf{G}}{dt} = \mathbf{K} \tag{463}$$

(where $\mathbf{G}$ = momentum of the electron, $\mathbf{K}$ = *external* force) is interpreted as follows[374]: It is postulated, first, that all forces acting on an electron are electromagnetic in origin, i.e. they are given by the Lorentz expression (212 a); and secondly, that the total force acting on an electron should always vanish,

$$\int \rho \left\{ \mathbf{E} + \frac{1}{c}(\mathbf{v} \wedge \mathbf{H}) \right\} dV = 0, \tag{464}$$

where the integration is to be carried out over the volume of the electron. This total force can be split into two parts. One is a force due to the external field

$$\int \rho \left\{ \mathbf{E}^{\text{ext}} + \frac{1}{c}(\mathbf{v} \wedge \mathbf{H}^{\text{ext}}) \right\} dV = \mathbf{K}$$

which is on the right-hand side of Eq. (463). The other is the force exerted by the electron on itself, which can be put equal to

$$\int \rho \left\{ \mathbf{E}^{\text{int}} + \frac{1}{c}(\mathbf{v} \wedge \mathbf{H}^{\text{int}}) \right\} dV = -\frac{d\mathbf{G}}{dt}.$$

In this expression, $\mathbf{G}$ is the electromagnetic momentum of the self-field of the electron. For accelerations which are not too large (quasi-stationary motion), it can be regarded as the momentum which corresponds to a uniform motion of the electron with the relevant instantaneous velocity. It depends, of course, on the charge distribution inside the electron.

The most natural assumption to make was that of a perfectly rigid electron. The theory for this case was worked out completely by Abraham.[375] In 1904, however, it was shown by H. A. Lorentz[376] that the deductions which can be made from the velocity dependence of the electron momentum will only be in agreement with the principle of relativity when it is assumed that a contraction of the electron takes place in the direction of motion, in the ratio $\sqrt{(1-\beta^2)}:1$. And Einstein[377]

[374] Cf. footnote 4, p. 1, H. A. Lorentz, *ibid.*, § 21.
[375] M. Abraham, *Ann. Phys., Lpz.*, **10** (1903) 105. See also footnote 4, p. 1, H. A. Lorentz, *ibid.*, § 21.
[376] Cf. footnote 10, p. 2, *Versl. gewone Vergad. Akad. Amst.*, *loc. cit.*
[377] Cf. footnote 12, p. 2, *ibid.*, § 10.

showed afterwards that the velocity dependence of the mass, energy and momentum follow from the principle of relativity alone, without the need for any assumptions about the nature of the electron (cf. § 29). Conversely, therefore, no information on the nature of the electron can be obtained from observations on the changes in mass.

It is however easy to see that the principle of relativity must necessarily lead to the existence of an energy of a non-electromagnetic kind for the electron, at least so long as the Maxwell–Lorentz theory is retained. This was first pointed out by Abraham.[378] Let us first assume that the charge distribution in a stationary electron is spherically symmetrical. We then obtain the energy and momentum of a moving electron from (351), in as far as they are electromagnetic in origin and given by the Maxwell–Lorentz expressions,

$$\mathbf{G} = \mathbf{u}\,\frac{\frac{4}{3}(E_0/c^2)}{\sqrt{(1-\beta^2)}}, \qquad E = \frac{E_0(1 + \frac{1}{3}u^2/c^2)}{\sqrt{(1-\beta^2)}}. \tag{351}$$

If these expressions represented the total energy and momentum as well, we should have, from (317), (318),

$$E = \int \left(\mathbf{u}\cdot\frac{d\mathbf{G}}{dt}\right) dt = \int \left(\mathbf{u}\cdot\frac{d\mathbf{G}}{d\beta}\right) d\beta.$$

This is however not the case. Instead, the integral on the right-hand side has the value

$$\frac{\frac{4}{3}E_0}{\sqrt{(1-\beta^2)}} + \text{const.}$$

If, in contrast to the energy, the *momentum* is assumed to be purely electromagnetic, the total energies E_0 and E of the stationary and moving electron, respectively, and its rest mass, become

$$\bar{E} = \frac{\bar{E}_0}{\sqrt{(1-\beta^2)}}, \qquad \bar{E}_0 = \tfrac{4}{3}E_0, \qquad m_0 = \frac{\bar{E}_0}{c^2} = \frac{4}{3}\frac{E_0}{c^2}. \tag{465}$$

The rest mass m_0 is here defined by

$$\mathbf{G} = \frac{m_0\,\mathbf{u}}{\sqrt{(1-\beta^2)}}.$$

These relations are in agreement with the theorem of the inertia of energy, as required (with the additive constant in $\bar{E}$ already fixed so as to fit in with this theorem). *The total energy of the stationary electron is equal to four-thirds of its electromagnetic energy as given by Lorentz.*

It would seem from the above arguments as if the rigid electron of the "absolute" theory were compatible with a purely electromagnetic world picture (or rather, with the particular electromagnetic world picture which is based on the Maxwell–Lorentz theory), in contrast to the electron

[378] M. Abraham, *Phys. Z.*, 5 (1904) 576: *Theorie der Elektrizität*, Vol. 2, (1st edn., Leipzig 1905), p. 205.

as demanded by the theory of relativity. But this is not correct, and for the following reason: The hypothesis of a rigid electron is a concept which is quite foreign to electrodynamics. If it had not been introduced, we would have had to postulate that not only the *total* force acting on the electron (Eq. (464)) should vanish, but even the force acting at each individual point,

$$\rho \left\{ \mathbf{E} + \frac{1}{c}(\mathbf{v} \wedge \mathbf{H}) \right\} = 0.$$

It is clear that a stationary charge ($\mathbf{v} = 0$) is incompatible with this postulate, for it follows that $\rho = 0$ (note that div $\mathbf{E} = \rho$). We therefore see that *the Maxwell–Lorentz electrodynamics is quite incompatible with the existence of charges, unless it is supplemented by extraneous theoretical concepts.* From a purely electromagnetic point of view, therefore, the electron of the "absolute" theory is really no better then the electron in relativity. It will, in any case, be necessary to introduce forces which hold the Coulomb repulsive forces of the electron charge on itself in equilibrium, and such forces are not derivable from Maxwell–Lorentz electrodynamics. Poincaré[379] already recognized the need for this and purely formally introduced a scalar cohesive pressure p, on whose nature he could not make any statement. Generally speaking, the problem of the electron has to be formulated as follows: The energy–momentum tensor S_{ik} of Maxwell–Lorentz electrodynamics has to have terms added to it in such a way that the conservation laws

$$\frac{\partial T_i{}^k}{\partial x^k} = 0 \qquad\qquad (341)$$

for the total energy–momentum tensor become compatible with the existence of charges. These additional terms will certainly have to depend on physical quantities which are causally determined by differential equations. (In § 42 we had made the phenomenological "Ansatz" $\mu_0 u_i u_k$ for the energy tensor of an isolated electron.) How far this formulation will have to be modified from the point of view of the *general* theory of relativity, will be discussed in §§ 65 and 66.

We are now in a position to answer the question raised by Ehrenfest[380], whether a uniform translational motion in all directions, in the absence of forces, is possible for an electron which is not spherically symmetric, even when at rest. For in such a case the electromagnetic momentum of the moving electron will not always be in the direction of its velocity, so that the electromagnetic forces will exert a rotational couple on the electron. It was stressed by Laue[381] that the situation is quite analogous to that in Trouton and Noble's experiment. In this experiment, the electromagnetic couple is compensated by a couple generated by the elastic

[379] Cf. footnote 11, p. 2, *R.C. Circ. mat. Palermo., loc. cit.*
[380] P. Ehrenfest, *Ann. Phys., Lpz.,* **23** (1907) 204; remarks on this by A. Einstein, *Ann. Phys., Lpz.,* **23** (1907) 206.
[381] Cf. footnote 240, p. 129, *ibid.*

energy current. In our case, the compensatory effect is similarly due to an energy current which is represented by the above-mentioned additional terms in the energy–momentum tensor. The introduction of such additional terms becomes necessary not just for the case of a moving electron, but even for that of a stationary one. Ehrenfest's question must therefore be answered in the affirmative.

We still have to discuss what can be said about the *dimensions* of the electron, both from the above theoretical and from the experimental points of view. *Experimentally* it is known with a high degree of probability that, in the last analysis, all matter consists of hydrogen nuclei and electrons. *Naturally, all our previous statements about the electron hold equally for the hydrogen nucleus.* Experiment has only told us about the dimensions of these particles that they are certainly not larger than 10^{-13} cm. In other words, two such particles will behave practically like point charges over this distance, as far as the forces which they exert on each other are concerned. The possibility that the dimensions of the particles could be very much smaller than 10^{-13} cm is not excluded by the experiments hitherto carried out. *Theoretically*, definite statements can only be made from Lorentz's point of view, as follows: A sphere of radius a and continuous surface charge distribution has an energy

$$E = \frac{e^2}{8\pi a},$$

where e is the total charge, measured in Heaviside units. It follows from (465) that

$$m_0 = \frac{e^2}{6\pi a c^2}, \qquad a = \frac{e^2}{6\pi m_0 c^2}. \tag{466}$$

If a different charge distribution were assumed, this would only modify the numerical coefficient, but not the order of magnitude of a. The value for a is obtained from the known rest masses of the electron and hydrogen nucleus. For the former it is of order 10^{-13} cm, for the latter about 1800 times smaller, corresponding to its greater mass. But it must be remarked that these considerations rest on very weak theoretical foundations. For we have seen that they are based on the following hypotheses:

(i) The charge distribution of a stationary electron (hydrogen nucleus) is spherically symmetric.

(ii) The total momentum of a moving electron (hydrogen nucleus) is given by the expression

$$\mathbf{G} = \frac{1}{c} \int \mathbf{E} \wedge \mathbf{H} \, dV$$

of the Maxwell–Lorentz theory; this is therefore assumed to be valid also for extreme concentrations of the charges and fields.

The second hypothesis, in particular, appears doubtful. From the experimental data available so far, no experimental support whatever can be found for the dimensions calculated in this way, in particular not for the theoretical requirement that the radius of the hydrogen nucleus should be very much smaller than that of the electron.[381a]

64. Mie's theory

The first attempt to set up a theory which could account for the existence of electrically charged elementary particles, was made by Mie[382]. He set himself the task to generalize the field equations and the energy–momentum tensor in the Maxwell–Lorentz theory in such a way that the Coulomb repulsive forces in the interior of the electrical elementary particles are held in equilibrium by other, *equally electrical*, forces, whereas the deviations from ordinary electrodynamics remain undetectable in regions outside the particles.

Mie retains the first set of Maxwell's equations

$$\frac{\partial F_{ik}}{\partial x^l} + \frac{\partial F_{li}}{\partial x^k} + \frac{\partial F_{kl}}{\partial x^i} = 0, \tag{203}$$

from which the existence of a four-vector potential follows,

$$F_{ik} = \frac{\partial \phi_k}{\partial x^i} - \frac{\partial \phi_i}{\partial x^k} \tag{206}$$

(cf. § 28). Furthermore, the four-current density will certainly have to satisfy the continuity equation

$$\frac{\partial s^k}{\partial x^k} = 0. \tag{197}$$

From this follows the existence of an antisymmetrical tensor of rank 2, $H^{ik} = -H^{ki}$, which satisfies the equation

$$s^i = \frac{\partial H^{ik}}{\partial x^k}. \tag{467}$$

Here, H_{ik} combines the two vectors **D** and **H**, just as F_{ik} combines the vectors **E** and **B**. It is thus seen that, for $H^{ik} = F^{ik}$, the equations reduce to those of ordinary electrodynamics and that they agree formally with those of the phenomenological electrodynamics of ponderable bodies.

Now however these field equations acquire a new physical content by virtue of the following, decisive, assumption: *The vectors H^{ik} and s^k are to be universal functions of F_{ik} and ϕ_i,*

$$H^{ik} = u_{ik}(\mathbf{F}, \mathbf{\varphi}), \qquad s^k = v_k(\mathbf{F}, \mathbf{\varphi}). \tag{467a}$$

[381a] On this point, we cannot agree with the view presented in M. Born's book, *Die Relativitätstheorie Einsteins* (Berlin 1920), p. 192.

[382] G. Mie, 'Grundlagen einer Theorie der Materie', *Ann. Phys., Lpz.*, 37 (1912) 511; 39 (1912) 1; 40 (1913) 1. Cf. also the version by M. Born, *Nachr. Ges. Wiss. Göttingen*, (1914) 23, which brings out the analogy between the derivation of the energy–momentum law from the variational principle of Mie's theory, and the derivation of the energy conservation law from Hamilton's principle, in ordinary mechanics. See also H. Weyl, *Raum-Zeit-Materie* (1st edn., 1918) § 25, p. 165; (3rd edn., 1920) § 25, p. 175.

The first six relations differ inherently from those of phenomenological electrodynamics, in that H^{ik} also depends on ϕ_i explicitly. In Mie's theory, not only the potential difference, but also the absolute value of the potential acquires a real meaning. The equations do not remain unchanged if one replaces φ by $\varphi + const$. We shall see later that this feature of Mie's theory leads to a serious difficulty. The last four equations (467 a) are essential for the existence and for the equations of motion of the material particles (electron and hydrogen nucleus). Mie calls, more or less arbitrarily, ϕ_i and F_{ik} "quantities of intensity", s^k and H^{ik} "quantities of magnitude".

By means of (467a), no less than ten universal functions are introduced into the theory. As Mie found, however, the energy principle brings about a great simplification, reducing the ten unknown universal functions to a single one. For it is seen that an equation of the form

$$\frac{\partial W}{\partial t} + \operatorname{div} \mathbf{S} = 0$$

(W = energy density, $\mathbf{S}$ = energy current)

can only be derived from (206) and (457), when an invariant $L(\mathbf{F}, \varphi)$ (in the first place, relative to the Lorentz group) exists, from which H^{ik} and s^i can be obtained by differentiation,

$$H^{ik} = \frac{\partial L}{\partial F_{ik}}, \qquad s^i = -\frac{1}{2}\frac{\partial L}{\partial \phi_i}, \tag{468}$$

so that

$$\delta L = H^{ik}\,\delta F_{ik} - 2s^i\,\delta\phi_i. \tag{468a}$$

A simple calculation then shows that equations (467) follow from the action principle

$$\delta \int L\, d\Sigma = 0, \tag{469}$$

provided the variation fulfils the condition that the relations (206) also remain valid for the field which is to be varied.

One can make a number of general statements about the invariant L, which is often called the world function. First of all, the only independent invariants which can be formed from the antisymmetric tensor F_{ik} and the vector ϕ_i, are the following:

(1) The square of the tensor $F_{ik} : \frac{1}{2}F_{ik}F^{ik}$.
(2) The square of the vector $\phi_i : \phi_i\phi^i$.
(3) The square of the vector $F_{ik}\phi^k : F_{ir}\phi_s F^{is}\phi^r$.
(4) The square of the vector $F^*_{ik}\phi^k$ or, which amounts to the same, the square of the tensor $F_{ik}\phi_l + F_{li}\phi_k + F_{kl}\phi_i$.

L must therefore be a function of these four invariants.† If L becomes equal to the first of the above invariants, the field equations of Mie's theory

† See suppl. note 20.

degenerate into the ordinary equations of electron theory for charge-free space. *Thus, L can only be markedly different from $\frac{1}{2}F_{ik}F^{ik}$ in the interior of the material particles.* Nothing further can be said about the world function L. It is impossible to narrow down the number of alternatives to such an extent that one would be led unambiguously to a quite definite world function. Rather, there remains an infinite manifold of possibilities.

We next have to determine the energy–momentum tensor T_{ik} as a function of the field quantities. Hilbert[383] and Weyl[384] showed that the corresponding calculations can be simplified considerably by writing the equations in the notation of the general theory of relativity and then using the method for varying the g_{ik} which was introduced in § 55. Only then do the formal connections become clearly apparent. We prepared the ground for this here by our notation for the above formulae and we differ in this respect from Mie, who of course based his papers of 1912 and 1913 on the special theory of relativity. First of all, just as for the case of ordinary electrodynamics dealt with in § 54, the set of equations (203), (208) remains valid also for an arbitrary G-field. On the other hand, (197) and (467) have to be replaced by

$$\frac{\partial \mathfrak{s}^i}{\partial x^i} = 0 \qquad\qquad (197\,\mathrm{a})$$

$$\mathfrak{s}^i = \frac{\partial \mathfrak{H}^{ik}}{\partial x^k}. \qquad\qquad (467\,\mathrm{b})$$

As was noticed by Weyl, the "quantities of magnitude" appear here as tensor *densities* (i.e. multiplied by $\sqrt{(-g)}$), whereas the "quantities of intensity" remain ordinary tensors. Equally, the relations (468), (468a) and Hamilton's principle (469) remain valid, and the latter can of course also be written in the form

$$\delta \int \mathfrak{L} \, dx = 0. \qquad\qquad (469\,\mathrm{a})$$

In order to find the energy tensor T_{ik} we need only determine the variation of the action function which is obtained by varying the G-field. Since, in this case, L is independent of the derivatives of the g_{ik}, the following relation holds if the electromagnetic field is kept constant,

$$\delta \mathfrak{L} = \mathfrak{T}_{ik}\,\delta g^{ik},$$

so that

$$T_{ik} = \frac{\partial L}{\partial g^{ik}} - \tfrac{1}{2}L g_{ik}, \qquad T_i{}^k = \frac{\partial L}{\partial g^{ir}}g^{rk} - \tfrac{1}{2}L\delta_i{}^k. \qquad (470)$$

If, on the other hand, we substitute into the general expression

$$\delta L = \frac{\partial L}{\partial g^{ik}}\delta g^{ik} + H^{ik}\,\delta F_{ik} - 2s^i\,\delta\phi_i,$$

[383] Cf. footnote 99, p. 63, *ibid.*, I.
[384] H. Weyl, *Raum-Zeit-Materie* (1st edn., 1918), p. 184 *et seq.*; (3rd edn., 1920), p. 199.

resulting from (468 a), a variation of the field quantities produced by an infinitesimal coordinate transformation (as given by (163), (164)), δL must vanish identically. In other words,

$$2\frac{\partial \xi^i}{\partial x^k}\left(\frac{\partial L}{\partial g^{ir}}g^{rk} - H^{kr}F_{ir} + s^k\phi_i\right) \equiv 0,$$

which is only possible if the bracket itself vanishes identically. Finally, we can take the value of $(\partial L/\partial g_{ir})g^{rk}$ from there and substitute it in (470). We then obtain

$$T_i{}^k = H^{kr}F_{ir} - s^k\phi_i - \tfrac{1}{2}L\,\delta_i{}^k. \tag{470a}$$

It is seen from their derivation that the corresponding covariant components T_{ik} are symmetrical. On the basis of the results of § 55 we can see, moreover, that the energy–momentum law is a consequence of the field equations. In the absence of gravitational fields it is of the form

$$\frac{\partial T_i{}^k}{\partial x^k} = 0, \tag{341}$$

and in the presence of gravitational fields,

$$\frac{\partial \mathfrak{T}_i{}^k}{\partial x^k} - \tfrac{1}{2}\mathfrak{T}^{rs}\frac{\partial g_{rs}}{\partial x^i} = 0. \tag{341a}$$

The expression (470 a) for the energy–momentum tensor is identical with that obtained by Mie by direct calculation.

Let us now consider once more the problem of the equation of motion, and whether it is possible for material particles to exist. In ordinary electrodynamics the electric field strength is defined as the force acting on a (stationary) charge. This simple meaning of the field strength no longer applies to the interior of a material particle in Mie's theory. Indeed, the ponderomotive force is everywhere equal to zero. Nevertheless, the practical meaning of the total charge of a particle still remains. For, let us consider a charged particle in an external field. In this case, it follows from (341) that

$$\frac{d}{dx^4}\int T_i{}^4\,dx^1\,dx^2\,dx^3 = -\int T_i{}^k n_k\,d\sigma. \qquad (i, k = 1, 2, 3)$$

$$(n_k = \text{unit normal to the surface})$$

The second integral is taken over a surface sufficiently far removed from the material particle. Since ordinary electrodynamics is valid on this surface, the surface integral has the same value as there, i.e. it represents the Lorentz force. We have thus given an electrodynamical proof of the equation of motion (210) of an electron, within the framework of Mie's theory. At the same time, it is seen that the rest mass of the particle is determined by

$$m_0 = \frac{E_0}{c^2} = -\int T_4{}^4\,dx^1\,dx^2\,dx^3, \tag{471}$$

according to the theorem of the inertia of energy. For $T_4{}^4$ we have to substitute the expression obtained from (470).

Mie assumes the field of the stationary electron to be static and spherically symmetric. As was explained in the previous section, the latter assumption is admittedly not justified by our experimental knowledge alone, but recommends itself for its simplicity. We will then have to look for those solutions of the field equations which are regular everywhere—for $r = 0$ as well as for $r = \infty$. A world function which is to correspond to reality will have to lead to one, and one only, solution for every kind of electricity. *It has not been possible, so far, to find a world function which satisfies this condition.* On the contrary, the trial expressions for L which have so far been discussed, lead to the conclusion, in contradiction with experiment, that elementary particles with arbitrary values of their total charge can exist. This in itself is not sufficient reason for abandoning Mie's electrodynamics, since it has not been proved that a world function *cannot* exist which is compatible with the existence of *certain* elementary particles.

A much more serious difficulty, it seems to us, is caused by a fact already noticed by Mie. Once we have found a solution for the electrostatic potential φ of a material particle of the required kind, $\varphi + const.$ will not be another solution, because the field equations of Mie's theory contain the absolute value of the potential. *A material particle will therefore not be able to exist in a constant external potential field.* This, to us, seems to constitute a very weighty argument against Mie's theory. In the theories which we are going to discuss in the following sections, this kind of difficulty does not arise.

An attempt by Weyl should be mentioned here, in which he tries to make the asymmetry between the two kinds of electricity understandable from the point of view of Mie's theory.[†] If the world function L is not a rational function of $\sqrt{(-\phi_i \phi^i)}$, we can put

and
$$L = \tfrac{1}{2}F_{ik}F^{ik} + w(+ \sqrt{(-\phi_i \phi^i)}) \quad \text{for } \phi_i \phi^i < 0, \phi_4 > 0$$
$$L = \tfrac{1}{2}F_{ik}F^{ik} + w(- \sqrt{(-\phi_i \phi^i)}) \quad \text{for } \phi_i \phi^i < 0, \phi_4 < 0$$

where w denotes any function which is not even. For the statical case the field equations will not remain invariant for an interchange of φ with $-\varphi$ (positive and negative electricity). Quite generally, if L is a many-valued function of the four above-mentioned fundamental invariants, it is possible to choose one of its branches as world function for positive electricity, and another for negative electricity. We shall come back to this possibility in § 67.

65. Weyl's theory[‡]

In a series of papers[385] Weyl has developed an extremely profound theory, which is based on a generalization of Riemannian geometry and

† See suppl. note 21.

‡ See suppl. note 22.

[385] H. Weyl, *S.B. preuss. Akad. Wiss.* (1918) 465; *Math. Z.* **2** (1918) 384; *Ann. Phys., Lpz.,* **59** (1919) 101; *Raum-Zeit-Materie*, (3rd edn., 1920), Chaps. II and IV, §§ 34 and 35, p. 242 *et seq.*; *Phys. Z.*, **21** (1920) 649.

which claims to interpret all physical events in terms of gravitation and electromagnetism, and these, in turn, in terms of the world metric. The theory's basic structure and results obtained to date will be discussed at *this* juncture, because the theory also makes certain statements on the nature of material particles.

(α) *Pure infinitesimal geometry. Gauge invariance.* The transition from Euclidean to Riemannian geometry was seen in Part II to be effected by no longer assuming that the transference of the *direction* of a vector from a point P to a point P' is independent of the path taken. Weyl takes this a step further by permitting a corresponding path dependence for the transference of length. It will then only be possible to compare lengths measured *at one and the same* world point, but not those at different world points. In other words, only the relations between the g_{ik} can be determined by measurements, but not these quantities themselves. Let us first assume some arbitrary (continuous) absolute values for the g_{ik} and define

$$ds^2 = g_{ik}\, dx^i\, dx^k$$

to be the square of the length of a measuring rod, where dx^i are the co-ordinate differences of its end points. (For brevity, we speak here, and in what follows, of the length of a measuring rod, but the same arguments apply of course to the period of a clock in the case of a time-like line element.) If now the measuring rod is displaced along a definite curve $x^i = x^i(t)$ from the point $P(t)$ to the point $P'(t+dt)$, the square of the length $ds^2 = l$, will then be changed. We shall assume axiomatically that it will always be changed by a definite fraction of l,

$$\frac{dl}{dt} = -\, l\, \frac{d\varphi}{dt} \qquad (472)$$

where φ is a certain function of t and is independent of l. We introduce as a second axiom that $d\varphi/dt$ only depends on the first derivatives of the coordinates, dx^i/dt. Since, moreover, equation (472) has to hold for an arbitrary choice of the parameter t, $d\varphi/dt$ *must be a homogeneous function of first degree of the dx^i/dt*. This function can be specified still more by making use of the concept of a parallel displacement, which had been discussed in § 14. This concept had been defined there by two conditions: The first stated that the components of a vector would remain unchanged for an *infinitesimal* parallel displacement *in a suitably chosen coordinate system*, whereas the second expressed the fact that the length of a vector would remain unaltered during a parallel displacement. The first assumption can be retained unchanged; it leads to the expression (64) for the change in the vector components,

$$\frac{d\xi^i}{dt} = -\, \Gamma^i{}_{rs}\frac{dx^s}{dt}\xi^r \qquad (64)$$

with

$$\Gamma^i{}_{rs} = \Gamma^i{}_{sr}. \qquad (65)$$

The second assumption, however, loses its meaning here, because the lengths of two vectors at different points can no longer be compared. Instead, the assumption has to be replaced by the condition that the length should change according to (472),

$$\frac{d}{dt}(g_{ik}\xi^i\xi^k) = \frac{d}{dt}(\xi_i\xi^i) = -g_{ik}\xi^i\xi^k\frac{d\varphi}{dt}, \qquad (473)$$

for a parallel displacement. Substituting (64), it first of all follows that $d\varphi/dt$ must be a linear form of the dx^i/dt,

$$d\varphi = \phi_i\,dx^i. \qquad (474)$$

Only in this case, therefore, is a parallel displacement possible. Furthermore, we have from

$$\Gamma_{i,rs} = g_{ik}\Gamma^k{}_{rs}, \qquad \Gamma^i{}_{rs} = g^{ik}\Gamma_{k,rs}, \qquad (66)$$

that

$$\frac{\partial g_{ir}}{\partial x^s} + g_{ir}\phi_s = \Gamma_{i,rs} + \Gamma_{r,is}. \qquad (475)$$

The geodesic components of Weyl's geometry are thus different from those of Riemannian geometry. We shall always denote by an asterisk expressions in Riemannian geometry. They reduce to those of Weyl's geometry for the case $\phi_i = 0$. If therefore the quantities (69) are denoted by $\Gamma^*{}_{i,rs}$, then

$$\Gamma_{i,rs} = \Gamma^*{}_{i,rs} + \tfrac{1}{2}(g_{ir}\phi_s + g_{is}\phi_r - g_{rs}\phi_i). \qquad (476)$$

We had fixed the absolute values of the g_{ik} quite arbitrarily. Instead of the set of values g_{ik} we could equally well have used the set λg_{ik}, where λ is an arbitrary function of position. All elements of length would in that case have to be multiplied by λ and, because of (472), we would have found the set of values

$$\phi_i - \frac{\partial \log\lambda}{\partial x^i} = \phi_i - \frac{1}{\lambda}\frac{\partial\lambda}{\partial x^i},$$

instead of ϕ_i. Fixing the factor λ, the gauge, in Weyl's geometry is thus on a par with the choice of coordinates in Riemannian geometry. *Just as there we had postulated the invariance of all geometrical relations and physical laws under arbitrary coordinate transformations, so we have now to postulate, in addition, their invariance with respect to the substitutions*

$$\bar{g}_{ik} = \lambda_{ik}, \qquad \bar{\phi}_i = \phi_i - \frac{1}{\lambda}\frac{\partial\lambda}{\partial x^i}, \qquad (477)$$

i.e. with respect to changes in the gauge (gauge invariance).

  (β) *Electromagnetic field and world metric.* From (472) it follows by integration that

$$\log l \Big|_P^{P'} = - \int_P^{P'} \phi_i \, dx^i,$$

(478)

$$l_{P'} = l_P \exp \left\{ - \int_P^{P'} \phi_i \, dx^i \right\}.$$

When the linear form $\phi_i \, dx^i$ is a total differential, the length of a vector becomes independent of the path along which it is transferred, and we revert to the Riemannian case. The necessary and sufficient condition for this is the vanishing of the expressions

$$F_{ik} = \frac{\partial \phi_k}{\partial x^i} - \frac{\partial \phi_i}{\partial x^k}.$$

(479)

We see from (477) that, in this case, the vector ϕ_i can always be made to vanish by a suitable choice of the gauge. In the general case, however, the quantities F_{ik} will be different from zero. They then form the covariant components of an antisymmetric tensor of rank 2, which moreover remain unchanged for changes in the gauge, because of (477). In addition, they satisfy the equations

$$\frac{\partial F_{ik}}{\partial x^l} + \frac{\partial F_{li}}{\partial x^k} + \frac{\partial F_{kl}}{\partial x^i} = 0,$$

(480)

which follow from (479). We see that relations (479) and (480) are of exactly the same form as equations (206) and (203) of electron theory. But the analogy can be carried still further. If we take the view (contrary to the assumptions of Mie's theory) that electromagnetic phenomena are primarily determined by the space and time changes of the field strengths alone, and that the potentials, on the other hand, are only looked upon as mathematical auxiliary quantities, then all potential values ϕ_i which lead to the same field strengths are physically completely equivalent; thus a gradient $\partial \psi / \partial x^i$ remains undetermined in the ϕ_i. But as we have seen, exactly the same applies to the metric vector ϕ_i. Following Weyl, we are thus led to identify both sets of quantities ϕ_i, F_{ik}: *The metric vector ϕ_i which determines the behaviour of lengths according to (478), is to be identical with the electromagnetic four-vector potential (apart from a numerical factor).* Just as in Einstein's theory the gravitational effects are closely linked with the behaviour of measuring rods and clocks, such that they follow from it unambiguously, so the same holds in Weyl's theory for electromagnetic effects. In this sense, both gravitation and electricity appear as a consequence of the world metric in this theory.

This point of view had to modified subsequently by Weyl. For, the basic assumptions of the theory in its original form lead to deductions which seem to be in contradiction with experiment. This was stressed by Einstein.[386] Let us think of an electrostatic field connected with a static

[386] A. Einstein, *S.B. preuss. Akad. Wiss.* (1918) 478, including Weyl's reply.

G-field. The spatial components ϕ_i $(i = 1, 2, 3)$ then vanish, and the time component $\phi_4 = \varphi$ as well as the g_{ik} are time independent. The gauge is thus fixed to within a *constant* factor. If we apply relation (478) to the period of a stationary clock, it follows directly that

$$\tau = \tau_0\, e^{\alpha \varphi t}, \tag{481}$$

where α is a factor of proportionality. The meaning of this equation is the following: Let two identical clocks C_1, C_2, going at the same rate, be placed at first at the point P_1, at an electrostatic potential φ_1. Let clock C_2 then be taken to a point P_2, at potential φ_2, for t secs. and finally returned to P_1. The result will be that the rate of clock C_2, compared with that of clock C_1, will be increased or decreased, respectively, by a factor $\exp[-\alpha(\varphi_2 - \varphi_1)t]$ (depending on the sign of α and of $(\varphi_2 - \varphi_1)$). In particular, this effect should be noticeable in the spectral lines of a given substance, and spectral lines of definite frequencies could not exist at all. For, however small α is chosen, the differences would increase indefinitely in the course of time, according to (481). Weyl's present attitude to this problem is the following: *The ideal process of the congruent transference of world lengths, as determined by (472), has nothing to do with the real behaviour of measuring rods and clocks; the metric field must not be defined by means of information taken from these measuring instruments.* In this case the quantities g_{ik} and ϕ_i are, by definition, no longer observable, in contrast to the line element ds^2 of Einstein's theory. This relinquishment seems to have very serious consequences. While there now no longer exists a direct contradiction with experiment, the theory appears nevertheless to have been robbed of its inherent convincing power, from a physical point of view.[387] For instance, the connexion between electromagnetism and world metric is not now essentially physical, but purely formal. For there is no longer an immediate connection between the electromagnetic phenomena and the behaviour of measuring rods and clocks. There is only an interrelation between the former and the ideal process which is mathematically defined as congruent transference of vectors. Besides, there exists only formal, and not physical, evidence for a connection between world metric and electricity. This is quite in contrast to the connexion between world metric and gravitation, for which strong empirical support can be found in the equality of gravitational and inertial mass, and which is a rigorous consequence of the principle of equivalence and of special relativity.

(γ) *The tensor calculus in Weyl's geometry.* Before writing down the field equations, we still have to describe briefly the formal rules for the setting up of gauge-invariant equations. It is clear that in Weyl's theory the concept of a tensor has to be modified in such a way that a system of equations which expresses the vanishing of all the components of a tensor must not only remain invariant for an arbitrary change in the coordinates, but also for an arbitrary change in the gauge (477). In fact, it is found convenient to denote only those quantities as tensors, which are merely

---

[387] A. Einstein believes that, even in this version, the theory will not stand up to a comparison with reality (*Phys. Z.*, **21** (1920) 651; see also footnote 18, p. 4, *ibid.*).

multiplied by a power of λ, λ^e, for a transformation (477); e is called the weight of the tensor. Thus, g_{ik} is of weight $+1$, g^{ik} of weight -1, $\sqrt{(-g)}$ in a four-dimensional world of weight 2, $\Gamma^r{}_{ik}$ is, according to (64) or (476), *absolutely* gauge-invariant, i.e. of weight 0.

All those operations which are solely based on the concept of a parallel displacement can naturally be directly carried over into Weyl's geometry; only, we have to use expressions (66), (476) in place of (66), (69) for $\Gamma^r{}_{ik}$. Thus, the geodesic lines can be defined here, too, by the condition that their tangents should always remain parallel to themselves; they again satisfy equations (80). Eqs. (77 a) $(u_i u^i = \text{const.})$, however, have to be replaced by

$$\frac{d}{d\tau}(u_i u^i) = -(u_i u^i)(\phi_k u^k)$$

because of (472) and (474). If, in particular, $u_i u^i = 0$ at a given point on the geodesic line, this relation will always remain valid. This makes it possible to determine geodesic null lines. The property of geodesic lines, that they are also shortest lines, is dropped in Weyl's geometry, because the concept of a curve length becomes meaningless here. As in § 16 we arrive at the curvature tensor by means of a parallel displacement of a vector along a closed curve,

$$R^h{}_{ijk} = \frac{\partial \Gamma^h{}_{ij}}{\partial x^k} - \frac{\partial \Gamma^h{}_{ik}}{\partial x^j} + \Gamma^h{}_{k\alpha}\Gamma^\alpha{}_{ij} - \Gamma^h{}_{j\alpha}\Gamma^\alpha{}_{ik}. \tag{86}$$

The components in this equation are of weight 0, and the components of R_{hijk} therefore of weight 1. The symmetry relations for this curvature tensor are however different from those for the Riemannian tensor, which are determined by (92). Weyl has discussed this in more detail and has also calculated explicitly the expression (86) for the curvature tensor by substituting (476). As in § 17, we obtain the contracted curvature tensor R_{ik}, (94), whose covariant components are of weight 0, and the invariant R, (95), of weight -1. Finally, all operations of §§ 19 and 20 also remain valid in Weyl's theory, provided, first, the differentiated components of tensors or tensor densities are of weight 0, and secondly, expressions (66) and (476) are used for the quantities $\Gamma^r{}_{ik}$. It will be noticed that it is quite sufficient for the proofs of most of the theorems to realize that the concept of a parallel displacement, according to (64), is determined in an invariant way with the help of the $\Gamma^r{}_{ik}$, without having to know the connection between the metric quantities g_{ik}, ϕ_i. In the latest versions of his theory, Weyl stressed this point very strongly, by developing his geometry in three stages. In the first, he derived those theorems which are valid in arbitrary manifolds; in the second, the relations which are based on the concept of parallel displacement (called "affine connexion" by Weyl); and finally, the conclusions resulting from the existence of the two fundamental metric forms: the quadratic $g_{ik}\, dx^i\, dx^k$ (gravitation) and the linear $\phi_i\, dx^i$ (electricity). The linking of these two ranges of phenomena, which had been separate in the earlier theories, finds expression in the formal fact that the g_{ik} and ϕ_i occur simultaneously in the geodesic components Γ^r_{ik}, and thus also in most of the other gauge-invariant equations.

Of particular importance for the physical applications are the modifications and extentions, in Weyl's theory, of the statements on infinitesimal coordinate transformations and invariant integrals which we made in § 23. First of all, the infinitesimal changes in the gauge appear on an equal footing with the infinitesimal coordinate transformations. From (477) we have, with $\lambda = 1 + \epsilon\pi(x)$,

$$\delta g_{ik} = \epsilon\pi g_{ik}. \quad (\delta g^{ik} = -\epsilon\pi g^{ik}), \quad \delta\phi_i = -\epsilon\frac{\partial\pi}{\partial x^i}. \quad (482)$$

Furthermore, it is evident that, in Weyl's theory, only scalar densities $\mathfrak{W}$ of weight 0 lead to invariant integrals $\int \mathfrak{W}\, dx$. The corresponding scalars are then of weight -2, because of the factor $\sqrt{(-g)}$ in a four-dimensional world. Scalars of this kind will therefore play an important rôle in what follows. Among them, there are four which are rational combinations of the components of the curvature tensor,[388]

$$\tfrac{1}{2}F_{ik}F^{ik}, \quad R_{hijk}R^{hijk}, \quad R_{ik}R^{ik}, \quad R^2. \quad (483)$$

The invariant R which appears in the action principle of Einstein's theory is however of weight -1. It was pointed out by Weyl that, because the scalar densities belonging to (483) have the weight 0, a four-dimensional world has a distinct advantage over a metrical manifold with a different number of dimensions. In fact, in such a manifold no scalar densities can be constructed which are of weight 0 and would have such a simple structure.

(δ) *Field equations and action principle. Physical deductions.* We now have to look for gauge-invariant physical laws. According to Weyl, all processes must be reducible to electromagnetic and gravitational effects. There are thus the fourteen independent quantities of state ϕ_i, g_{ik} available. But since gauge invariance has now been added to the invariance under coordinate transformations, the general solution of the field equations must contain five, instead of four, arbitrary functions; there must therefore also be five identities between the fourteen field equations. We shall see that four of the identities express the energy–momentum law, analogously to those in Einstein's theory, and that the fifth identity expresses the conservation of charge.

We shall first try to retain Maxwell's equations and to identify the material energy tensor with that of Maxwell, and shall only replace the curvature tensor of Riemannian geometry in Einstein's equations by that of Weyl's theory. It will however be seen that only the former "prescription" is possible, and not the latter. Let us first of all investigate the Maxwell theory. The first set of Maxwell's equations are automatically satisfied, as has already been mentioned. But since the field strengths F_{ik} are of weight 0, the same will be true, in a four-dimensional world,

---

[388] It was shown by R. Weitzenböck, *Wiener Ber., math.-nat. Kl.*, IIa, **129** (1920) 683 and 697, that the invariants given here are the only ones of this kind.

of the contravariant components $\mathfrak{F}^{ik}$ of the corresponding tensor density. The equations

$$\frac{\partial \mathfrak{F}^{ik}}{\partial x^k} = \mathfrak{s}^i$$

are therefore gauge-invariant: *Maxwell's equations remain invariant when g_{ik} is replaced by λg_{ik}.* Bateman's theorem, that Maxwell's equations are invariant under conformal transformations (§ 28), is contained in this statement as a special case. Such transformations do indeed change the normal values $\delta_i{}^k$, which the g_{ik} have in special relativity, into $\lambda \delta_i{}^k$. The gauge invariance of Maxwell's equations is connected with the fact that the action integral

$$J = \int \tfrac{1}{4} F_{ik} \mathfrak{F}^{ik} \, dx$$

from which they are produced is itself gauge-invariant. We wish to remark at this point that the reason for the, seemingly accidental, vanishing of the scalar of the Maxwell energy tensor (§ 30, Eq. (223)) is also to be found in the gauge invariance of this action integral. For, a variation of the action integral, keeping the F_{ik} constant, leads to

$$\delta J = \int \mathfrak{S}_{ik} \, \delta g^{ik} \, dx,$$

according to § 55. If we now look for the condition that J should remain unchanged for the infinitesimal change in gauge, $\lambda = 1 + \epsilon \pi(x)$, it follows directly from (482) that $\mathfrak{S}_i{}^i = 0$, as required.[389]

The situation is quite different when we come to deal with Einstein's theory, as opposed to Maxwell's. To start with, the law that the world lines of point-masses and light rays are geodesics is no longer generally valid in Weyl's theory. A point-mass only moves along a geodesic world line in the absence of electromagnetic fields, and for a light ray the equation of the geodesic line loses its meaning for the following reason. Even in the absence of gravitational fields, the terms which contain the four-vector potential ϕ_i will introduce into the equation of the geodesic line oscillatory functions of the light period. Only the gauge-invariant equation

$$g_{ik} \, dx^i \, dx^k = 0$$

of the null cone remains valid for the world lines of the light rays. The

[389] The ϕ_i need not be varied here, since J contains them only in the form of the gauge-invariant F_{ik}. This connexion admits of an interesting application to Nordström's gravitational theory. As was mentioned in § 56, the line element in this theory is of the form

$$ds^2 = \Phi \sum_i (dx^i)^2.$$

It follows first of all from the gauge invariance of Maxwell's equations that they also remain valid in the presence of gravitational fields in Nordström's theory, so that gravitational fields have no influence on electromagnetic processes (e.g. no bending of light rays). Conversely, because of the vanishing of Maxwell's energy scalar, the electromagnetic energy does not generate gravitational fields in Nordström's theory, since the gravitational field equations only contain the energy scalar. It is seen from the above that this circumstance, too, has as its formal basis the gauge invariance of Maxwell's equations.

attempt to make use of the field equations of Einstein's theory in Weyl's theory (by replacing the Riemannian curvature quantities by the more general ones of Weyl) is finally wrecked by the fact that in the equation

$$G_{ik} = - \kappa T_{ik}$$

the left-hand side would be of weight 0, the right-hand side of weight -1; the latter is easily seen by taking Maxwell's energy tensor as an example. This failure is due to the fact that the action integral $\int \Re \, dx$ from which Einstein's field equations are produced is not gauge-invariant, since the integral is of weight 1 instead of 0. If, therefore, we wish to retain the principle of gauge invariance, we have to abandon the Einstein field equations. This last remark, however, already points the way to a method of obtaining gauge-invariant field equations. An action principle

$$\delta \int \mathfrak{W} \, dx = 0 \qquad (484)$$

has to be set up, in which the integral is invariant with respect to changes in the gauge, too. If the variations vanish at the boundary, and if in general for varying the ϕ_i and g_{ik}

$$\delta \int \mathfrak{W} \, dx = \int (\mathfrak{w}^i \, \delta\phi_i + \mathfrak{W}^{ik} \, \delta g_{ik}) \, dx, \qquad (485)$$

then the physical laws are expressed by

$$\mathfrak{w}_i = 0, \qquad \mathfrak{W}^{ik} = 0. \qquad (486)$$

By looking for conditions that the integral $\int \mathfrak{W} \, dx$ should be invariant with respect to infinitesimal coordinate transformations and infinitesimal changes in the gauge, we obtain five identities between these fourteen equations, as had been postulated above on the grounds of causality. These are

$$\frac{\partial \mathfrak{w}^i}{\partial x^i} + \mathfrak{W}_i{}^i \equiv 0 \qquad (487)$$

and

$$\frac{\partial \mathfrak{W}_i{}^k}{\partial x^k} - \Gamma^r{}_{si} \mathfrak{W}_r{}^s + \tfrac{1}{2} F_{ik} \mathfrak{w}^k \equiv 0. \qquad (488)$$

We can, next, consider variations of the action integral which do not vanish at the boundary. This will enable us to construct unambiguously from the action invariant a vector density $\mathfrak{s}^i$ and the density of an *affine* tensor $\mathfrak{S}_i{}^k$, which satisfy the relations

$$\frac{\partial \mathfrak{s}^i}{\partial x^i} \equiv \frac{\partial \mathfrak{w}^i}{\partial x^i} \quad \text{and} \quad \frac{\partial \mathfrak{S}_i{}^k}{\partial x^k} \equiv \frac{\partial \mathfrak{W}_i{}^k}{\partial x^k} \qquad (489)$$

identically, without vanishing themselves by virtue of the physical laws. Weyl therefore calls $\mathfrak{s}^i$ the four-current, and $\mathfrak{S}_i{}^k$ the energy components. We thus see: *In Weyl's theory the charge conservation law is formally on exactly the same footing as the energy conservation law. Both are doubly derived from the physical laws, thus producing the necessary five identities between them.* The components of the total energy, which only form an affine tensor even in Einstein's theory (i.e. which are only covariant under *linear* transformations), can now no longer be split into a part due to gravitation and one due to matter itself; thus, an energy–momentum tensor $T_i{}^k$ of matter does not exist at all in this theory. It has to be admitted that the action principle enables us to recognize these relationships in a particularly simple and transparent manner. But it should be added that it is not at all self-evident, from a physical point of view, that the physical laws should be derivable from an action principle. It would, on the contrary, seem far more natural to derive the physical laws from purely physical requirements, as was done for Einstein's theory in § 56.

In order to be able to make further deductions from the theory, we must now make specific assumptions about the shape of the action function. The number of possibilities is not as great here as in Mie's theory. There, a new invariant could be derived from an *arbitrary* function $f(J_1, J_2, ...)$ of arbitrary invariants $J_1, J_2, ...$. This is no longer the case here, because the invariants have to be of weight -2, in order that the corresponding scalar densities should be of weight 0. A new, admissible, action function can therefore only be produced by a function which is at most *homogeneous of first degree* in these invariants. Even so, the manifold of admissible action functions still remains fairly considerable. The most natural assumption is that the action invariant should be a rational function of the curvature components. According to what was said in (γ), the action function will then have to be a linear combination of the invariants (483).[390] The calculation first of all leads to the *validity of Maxwell's equations*

$$\frac{\partial \mathfrak{F}^{ik}}{\partial x^k} = \mathfrak{s}^i \tag{208a}$$

and also to the expression

$$s_i = k\left(\frac{\partial R}{\partial x^i} + R\phi_i \right) \tag{490}$$

for the four-current (R is the curvature invariant in Weyl's geometry, k a constant). For the statical case we obtain from this that

$$R = \text{const.} \tag{491}$$

If charges are present at all, this constant cannot vanish. If, in addition, it is assumed to be positive, *it follows automatically that the curvature of space is positive and that the universe is finite,* so that it is unnecessary to add a special λ-term to the gravitational equations. This is a particular merit of Weyl's theory. As for the gravitational equations themselves, finally, these are not identical with Einstein's equations, even in the absence of an electromagnetic field ($\phi_i = 0$), as might have been expected from earlier arguments, and they are of higher order than the second.

[390] H. Weyl (Cf. footnote 385, p. 192, *Ann. Phys., Lpz., loc. cit.,* and *Raum-Zeit-Materie, loc. cit.*) thinks it probable that in particular the assumption $W = \frac{1}{2} F_{ik} F^{ik} + c R_{hijk} R^{hijk}$ corresponds to reality.

But it can be shown that for the case of a static, spherically symmetric, field in the space outside a "material particle", the gravitational field (421) of Einstein's theory is at the same time a solution of the gravitational equations of Weyl's theory. This case is, in practice, the only important one and is decisive for the perihelion precession of Mercury and the bending of light rays. *Weyl's theory is thus capable of explaining the perihelion precession of Mercury and the bending of light rays just as well as Einstein's theory.*[391]

We still have to discuss what deductions can be made relating to the problem of the structure of matter. Once more, the problem consists in determining those static, spherically symmetric, solutions of the field equations which are nowhere singular. It has again to be demanded of an action function which is to correspond to reality, that it should admit only of one solution each for the two kinds of electricity. An aspect which is essentially new, compared with Mie's theory, appears in the condition for regularity, not at infinity, but on the "equator" of the universe, because of the finiteness of the universe. We are thus led to conjecture that a connection exists between the size of the universe and that of the electron, which might seem somewhat fantastic. The forces which keep the electron together are, in this theory, only partly electrical in nature, and are partly gravitational forces. Even with the special assumptions for the action function, which were discussed above in detail, the differential equations become so complicated that it has not so far been possible to integrate them. Apart from this, the differential equations are the same for positive and negative electricity (cf. § 67), so that the completely asymmetric conditions which obtain in reality are certainly not represented correctly.† *Summarizing, we can say that Weyl's theory has not succeeded in getting any nearer to solving the problem of the structure of matter.* As will be argued in more detail in § 67, there is, on the contrary, something to be said for the view that a solution of this problem cannot at all be found in this way.

66. Einstein's theory

Einstein[392] tried to approach the inquiry into the structure of material particles from a completely different angle. The field equations (401) and (452), respectively, were based on the assumption of a material energy–momentum tensor $\mathfrak{T}_i{}^k$ which satisfies the equation

$$\frac{\partial \mathfrak{T}_i{}^k}{\partial x^k} - \tfrac{1}{2}\mathfrak{T}^{rs}\frac{\partial g_{rs}}{\partial x^i} = 0. \tag{341a}$$

This assumption will be retained here. Since Maxwell's energy tensor

$$\mathfrak{S}_i{}^k = F_{ir}\mathfrak{F}^{ir} - \tfrac{1}{4}F_{rs}\mathfrak{F}^{rs}\delta_i{}^k \tag{222a}$$

(cf. § 54) only satisfies this condition in charge-free space, some further

[391] Apart from the papers by Weyl, quoted in footnote 385, p. 192, see also W. Pauli, jr., *Verh. dtsch. phys. Ges.*, **21** (1919) 742, where, specifically, the action principle mentioned in footnote 390, p. 201, is used as a basis.

† See suppl. note 21.

[392] A. Einstein, *S.B. preuss. Akad. Wiss.* (1919) 349; also in the collection Lorentz–Einstein–Minkowski, *Das Relativitätsprinzip*, (5th edn., Berlin 1920).

terms have to be added to $\mathfrak{S}_i{}^k$. Mie had assumed that these terms are of an electrical nature, i.e. that they are functions of the electrical quantities F_{ik}, ϕ_i. Einstein, on the contrary, assumes that *the material particles are held together solely by gravitational forces*, so that these additional terms have to depend on the g_{ik} and their derivatives. Although the Maxwell tensor $\mathfrak{S}_i{}^k$ cannot now be called the total energy tensor of matter and does not satisfy equation (341 a), Einstein starts quite analogously as in § 56 with the hypothesis that *this energy tensor $\mathfrak{S}_i{}^k$ of Maxwell should be proportional to a differential expression of second order which is formed from the g_{ik} alone.* This simple assumption is decisive for Einstein's theory. We conclude from this and from the general covariance requirement, as in § 56, that the field equations must be of the form

$$R_{ik} + \bar{c}R g_{ik} = - \kappa S_{ik}.$$

The addition of another term proportional to g_{ik} will be seen to be superfluous here. But since equation (341 a) for $\mathfrak{S}_{ik}$ is not valid, we are no longer entitled to put, as previously, $\bar{c} = -\frac{1}{2}$. Instead, $\bar{c}$ is determined by another condition. According to (223), the scalar $S_i{}^i$ vanishes; we must put $\bar{c} = -\frac{1}{4}$, so that the scalar of the left-hand side of the field equations also vanishes identically. For this case, the field equations read

$$R_{ik} - \tfrac{1}{4}g_{ik}R = - \kappa S_{ik}. \tag{492}$$

In addition, the equations of electron theory,

$$\frac{\partial F_{ik}}{\partial x^l} + \frac{\partial F_{li}}{\partial x^k} + \frac{\partial F_{kl}}{\partial x^i} = 0 \tag{203}$$

and

$$\frac{\partial \mathfrak{F}^{ik}}{\partial x^k} = \mathfrak{s}^i, \tag{208}$$

are to remain valid. A simple enumeration shows that (203) and (492) contain just four independent equations less than there are unknowns, as required by a general relativistic theory. It should also be mentioned that the field equations do not appear to be derivable from an action principle, in this case. It follows from § 54 (203) and (208), that the divergence of S_{ik} has the value

$$- F_{ik}\mathfrak{s}^k$$

of the (negative) force vector of Lorentz. Therefore the divergence of the field equations (492) leads to the relation

$$F_{ik}\mathfrak{s}^k - \frac{1}{4\kappa}\frac{\partial R}{\partial x^i} = 0, \tag{493}$$

since the divergence of $R_{ik} - \tfrac{1}{2}g_{ik}R$ vanishes. *This relation shows that for these field equations the Coulomb repulsive forces are indeed held in equilibrium by a gravitational pressure.* If we put $\mathfrak{s}^k = \rho_0 u^k$, it also follows that

$$\frac{\partial R}{\partial x^i}u^i = \frac{dR}{d\tau} = 0, \tag{494}$$

i.e. R remains constant on the world line of one and the same element of matter. In charge-free space

$$\frac{\partial R}{\partial x^i} = 0,$$

from (493), so that

$$R = \text{const.} = R_0. \tag{495}$$

In the interior of a material particle, R decreases continuously from the value R_0 to smaller and smaller values, up to the centre of the particle. (493) shows that $(1/4\kappa)R$ represents directly the potential energy of the gravitational forces which hold the particle together.

We now have to look for the material energy–momentum tensor T_{ik}. *For this, the Eq. (452), which contains the λ-term, is to remain valid.* From (453), we have $R = -4\lambda$ for matter-free space. Comparison with (495) shows that we have to put

$$R_0 = -4\lambda, \qquad \lambda = -\frac{R_0}{4}. \tag{496}$$

It constitutes one of the main advantages of this new formulation that the constant λ is not characteristic of the fundamental law itself, but has the meaning of an *integration constant*. Eq. (452) can now be written

$$G_{ik} + \tfrac{1}{4}R_0 g_{ik} = -\kappa T_{ik},$$

whereas (492) leads to

$$G_{ik} + \tfrac{1}{4}R g_{ik} = -\kappa S_{ik}.$$

Comparison then shows that

$$T_{ik} = S_{ik} + \frac{1}{4\kappa}(R - R_0)g_{ik}. \tag{497}$$

This tensor therefore satisfies automatically the earlier equation (452), because of (492), and thus also Eq. (341a). In addition it vanishes in matter-free space. We are thus quite justified, from a physical point of view too, to call it the energy tensor of matter. The material energy density $-\mathfrak{T}_4{}^4$ is composed of two parts, one originating from the electromagnetic and one from the gravitational field, both of which are positive. It is easy to see that the spatially finite universe with constant, stationary, mass density $(T_1{}^1 = T_2{}^2 = T_3{}^3 = 0, T_4{}^4 = -\mu_0 c^2)$ is a solution of the new field equations. All the relations of § 62(β) remain unaltered. The electromagnetic tensor $S_i{}^k$ is generally calculated from (497) to be

$$S_i{}^k = T_i{}^k - \tfrac{1}{4}T\delta_i{}^k \tag{498}$$

so that, for our case,

$$S_1{}^1 = S_2{}^2 = S_3{}^3 = \tfrac{1}{4}\mu_0 c^2, \qquad S_4{}^4 = -\tfrac{3}{4}\mu_0 c^2. \tag{499}$$

The energy of a spatially finite universe is three-quarters electromagnetic and one-quarter gravitational in origin. This contribution of the electromagnetic to the total energy is exactly the same as that derived in § 63

on the basis of specific (not necessarily correct) assumptions about the
electron.

If we now try to determine the field of a material particle from the
differential equations (203) or (206), (208) and (492), respectively, we
find that we are one equation short for the determination of the unknowns
in the static spherically symmetrical case. *According to the theory of Einstein
as developed here, every static spherically symmetrical distribution of electricity is in equilibrium.* However satisfactory the foundations of this
theory may be, it, too, is not capable of providing an answer to the
problem of the structure of matter.

67. General remarks on the present state of the problem of matter†

Each of the theories which we discussed has its particular advantages
and drawbacks. Their joint failure prompts us, however, to summarize
specifically those shortcomings and difficulties which are common to them
all.

It is the aim of all continuum theories to derive the atomic nature of
electricity from the property that the differential equations expressing
the physical laws have only a discrete number of solutions which are
everywhere regular, static, and spherically symmetric. In particular, one
such solution should exist for each of the positive and negative kinds of
electricity. It is clear that differential equations which have this property
must be of a particularly complicated structure. It seems to us that this
complexity of the physical laws in itself already speaks against the continuum theories. For it should be required, from a physical point of view,
that the existence of atomicity, in itself so simple and basic, should
also be interpreted in a simple and elementary manner by theory and
should not, so to speak, appear as a trick in analysis.

Furthermore, we have seen that the continuum theories are forced to
introduce special forces which keep the Coulomb repulsive forces in the
interior of the electrical elementary particles in equilibrium. If we assume
that these forces are *electrical in nature*, we have to assign an absolute
meaning to the four-vector potential, which leads to the difficulties
discussed in § 64. The other alternative, that the electrical elementary
particles are held together by *gravitational forces*, is however countered
by a very weighty, empirical, argument. For one would expect, in such a
case, that a simple numerical relation would exist between the gravitational mass of the electron and its charge. Actually, the relevant dimen-
sionless number $e/(m\sqrt{k})$ (k = ordinary gravitational constant) is of the
order of 10^{20} (see also § 59.)!

It must also be required of the field equations that they should account
for the asymmetry (difference in mass) in the two kinds of electricity.‡
It is however easy to see that this is in contradiction with their general
covariance,[393] from a formal point of view. For the statical case, the field

† See suppl. note 23.
‡ See suppl. note 21.
[393] W. Pauli, jr., *Phys. Z.*, **20** (1919) 457.

equations only contain the electrostatic potential φ as a variable, apart
from the g_{ik} (i, $k = 1, 2, 3$ or $i = k = 4$). As a special case of general
covariance, the differential equations must, in particular, be covariant
under time reversal $x'^4 = -x^4$. But then φ goes over into $-\varphi$, whereas
the g_{ik} remain unchanged (in our case, $g_{ik} = 0$ for $i = 1, 2, 3$). If there-
fore φ, $g_{ik}(g_{i4} = 0)$ is a solution of the field equations, then $-\varphi$, $g_{ik}(g_{i4} = 0)$
is also a solution, in contradiction to the asymmetry of the two kinds of
electricity. One could try to avoid such a conclusion by introducing non-
rational action functions, as was indicated at the end of § 64. But, first of
all, the field equations would then become even more complicated, and,
secondly, the selection of the unambiguous branch of the action function
is not carried out in a generally covariant manner: covariance no longer
exists, e.g. for the time reversal $x'^4 = -x^4$.

Finally, a conceptual doubt should be mentioned.[394] The continuum
theories make direct use of the ordinary concept of electric field strength,
even for the fields in the interior of the electron. This field strength is
however defined as the force acting on a test particle, and since there are
no test particles smaller than an electron or a hydrogen nucleus, the field
strength at a given point in the interior of such a particle would seem to
be unobservable, by definition, and thus be fictitious and without physical
meaning.

Whatever may be one's attitude in detail towards these arguments,
this much seems fairly certain: new elements which are foreign to
the continuum concept of the field will have to be added to the basic
structure of the theories developed so far, before one can arrive at a
satisfactory solution of the problem of matter.

[394] Cf. footnote 391, p. 202, W. Pauli, jr., *ibid.*, and the 'Nauheimer Diskussion', *Phys
Z.*, **21** (1920) 650.

SUPPLEMENTARY NOTES

Note 1. (p. xiii) **Selection of books concerned with the later development of relativity theory**

A. Einstein, *The Meaning of Relativity* (5th edn., Princeton 1956).
M. v. Laue, *Die Relativitätstheorie* (Braunschweig):
Vol. 1, *Spezielle Relativitätstheorie* (6th edn., 1955).
Vol. 2, *Allgemeine Relativitätstheorie* (3rd edn., 1953).
H. Weyl, *Raum-Zeit-Materie* (5th edn., Berlin 1923). English translation (with new preface) *Space–Time–Matter* (New York 1950).
A. S. Eddington, *The Mathematical Theory of Relativity* (2nd edn., Cambridge 1924; reprint 1953). German translation with an appendix by A. Einstein (Berlin 1925).
Richard C. Tolman, *Relativity, Thermodynamics and Cosmology* (Oxford 1934).
P. G. Bergmann, *An Introduction to the Theory of Relativity* (New York 1942).
E. Schrödinger, *Space-Time Structure* (Cambridge 1950).
A. Lichnerowicz, *Théories rélativistes de la gravitation et de l'électromagnétisme* (Paris 1955).
P. Jordan, *Schwerkraft und Weltall* (2nd edn., Braunschweig 1955).
Cinquant' anni di relatività, 1905–1955 (Florence 1955).
Proceedings of the Congress 'Jubilee of Theory of Relativity,' Berne, July 1955, *Helv. phys. acta*, Suppl. IV, 1956.
Volume 'Einstein' in the *Library of living Philosophers* (Evanston 1949; 2nd edn., 1951; German translation, Stuttgart 1955).

Note 2. (p. 4)

R. J. Kennedy and E. M. Thorndike, *Phys. Rev.*, **42** (1932) 400, performed an important variation of the Michelson experiment, in which the difference in length of the two arms in the interferometer was kept large. The negative result of this experiment excludes the possibility of a dependence on the velocity of the earth of the time needed for the light to traverse any closed path in a terrestrial laboratory. See also the theoretical discussion of this experiment by H. P. Robertson, *Rev. Mod. Phys.*, **21** (1949) 378.

Note 3. (p. 8)

Meanwhile, the Michelson experiment with extra-terrestrial light (sun and stars) has actually been carried out with negative result by R. Tomaschek, *Ann. Phys., Lpz.*, **73** (1924) 105.

Note 4. (p. 19)

The experiments of Harress and Sagnac have been repeated by B. Pogany, *Ann. Phys., Lpz.*, **80** (1926) 217; *Naturwissenschaften*, **15** (1927) 177; *Ann. Phys., Lpz.*, **85** (1928) 244. The proposed optical experiment to test the rotation of the earth has actually been carried out: A. A. Michelson, *Astrophys. J.* **61** (1925) 137 and A. A. Michelson and H. G. Gale, *Astrophys. J.* **61** (1925) 1401.

Note 5. (p. 20)

The second-order Doppler effect has actually been verified experimentally by comparing the arithmetical mean of the frequency shift of two light beams of exactly opposite directions with the unshifted light frequency emitted by atoms at rest. The experiment has been made by H. E. Ives and C. R. Stilwell, *J. opt. Soc. Amer.*, **28** (1938) 215; **31** (1941) 369, and repeated by G. Otting, *Phys. Z.*, **40** (1939) 681. The prediction of special relativity theory is confirmed with great accuracy. (Hence the proposal of Abraham mentioned on p. 14 of the text is also disproved experimentally.)

A good means of verifying experimentally the time dilatation of special relativity is the dependence of the life time of decaying mesons on their energy. Theoretically, the life time should be proportional to the relativistically defined energy of the particles (see § 37). Qualitatively, the effect is well checked both in cosmic rays† and with artificially produced mesons,‡ but the accuracy is not very good at present, as until now the experiments have not been made with the particular purpose of checking the theoretical formulas for time dilatation.

Note 5a. (pp. 33 and 43-46)

The systematic nomenclature introduced here has not found its way into the literature. It will strike the reader of today as particularly unusual that the rank of a "surface" tensor should be different from the number of tensor indices. The curvature tensor, for instance, is called here a "surface tensor of rank 2", while the number of its indices is four.

Note 6. (p. 38)

It is logically possible to abandon any derivation of the "geodesic components" Γ^l_{ik} from the metric of the space, defined by the tensor g_{ik}. The "pseudotensor field" Γ^l_{ik} is then axiomatically assumed to satisfy the transformation law

$$
\begin{aligned}
\Gamma'^l_{ik} &= \frac{\partial x^r}{\partial x'^i} \frac{\partial x^s}{\partial x'^k} \left(\frac{\partial x'^l}{\partial x^t} \Gamma^t_{rs} - \frac{\partial^2 x'^l}{\partial x^r \partial x^s} \right) \\
&= \frac{\partial x'^l}{\partial x^t} \left(\frac{\partial x^r}{\partial x'^i} \frac{\partial x^s}{\partial x'^k} \Gamma^t_{rs} + \frac{\partial^2 x^t}{\partial x'^i \partial x'^k} \right),
\end{aligned}
\tag{I}
$$

which is equivalent to the general covariance of the definition (64) for the parallel displacement of contravariant vectors.

Instead of Eq. (67) of the text, which contains explicitly the metric,

† See, for instance, B. Rossi and D. B. Hall, *Phys. Rev.*, **59** (1941) 223.

‡ R. Durbin, H. H. Loar and W. W. Havens, *Phys. Rev.*, **88** (1952) 179 (Especially p. 183) measured the life time of π-mesons of kinetic energy of 73 MeV, so that the time-dilatation factor was about 1·5. The measured dilatation was in agreement with this to 10 per cent.

the invariance of any scalar product $a^i b_i$ of a contravariant vector a^i and covariant vector b_i under parallel displacement has to be postulated here. With this condition,

$$\frac{d}{dt}(a^i b_i) = 0,$$

we derive from (64) the displacement law

$$\frac{db_i}{dt} = + \Gamma^r_{is}\frac{dx^s}{dt}b_r \tag{2}$$

for covariant vectors.

This generalization of the Riemannian geometry is called "affine geometry", the parallel displacement of vectors is called "affine connexion" and a space in which it is defined an "affinely connected space".

At first it would seem natural to maintain the symmetry condition (65) for the Γ's, because the antisymmetric part of Γ^l_{ik} would be a new independent tensor in the proper sense, according to the transformation law (1). Only the symmetric part of Γ^l_{ik} can be locally transformed away at a given point. In Note 23 we shall, however, discuss also non-symmetric Γ's, in connexion with attempts to apply this generalized affine geometry to physics.

In the following Note 7, however, we restrict ourself to symmetric Γ's.

Note 7. (pp. 60 and 69)

(a) *Covariant differentiation in an affinely connected space.*† With the help of geodesic components $\Gamma^i_{kl} = \Gamma^i_{lk}$ without any metric, the covariant differentiation of Ricci and Levi-Civita, mentioned on p. 59 of the text, can be uniquely defined. We denote here, differing slightly from the text, the covariant differentiation with a semicolon so that, for instance [see (148 a) and (148 b)],

$$a^i_{;k} = \frac{\partial a^i}{\partial x^k} + \Gamma^i_{rk}a^r \tag{1}$$

$$b_{i;k} = \frac{\partial b_i}{\partial x^k} - \Gamma^r_{ik}b_r. \tag{2}$$

One postulates for a scalar c

$$c_{;k} \equiv \frac{\partial c}{\partial x^k} \tag{3}$$

and for a product, with or without contraction, the general rule

$$(a^{\cdots}_{\cdots}b^{\cdots}_{\cdots})_{;k} = a^{\cdots}_{\cdots;k}b^{\cdots}_{\cdots} + a^{\cdots}_{\cdots}b^{\cdots}_{\cdots;k}, \tag{4}$$

† For a standard work on the subject dealt with in this note, cf. J. A. Schouten, *Ricci-Calculus* (2nd edn., Berlin and London 1954).

which says that the covariant differentiation of a product is analogous to the ordinary differentiation of a product. With the rules (3) and (4) one can derive each of the formulae (1) or (2) from the other one, for in

$$(a^s b_s)_{;k} = a^s_{,k} b_s + a^s b_{s;k}$$

the terms with the Γ's cancel.

Moreover, the general formula (152) of the text for the covariant differentiation of a tensor is in agreement with the product rule (4) and follows from this rule if the particular cases (1), (3), or (2), (3) are assumed.

It is easy to generalize the concept of parallel displacement from vectors to all tensors. The condition that an arbitrary tensor field $a^{...}_{...}$ is invariant under parallel displacement along a given curve can be written

$$a^{...}_{...;r} \frac{dx^r}{dt} = 0. \tag{5}$$

If $a^{...}_{...}$ is not given as a field, one has to replace

$$\frac{\partial a^{...}_{...}}{\partial x^r} \frac{dx^r}{dt} \quad \text{by} \quad \frac{da^{...}_{...}}{dt},$$

so that (5) determines the dependence of $a^{...}_{...}$ along the given path.

We further note, that (152) is in agreement with the vanishing of the covariant differentiation of the trivial mixed tensor which is given by the Kronecker symbol $\delta_i{}^k$.

Applying the product rule (4) to a determinant $D = \det | a_{ik} |$ of a tensor a_{ik} of the second rank (which transforms like g) one obtains

$$D_{;k} = \frac{\partial D}{\partial x^k} - 2\Gamma^\alpha_{\alpha k} D.$$

Hence

$$(\sqrt{D})_{;k} = \frac{\partial \sqrt{D}}{\partial x_k} - \Gamma^\alpha_{\alpha k} D.$$

From this and (3) it follows for any scalar density $\mathfrak{A} = a \sqrt{D}$ that

$$\mathfrak{A}_{;k} = \frac{\partial \mathfrak{A}}{\partial x^k} - \Gamma^\alpha_{\alpha k} \mathfrak{A}, \tag{6}$$

but for the divergence of a vector density one finds that the Γ's cancel

$$\mathfrak{A}^k{}_{;k} = \frac{\partial \mathfrak{A}^k}{\partial x^k} \tag{6a}$$

(see p. 55).

(b) *The curvature tensor in an affinely connected space.* The considerations of § 16, which lead to the equation (86) of the curvature tensor

$$R^h{}_{ijk} = \frac{\partial \Gamma^h{}_{ij}}{\partial x^k} - \frac{\partial \Gamma^h{}_{ik}}{\partial x^j} + \Gamma^h{}_{k\alpha} \Gamma^\alpha{}_{ij} - \Gamma^h{}_{j\alpha} \Gamma^\alpha{}_{ik} \tag{7}$$

lso hold in the affine space, as they do not contain the metric. This
ensor still has the symmetry properties

$$R^h{}_{ikj} = -R^h{}_{ijk}, \qquad R^h{}_{ijk} + R^h{}_{jki} + R^h{}_{kij} = 0. \qquad (7\,\mathrm{a})$$

However, there is now no simple means for lowering the first index and
herefore no analogy to the property of skew-symmetry in the first two
ndices of the Riemannian curvature tensor R_{hijk} [see (92)].

This has an important consequence for the contracted curvature tensor,
defined in analogy with (93) and (94) by

$$R_{ik} = R^\alpha{}_{iak} = \frac{\partial \Gamma^\alpha{}_{ia}}{\partial x^k} - \frac{\partial \Gamma^\alpha{}_{ik}}{\partial x^\alpha} + \Gamma^\beta{}_{ia}\Gamma^\alpha{}_{k\beta} - \Gamma^\alpha{}_{ik}\Gamma^\beta{}_{\alpha\beta}. \qquad (7\,\mathrm{b})$$

This tensor is no longer symmetric and therefore splits immediately into
ts two irreducible parts, the symmetric and the skew-symmetric one:

$$R_{ik} = R_{\underline{ik}} + R_{\underset{\smile}{ik}},$$

$$R_{\underline{ik}} = R_{\underline{ki}} = \frac{1}{2}\left(\frac{\partial \Gamma^\alpha{}_{ia}}{\partial x^k} + \frac{\partial \Gamma^\alpha{}_{ka}}{\partial x^i}\right) - \frac{\partial \Gamma^\alpha{}_{ik}}{\partial x^\alpha} + \Gamma^\beta{}_{ia}\Gamma^\alpha{}_{k\beta} - \Gamma^\alpha{}_{ik}\Gamma^\beta{}_{\alpha\beta} \qquad (8)$$

we use here the symmetry condition (65) for the Γ's and

$$R_{\underset{\smile}{ik}} = -R_{\underset{\smile}{ki}} = \frac{1}{2}\left(\frac{\partial \Gamma^\alpha{}_{ia}}{\partial x^k} - \frac{\partial \Gamma^\alpha{}_{ka}}{\partial x^i}\right). \qquad (9)$$

The latter fulfills the identity

$$\frac{\partial R_{\underset{\smile}{ik}}}{\partial x^l} + \frac{\partial R_{\underset{\smile}{li}}}{\partial x^k} + \frac{\partial R_{\underset{\smile}{kl}}}{\partial x^i} = 0. \qquad (10)$$

As the scalar densities (p. 32) are important for the variational principles
§ 23), we mention here that the simplest scalar density constructed
algebraically (without differentiations) from the curvature tensor is

$$\mathfrak{I}_1 = \sqrt{(|\det|R_{\underline{ik}}||)} \qquad (11)$$

Indeed, every tensor of the second rank can be used in this way to build
up an invariant volume element, as is usually done with the metric tensor.
One can also define the normalized subdeterminants $R^{\underline{ik}}$ by

$$R_{\underline{i\alpha}}R^{\underline{k\alpha}} = \delta_i{}^k \qquad (12)$$

and use them to raise the indices of $R_{\underset{\smile}{ik}}$

$$R^{\underset{\smile}{ik}} = R_{\underset{\smile}{lm}}R^{\underline{li}}R^{\underline{mk}}. \qquad (13)$$

The simplest invariant is then

$$I = \tfrac{1}{2}R_{\underset{\smile}{ik}}R^{\underset{\smile}{ik}} = \tfrac{1}{2}R_{\underset{\smile}{ik}}R_{\underset{\smile}{lm}}R^{\underline{li}}R^{\underline{mk}}. \qquad (14)$$

Combining (11) and (14) gives the scalar density

$$\mathfrak{I}_2 = (1 + \alpha\tfrac{1}{2}R_{\underset{\smile}{ik}}R_{\underset{\smile}{lm}}R^{\underline{li}}R^{\underline{mk}})\sqrt{(|\det|R_{\underline{ik}}|\cdot|)}, \qquad (15)$$

with an arbitrary numerical factor α. It consists of two parts added together with a plus sign, just what the wish for "unification" made various authors so eager to avoid. (Cf. below, Note 23).†

(c) *The identities of Bianchi.* The identity

$$a_{i;kl} - a_{i;l;k} = -R^h{}_{ikl}\,a_h \tag{16}$$

mentioned on p. 59 is useful to describe the differential identities of *Bianchi*, which hold both for the Riemannian and for the more general affine space.

From (16) it follows for an arbitrary tensor S_{ik} of the second rank that

$$S_{ik;l;m} - S_{ik;m\cdot l} = -(R^h{}_{klm}\,S_{ih} + R^h{}_{ilm}\,S_{hk}). \tag{17}$$

It follows, for instance, first for the particular tensor $S_{ik} = a_i\,b_k$; but as the relation is linear in the components of S_{ik}, it also holds for any sum of these particular tensors, and hence for a general tensor of the second rank.

Putting in (17) $S_{ik} = a_{i;k}$, and adding the three cyclic permutations of k, l, m one obtains, taking into account the cyclic symmetry (7 a) of $R^h{}_{klm}$

$$(a_{i;k;l;m} - a_{i;l;k;m}) + (a_{i;l;m;k} - a_{i;m;l;k}) + (a_{i;m;k;l} - a_{i;k;m;l}) = \\ -(R^h{}_{ilm}\,a_{h;k} + R^h{}_{imk}\,a_{h;l} + R^h{}_{ikl}\,a_{h;m}). \tag{18}$$

On the other hand, one obtains by covariant differentiation of (16) with respect to m_i followed by the cyclic permutation of k, l, m,

$$(a_{i;k;l;m} - a_{i;l;k;m}) + (a_{i;l;m;k} - a_{i;m;l;k}) + (a_{i;m;k;l} - a_{i;k;m;l}) = \\ -(R^h{}_{ilm}\,a_{h\cdot k} + R^h{}_{ikl}\,a_{h;m} + R^h{}_{imk}\,a_{h;l}) - (R^h{}_{ikl;m} + R^h{}_{ilm;k} + R^h{}_{imk;l})\,a_h \tag{19}$$

The left-hand side of (18) is identical with the left-hand side of (19). The same holds for the right-hand side of (18) and the first triplet on the right-hand side of (19) apart from the order of the terms. Hence the second triplet on the right-hand side of (19) must vanish for every vector a_i. This gives the famous identities of *Bianchi*

$$R^h{}_{ikl;m} + R^h{}_{ilm\cdot k} + R^h{}_{imk;l} = 0. \tag{20}$$

By contraction $h = k = \alpha$ one gets

$$R_{il\cdot m} - R_{im;l} + R^\alpha{}_{ilm;\alpha} = 0$$

or, with a change of notation in the indices

$$R_{hl\cdot k} - R_{hk;i} + R^\alpha{}_{hik;\alpha} = 0. \tag{21}$$

(d) *The simplification in the Riemannian space.* For a Riemannian space it is possible and also natural to link the affine connexion with the metric by the postulate that the *metric field* $g_{ik}(x)$ *is invariant under parallel displacement along any curve.* This gives immediately, according to (5)

$$g_{ik;r} = 0 \tag{22}$$

or

$$\frac{\partial g_{ik}}{\partial x^r} - g_{sk}\,\Gamma^s{}_{ir} - g_{is}\,\Gamma^s{}_{kr} = 0,$$

† A third invariant can be constructed in analogy with the one considered in Note 20

which is, in view of (66), identical with (68). The expression (69) for $\Gamma_{i,rs}$ follows from it, for symmetrical Γ's.

The condition (22) is equivalent to

$$g^{ik}{}_{;r} = 0, \qquad (22\,\text{a})$$

which is identical with Eq. (71) of the text. It enables one to raise and to lower indices of covariantly differentiated tensors be means of the metric tensor. If, as usual,

$$a_i = g_{ir}\,a^r, \qquad a^i = g^{ir}\,a_r,$$

one has also

$$a_{i;k} = g_{ir}\,a^r{}_{;k}\,, \qquad a^i{}_{;k} = g^{ir}\,a_{r;k}\,. \qquad (23)$$

We note, that Eq. (67) of the text is also equivalent to (22). In view of (3), the product rule (4), and Eq. (64) of the text, one has indeed,

$$\frac{d}{dt}(g_{ik}\,\xi^i\,\xi^k) = g_{ik;r}\,\xi^i\,\xi^k\frac{dx^r}{dt}.$$

We can now easily derive again the remaining symmetry property

$$R_{iklm} = -\,R_{kilm} \qquad (24)$$

or

$$g_{ih}\,R^h{}_{klm} + g_{kh}\,R^h{}_{ilm} = 0 \qquad (24\,\text{a})$$

of the metric curvature tensor, which is constructed with the help of the special $\Gamma^i{}_{kl}$, given by (66) and (69), which fulfil (22).

For this purpose we have only to insert g_{ik} for S_{ik} in the general formula (17). As the left-hand side vanishes in view of (22), the right-hand side must vanish too, which is identical with (24 a).

In § 16 it was shown, that from (24) there follows the symmetry of the contracted curvature tensor R_{ik}, which means the vanishing of its skew-symmetric part R_{ik}. It is this which makes the construction of invariants from the curvature tensor so unique for the metric space [see p. 48, Eq. (113)].

Finally, we derive from Eqs. (21) above, which followed from the Bianchi identities, the identities (182 a, b) for the tensor [see(109)]

$$G_{ik} = R_{ik} - \tfrac{1}{2}g_{ik}\,R,$$

which can be written

$$G_i{}^k{}_{;k} = G^k{}_{i;k} = 0 \qquad (25)$$

in the calculus of covariant differentiation.

For this purpose one has only to multiply (21) by g^{hk}, and to contract with respect to the indices h and k. Using (23) and $g^{hk}R^\alpha{}_{hik} = R_i{}^\alpha$ ([see (93)] one obtains

$$R^\alpha{}_{i;\alpha} - R_{;i} + R^\alpha{}_{i;\alpha} = 0$$

or

$$2(R^\alpha{}_i - \tfrac{1}{2}\delta^\alpha{}_i R)_{;\alpha} = 0$$

which is identical with (25). The importance of the simple connexion with the identities of Bianchi of the four identities (25) which are so funda-mental in general relativity, was also emphasized by Einstein in his later work.

Note 8. (pp. 69 and 162)

In the text, p. 69, it was stated: "An explicit evaluation now shown that

$$\frac{\partial}{\partial x^\sigma} \frac{\partial \mathfrak{G}}{\partial g^{ik}{}_\sigma} - \frac{\partial \mathfrak{G}}{\partial g^{ik}} = \mathfrak{G}_{ik} = \sqrt{(-g)}\, G_{ik},$$

where G_{ik} is the tensor defined in (109)."

Also, reference was made there to a paper by *Palatini* (see footnote (105) p. 69).

The method of Palatini is the following: he first notices that, according to (71), the variation $\delta\Gamma^r{}_{ik}$ of the Γ-field is a proper tensor, in contrast to $\Gamma^r{}_{ik}$ itself. Using the formula (94) for R_{ik} one obtains

$$\delta R_{ik} = (\delta\Gamma^\alpha{}_{i\alpha})_{;k} - (\delta\Gamma^r{}_{ik})_{;r} . \tag{1}$$

This can immediately be seen, first, in a geodesic coordinate system where at a particular point the Γ's themselves (but not necessarily their deriva-tives) vanish. Hence the equation holds generally by reason of the general covariance of both sides.

Using the product rule for covariant differentiation one gets

$$\sqrt{(-g)}\, g^{ik}\delta R_{ik} = [\sqrt{(-g)}(g^{ir}\delta\Gamma^\alpha{}_{i\alpha} - g^{ik}\delta\Gamma^r{}_{ik})]_{;r} + \\ + [(\sqrt{(-g)}g^{ik})_{;r} - (\sqrt{(-g)}g^{is})_{;s}\,\delta_r{}^k]\delta\Gamma^r{}_{ik}.$$

The first term is as a covariant divergence of a vector density, i.e. according to Note 7, Eq. (6 a), a common divergence, which, after integration over the volume, gives rise to a surface integral. The latter vanishes if the variation of the Γ-field vanishes at the boundary. We are therefore left with

$$\int \sqrt{(-g)}\, g^{ik}\delta R_{ik}\, dx = \int [(\sqrt{(-g)}g^{ik})_{;r} - (\sqrt{(-g)}g^{is})_{;s}\,\delta_r{}^k]\delta\Gamma^r{}_{ik}\, dx + \int_{\text{surface}} \tag{2}$$

Using

$$R_{ik}\delta(\sqrt{(-g)}g^{ik}) = \sqrt{(-g)}(R_{ik} - \tfrac{1}{2}g_{ik}R)\delta g^{ik} = \mathfrak{G}_{ik}\delta g^{ik}$$

we obtain therefore

$$\delta\int \mathfrak{R}\, dx = \int \mathfrak{G}_{ik}\delta g^{ik}\, dx + \int [(\sqrt{(-g)}g^{ik})_{;r} - (\sqrt{(-g)}g^{is})_{;s}\,\delta_r{}^k]\delta\Gamma^r{}_{ik}\, dx + \\ + \int_{\text{surface}} . \tag{3}$$

Until now we have not used the vanishing of the covariant derivative of the metric tensor. [Note 7, Eqs. (22) and (23).] If we do so, the second integral of the right-hand side of (3) vanishes and formula (180), p. 69

of the text is proved. Based on this development of Palatini, there exists a variant of the action principle (see § 57 of the text) where one treats the 10 functions g^{ik} and the 40 functions Γ^r_{ik} as independent variables.†

Provided the integrand $\mathfrak{M}$ of the matter part of the action integral [see Eq. (404) of the text] does not contain the Γ^r_{ik} explicitly, the variation of the Γ^r_{ik} gives, according to (3),

$$(\sqrt{(-g)}g^{ik})_{;r} - (\sqrt{(-g)}g^{is})_{;s}\,\delta_r{}^k = 0,$$

from which we can easily derive $g^{ik}{}_{;r} = 0$. These equations, just like the equation (401) of the gravitational field, are then, as a part of the field equations, a consequence of the variational principle.

The condition that the matter part $\mathfrak{M}$ of the action integral should not depend on the Γ^r_{ik} is certainly fulfilled for an electromagnetic field without electric currents [see Eq. (172)]. For the general representation of matter, however, the limitations in the applicability of the classical field concepts come into play and the condition in question does not seem to me trivial. Particularly the case in which $\mathfrak{M}$ contains spinor fields needs a closer investigation.

In a Riemannian space it seems to me simpler and more natural to assume the identities

$$g_{ik;r} = 0 \quad \text{or} \quad g^{ik}{}_{;r} = 0$$

from the beginning and to treat, in the variational principle, the 10 functions $g_{ik}(x)$ alone as independent variables.

For the use of Palatini's method in Einstein's field equations, see Note 23.

Note 8 a. (p. 70)‡

The identity of Eq. (184) of the text, i.e.

$$\frac{\partial}{\partial x^k}(\mathfrak{U}_i{}^k + \mathfrak{G}_i{}^k) \equiv 0, \tag{I}$$

becomes obvious on the basis of the relation

$$\mathfrak{U}_i{}^k + \mathfrak{G}_i{}^k \equiv \frac{\partial \mathfrak{B}_i{}^{kl}}{\partial x^l}, \tag{II}$$

first derived by P. Freud,‖ in which $\mathfrak{B}_i{}^{kl}$ is antisymmetric in k and l,

$$\mathfrak{B}_i{}^{kl} + \mathfrak{B}_i{}^{lk} \equiv 0. \tag{III}$$

† A. Einstein, *S.B. preuss. Akad. Wiss.* (1925), 414.

‡ I am indebted to Dr. V. Bargmann for having drawn my attention to the results dealt with in this note.

‖ P. Freud, *Ann. Math., Princeton,* **40** (1939) 417. Freud denotes by

$$\mathfrak{U}_i{}^k = \kappa(t_i{}^k + T_i{}^k)\sqrt{(-g)}$$

the expression which is denoted by $-(\mathfrak{U}_i{}^k + \mathfrak{G}_i{}^k)$ in this book; hence his $\mathfrak{A}_i{}^{kl}$ has the opposite sign to our $\mathfrak{B}_i{}^{kl}$.

This author also derived for the affine-tensor density $\mathfrak{B}_i{}^{kl}$ the expression

$$2\mathfrak{B}_i{}^{kl} = \sqrt{(-g)}[\delta_i{}^k(g^{rs}\,\Gamma^l{}_{rs} - g^{lr}\,\Gamma^s{}_{rs}) + \\ + \delta_i{}^l(g^{kr}\,\Gamma^s{}_{rs} - g^{rs}\,\Gamma^k{}_{rs}) + \\ + (g^{lr}\,\Gamma^k{}_{ir} - g^{kr}\,\Gamma^l{}_{ir})]. \tag{1}$$

The results of Freud can also be derived by a generalization of the result (181) of the text for the variation of $\int \mathfrak{R}\, dx$ with arbitrary functions ξ^i, where the contribution of the surface integral in (177) has to be taken into account.

In view of (182 a) one obtains, after some transformations,

$$\delta \int \mathfrak{R}\, dx = 2 \int \frac{\partial}{\partial x^k}\left[(\mathfrak{U}_i{}^k + \mathfrak{G}_i{}^k)\xi^i - \mathfrak{B}_i{}^{jk}\frac{\partial \xi^i}{\partial x^j}\right] dx, \tag{2}$$

where

$$\mathfrak{B}_i{}^{jk} = g^{jr}\frac{\partial \mathfrak{G}}{\partial g^{ir}{}_k} + \tfrac{1}{2}\delta_i{}^k\frac{\partial(\sqrt{(-g)}g^{jr})}{\partial x^r} - \frac{1}{2}\frac{\partial(\sqrt{(-g)}g^{jk})}{\partial x^i}. \tag{3}$$

The vanishing of the integrand of (2) for arbitrary functions ξ^i gives just the identities (I), (II), (III), while the expression (3) turns out to be identical with the expression (1) of Freud.

The identity (II) is useful because it permits one to compute the volume integrals of total energy and momentum as a flux through a surface.

Note 9. (p. 83)

Today the relativistic dependence of energy and momentum on velocity is taken as a matter of course in all experiments on high-energy particles, either occurring in cosmic rays or artificially produced with help of machines (cyclotrons, bevatrons, etc.) in which a high acceleration of charged particles takes place. For the computation of their orbits in these machines the relativistic formulae, the predictions of which have always been in agreement with experience, are also essential. A particular experiment, which checked the relativistic mass formula for electrons in a range of velocities up to nearly $0.8c$ was performed by M. M. Rogers, A. W. McReynolds and F. T. Rogers, Jr., *Phys. Rev.*, **57** (1940) 379.

Note 10. (p. 88)

The text here contains a historical error: the action principle in question had already been established by J. J. Larmor [see his book: *Aether and Matter*, (Cambridge 1900) Chap. 6].

Note 11. (pp. 111 and 115)

M. v. Laue [see his *Relativitätstheorie*, Vol. 1 (6th edn., 1955) § 19] has shown that only the unsymmetric energy–momentum tensor of Minkowski is correct for a phenomenological description of moving bodies (just as it is in crystals at rest). His argument also emphasizes the validity of the addition theorem of velocities for the ray-velocity (see Eq. (312) of the text), which is in agreement only with this unsymmetric tensor.

Note 12. (p. 119)

The considerations of Lewis and Tolman are simpler in the centre-of-gravity system of the colliding spheres.

Note 13. (p. 123)

A very striking example of the equivalence of mass and energy is the annihilation radiation of a pair of two electrons with opposite electric charge (positon and negaton), where the whole mass is transformed into radiation energy. [For a quantitative measurement of the wave-length of the emitted photon see J. Du Mond, D. A. Lind and B. B. Watson, *Phys. Rev.*, **75** (1949) 1226.]

The first quantitative verification of the mass–energy balance in nuclear reactions has been made by J. D. Cockcroft and E. T. Walton, *Proc. Roy. Soc.* A137 (1932) 229, in the reaction, where two α-particles are emitted by proton bombardment of Li-nuclei with mass number 7.

Today the equivalence of mass and energy (inertia of energy) postulated by Einstein is one of the most certain foundations of nuclear physics. It gives rise to the programme for interpreting the mass values of the fundamental particles as energy eigenvalues.

Note 14. (pp. 153–154)

While, due to different perturbing effects, no progress has been made in the meantime regarding the red shift of spectral lines in the sun, there is good agreement between theory and experiment for the red shift in the companion of Sirius. This is about 30 times as great as for the sun, due to the extreme density of the star. For details see the Proceedings of the Congress "Jubilee of Theory of Relativity", Berne, 1955, quoted in Note 1.

Note 15. (pp. 158–160)

The fact that the momentum–energy law (341 a), p. 157, for matter (including electromagnetic fields) is a consequence of the field equations (401) for gravitation alone leads one to expect that the law of motion for material particles (which can be described phenomenologically by the energy–momentum tensor Θ_{ik}, Eq. (322), p. 117) must also follow from these field equations without further assumptions.

This has indeed been proved in a series of papers by Einstein and collaborators, and later by Infeld and collaborators.† They considered particularly a point singularity (world line singularity in the four-dimensional space-time) in an external field and showed that the vanishing

---

† A. Einstein and J. Grommer, *S.B. preuss. Akad. Wiss.* (1927) 6 and 235.

A. Einstein, L. Infeld and B. Hoffmann, *Ann. Math., Princeton*, **39** (1938) 65.

L. Infeld, *Phys. Rev.*, **53** (1938) 836.

A. Einstein and L. Infeld, *Ann. Math., Princeton*, **41** (1940) 455; *Canad. J. Math.*, **1** (1949) 209.

L. Infeld and A. Schild, *Rev. Mod. Phys.* **21** (1949) 408 (limit of particle mass tending to zero).

L. Infeld and A. Scheidegger, *Canad. J. Math.*, **3** (1951) 195.

L. Infeld, *Acta phys. polon.* **13** (1954) 187.

See also Bergmann, *Introduction to the Theory of Relativity*, (New York 1942) Chap. XV.

of the covariant divergence of G_{ik} has the consequence that the assumption of a singularity of the metric field on a world line is only compatible with the validity of the field equations $G_{ik} = 0$ (or $R_{ik} = 0$) outside this world line, if the latter is a geodesic.

In order to prove this, approximation methods had to be applied. The most convenient one is a development in powers of c^{-2}, which means in powers of the operations of time derivatives (quasi-static fields, cf. § 58(α)). Another one, applied by Infeld and Schild, is the passing to the limit of a vanishing mass of the test body, which also applies in a rapidly varying field.

The result implies the non-existence of a solution of the gravitational field equations corresponding to two point-masses at rest. There is, however, a static solution with a singularity of the metric field on a line joining two points of space, which describes a one-dimensional density of matter.

In a similar way it can also be shown that, as a consequence of the gravitational equations alone, an electrically charged point-mass obeys a law of motion including the force (216) [see Eq. (225 a), p. 157 of the text].

The representation of matter by a point singularity may be of some formal mathematical interest and also convenient in some applications, but I wish to point out that it is not of basic significance in the investigation of the laws of motion in general relativity theory. The material energy tensor T_{ik} can also be introduced formally, without expressing it by other quantities, in the field equation (401) to characterize small, but finite, space-time regions in which the left-hand side is different from zero.[†] The divergence equations (341a), p. 157, which follow, can then be used to investigate the motion of the centre of these regions, much as for moving point singularities. In this way, N. Hu[‡] has investigated the small damping forces due to the emission of gravitational waves. Although they appear only in a very high and practically unobservable approximation in the development in powers of c^{-2}, they are of fundamental interest, as in this approximation it becomes less and less possible to distinguish sharply between the external field and the self-field of matter.

Note 16. (p. 169)

Most experts agree that the best determination of the light deflection by the sun is still the one by Campbell and Trumpler (Lick Observatory), which is in good agreement with the theoretical value. See the report by Trumpler and its discussion in the Proceedings of the Congress "Jubilee of Theory of Relativity," Berne 1955.

Note 17. (pp. 174, 178 and 179)

It is an interesting open question whether there exist rigorous solutions of the vacuum field equations $R_{ik} = 0$ which are everywhere regular and approach the line element of special relativity at the infinity of three-

---

† See H. Weyl, *Raum–Zeit–Materie*, (5th edn., 1923) § 38 and also V. Fock, *Theory of Space, Time and Gravitation* (London 1958).

‡ N. Hu, *Proc. Roy. Irish Acad.*, A **51** (1947) 87.

dimensional space. It has been proved that no static or stationary solutions of this kind exist (see the following Note), but one should expect such time-dependent solutions to correspond to standing gravitational waves (e.g. spherical ones). One can easily see that no such solutions exist for plane waves. For cylindrical waves, A. Einstein and N. Rosen, *J. Franklin Inst.*, **223** (1937) 43, have constructed such rigorous solutions which, however, do not approach the pseudo-Euclidean line element at the infinity of the space.

A decision of the more general mathematical problem of the existence of such rigorous solutions is desirable. If they exist, it would not be possible to formulate the "Mach principle" (p. 179) in such a way that it is a consequence of the relativistic field equations. This principle has to be reconsidered anyhow in the light of the more recent development of the cosmological problem (see Note 19).

Note 18. (p. 182, footnote 365)

After *Serini* had settled the non-existence of regular static solutions of the vacuum field equations $R_{ik} = 0$, the next question was to admit in the premise non-vanishing, but time-independent g_{a4} ($a = 1, 2, 3$). A first step in the proof of the non-existence of these more general stationary solutions was made by A. Einstein and W. Pauli, in *Ann. Math., Princeton*, (2), **44** (1943) 131. They showed that if such solutions exist, the deviations of the metric field from the pseudo-Euclidean at large distances r must decrease more rapidly than r^{-1}. (The old method of Serini is reprinted there in the Appendix.)

This restriction has been overcome by A. Lichnerowicz, *C. R. Acad. Sci., Paris*, **222** (1946) 432 [see also his book, *Théories rélativistes de la gravitation et de l'électromagnétisme*, (Paris 1955)] who proved in full generality the non-existence of stationary solutions of the equations $R_{ik} = 0$ which approach the pseudo-Euclidean line element at infinity. His method, like Serini's, is to show the vanishing of certain integrals with a positive-definite integrand.

Note 19. [pp. 179–183, (§ 62)] The cosmological problem

Since the first publication of this book an important new development has occurred. New solutions of Einstein's field equations describing a world, homogeneous in space, with a time-dependent metric have been found by A. Friedmann.† They also exist without Einstein's cosmological term [$-\lambda g_{ik}$ in Eq. (452)] for all three cases of positive, zero and negative constant curvature of the three-dimensional space. These solutions have been applied for the first time to the actual universe by G. Lemaître.‡ He also showed that Einstein's static solution is unstable with respect to a time variation of the density of matter. The application of these solutions to the actual universe became possible by Hubble's discovery of a red shift of the spectral lines emitted by nebulae, which is proportional to their distance. This cannot well be interpreted other than as a Doppler shift

---

† A. Friedmann, *Z. Phys.* **10** (1922) 377 and **21** (1924) 326.

‡ G. Lemaître, *Ann. Soc. Sci. Brux.* A **47** (1927) 49.

due to a velocity of the nebulae in the sense of an expansion of the whole system of matter.

Einstein[†] was soon aware of these new possibilities and *completely rejected the cosmological term* as superfluous and no longer justified. I fully accept this new standpoint of Einstein's.[‡]

The "Ansatz" of Friedmann for the metric is

$$ds^2 = R^2(t)\, d\sigma^2 - dx_4^2 \qquad \text{with} \quad x_4 = ct, \tag{1}$$

where $d\sigma$ is a three-dimensional time-independent line element corresponding to a space with constant curvature, which can be normalized to $\epsilon = +1, 0$ or -1 so that for $\epsilon \neq 0$, x^1, x^2, x^3 are measured in units of the radius of curvature $R(t)$. The time scale is normalized by the choice $g_{44} = -1$ in (1) and the x^a $(a = 1, 2, 3)$ are constant for matter following the expansion of the space. For $d\sigma^2 = \gamma_{ab}\, dx^a dx^b$ $(a, b, = 1, 2, 3)$ any of the expressions (122), (124) or (126) with $1/a^2 = \epsilon$ can be chosen, as for instance

$$d\sigma^2 = \frac{1}{[1 + (\epsilon/4)r^2]^2} \sum_a (dx^a)^2 \qquad \text{with} \quad r^2 = \sum_a (x^a)^2. \tag{2}$$

For the contracted curvature tensor P_{ab} which belongs to $d\sigma^2$ one has $P_{ab} = -2\epsilon\gamma_{ab}$, according to (117) and since $n = 3$. The equation for the geodesics give the result[||] that for a material particle

$$|p| \cdot R = \text{const.} \tag{3}$$

where $p = mv[1 - (v^2/c^2)]^{-\frac{1}{2}}$ is its momentum. If one passes to the de Broglie wavelength $\lambda = h/p$, one can write (3) also as

$$R/\lambda = \text{const.} \tag{3a}$$

which then also holds for light (photons). For the time scale of the line element, the square of which is given by (1) with $g_{44} = -1$, the velocity of light is constant, and for the frequency of light in this time scale it follows that

$$\nu \cdot R = \text{const.} \tag{3b}$$

This was shown by M. v. Laue[¶] without use of any quantum-theoretical concepts by pointing out that, due to the conformal invariance of the Maxwell equations, the frequency ν' corresponding to the line element $ds^2 = R^2(t')(d\sigma^2 - c^2 dt'^2)$ must be independent of time.

If μ is the mass density, $u = \mu c^2$ the corresponding energy density, p

---

[†] A. Einstein, *S.B. preuss. Akad. Wiss.* (1931) 235. Appendix to the 2nd edn. (1945) of *The Meaning of Relativity*, reprinted in all later editions.

[‡] Cf. for the following also the books: R. C. Tolman, *Relativity, Thermodynamics and Cosmology* (Oxford 1934), M. v. Laue, *Relativitätstheorie*, Vol. 2: *Allgemeine Relativitätstheorie* (3rd edn., 1953) p. 52; P. Jordan, *Schwerkraft und Weltall*, (2 edn., 1955).

[||] See literature, quoted in footnote[‡]. One can consider the special case

$$x^2 = x^3 = 0, \quad x^1 = r, \quad d\sigma = dr[1 + (\epsilon/4)r^2]^{-1}.$$

For a material particle one has $v = R\, d\sigma/dt$.

[¶] M. v. Laue, *S.B. preuss. Akad. Wiss.* (1931) 723.

the pressure, so that u and p are time-dependent, but constant in space, we have for the components of the energy tensor[†] T_{ik} with a, $b = 1, 2, 3$

$$T_{44} = u, \quad T_{4a} = 0, \quad T_{ab} = pg_{ab} = pR^2\gamma_{ab} \tag{4}$$

Hence

$$\left.\begin{array}{l} T = -u + 3p, \\ T_{44} - \tfrac{1}{2}g_{44}T = \tfrac{1}{2}(u + 3p), \\ T_{ab} - \tfrac{1}{2}g_{ab}T = \tfrac{1}{2}(u + 3p). \end{array}\right\} \tag{5}$$

The computation of the components of R_{ik} gives, if a dot denotes differentiation with respect to $x_4 = ct$,

$$R_{44} = \frac{3\ddot{R}}{R}, \quad R_{4a} = 0, \quad R_{ab} = -\gamma_{ab}(2\epsilon + \dot{R}^2 + R\ddot{R}). \tag{6}$$

The field equations without the cosmological λ-term in the form (401 a), p. 161 of the text,[‡]

$$R_{ik} = -\kappa(T_{ik} - \tfrac{1}{2}g_{ik}T),$$

give therefore

$$\left.\begin{array}{l} 3\dfrac{\ddot{R}}{R} = -\dfrac{\kappa}{2}(u + 3p), \\[2mm] 2\epsilon + 2\dot{R}^2 + R\ddot{R} = \dfrac{\kappa}{2}R^2(u - p), \end{array}\right\} \tag{7}$$

or also

$$\left.\begin{array}{l} 3\dfrac{\dot{R}^2 + \epsilon}{R^2} = \kappa u, \\[2mm] -\dfrac{2R\ddot{R} + \dot{R}^2 + \epsilon}{R^2} = \kappa p. \end{array}\right\} \tag{8}$$

The energy law (vanishing of the covariant divergence of T_{ik}) gives for $i = 4$ (the other three equations are identically fulfilled)

$$\dot{u} + \frac{3\,\dot{R}}{R}(u + p) = 0, \tag{9}$$

which also follows directly from (7) or (8). This equation can also be written

$$d(uR^3) + p\,d(R^3) = 0, \tag{9a}$$

which expresses the constancy of entropy in a material volume.

In the case of pure matter one obtains

$$p = 0, \quad uR^3 = \text{const.} = \tfrac{1}{3}A, \tag{9b}$$

in the case of pure radiation

$$p = \tfrac{1}{3}u, \quad uR^4 = \text{const.} \tag{9c}$$

[†] See Eq. (362), p. 133 and also pp. 163 and 170 of the text. One has to put here $u_4 = c$ and $u_a = 0$ for $a = 1, 2, 3$.

[‡] Compare footnote 320, p. 163 on our notation for κ.

The case $p = 0$ alone seems to be of practical interest, and will therefore be assumed in the following. Then (9 b) inserted in the first line of (8) gives immediately

$$R(\dot{R}^2 + \epsilon) = \kappa A,$$

or

$$\dot{R}^2 = \frac{\kappa A}{R} - \epsilon. \tag{10}$$

This equation can easily be integrated. For zero curvature one has, for instance,

$$\epsilon = 0, \qquad \tfrac{2}{3} R^{3/2} = \sqrt{(\kappa A)}\, c(t - t_0). \tag{11}$$

The measured *Hubble constant* is therefore

$$H \equiv \frac{1}{t_H} = c\,\frac{\dot{R}}{R}, \qquad \frac{1}{t_H} = \frac{c\sqrt{(\kappa A)}}{R^{3/2}} = \frac{2}{3}\,\frac{1}{t - t_0}, \tag{12}$$

or

$$t - t_0 = \tfrac{2}{3} t_H = \tfrac{2}{3} H^{-1}. \tag{13}$$

In this solution the time t_0 corresponds to $R = 0$, $u = \infty$ where the idealizations assumed in the model are no longer justified. In the sense that it is not possible theoretically to trace further back the extremely dense state of matter which existed at the time t_0 before the present, the time $t - t_0$ can be interpreted as the age of the universe.

For the other cases $\epsilon = +1$ and $\epsilon = -1$ we refer to the literature quoted in footnotes (†) and (‡), p. 220. If $H = 1/t_H$ is still defined by (12), and $R = 0$ for $t = t_0$, one finds for the "age of the universe" $t - t_0$ the inequalities

$$t - t_0 < \tfrac{2}{3} t_H \quad \text{for} \quad \epsilon = +1, \tag{13a}$$

$$t - t_0 > \tfrac{2}{3} t_H \quad \text{for} \quad \epsilon = -1. \tag{13b}$$

In the latter case the time length $t - t_0$ is also limited for known t_H by the possibilities for the values of R/kA.

An empirical lower bound for $t - t_0$ is given by the fact that the age of the firm crust of the earth is known to be about 3×10^9 years. For some time there seemed to be a discrepancy with the empirical value of the Hubble constant, which led to a value of the age of the universe which was too low.† More recently the astronomers found, however, a lower value of the Hubble constant H, leading to

$$t_H = \frac{1}{H} = (5 \cdot 6 \pm 2) \times 10^9 \text{ years.} ‡$$

There no longer seems to be any serious discrepancy between the empirical value of the Hubble constant, the age of the earth and the equations of general relativity without the cosmological term.‖

† See A. Einstein, *The Meaning of Relativity*, Appendix; and P. Jordan, *Schwerkraft und Weltall*, (2nd edn., 1955).
‡ A. R. Sandage, *Astr. J.*, **59** (1954) 180.
‖ See report H. P. Robertson in Proceedings of the Congress, "Jubilee of Theory of Relativity", Berne, 1955.

Note 20. (p. 189)

Here the invariant [see Eq. (54 a), p. 33)]

$$\frac{1}{\sqrt{(-g)}}(F_{23}F_{14} + F_{31}F_{24} + F_{12}F_{34}) = \tfrac{1}{4}F_{ik}F^{*ik}$$

has been erroneously omitted. The quantity is not invariant with respect to reflections of the space coordinates ($x'_a = -x_a$ for $a = 1, 2, 3$), but changes its sign under this transformation (pseudo-scalar). As the Lagrangians occurring in nature are invariant also under spatial reflections, only *the square of this quantity* can occur. This is actually the case in the application of the quantum theory of the positon to the polarization of the vacuum in homogeneous external electric and magnetic fields, as was shown by W. Heisenberg and H. Euler, *Z. Phys.*, **98** (1936) 714. It also plays a rôle in the non-linear electrodynamics of M. Born, *Proc. Roy. Soc.* A143 (1934) 410 and of M. Born and L. Infeld, *Proc. Roy. Soc.* A144 (1934) 425, **147** (1934) 522, **150** (1935) 141.

The other invariants mentioned under (2) to (4), p. 189 have, however, been entirely abandoned because of their lack of gauge invariance.

Note 21. (pp. 192, 202 and 205)

After the properties of negatons and positons had turned out to exactly symmetrical, the negative antiproton was found experimentally.[†]

All arguments in the text, which are based "on the asymmetry between the two kinds of electricity" have therefore to be discarded.

Note 22. (pp. 192–202) **Theory of Weyl**

While there has never been an empirical reason to believe that lengths and times of measuring rods and clocks depend on their prehistory (see § 65(β)), the theoretical situation has also changed very much since the establishment of wave mechanics. In this theory the complex wave equation which describes electrically charged matter (the wave function ψ may have one or several components) permits the group ($\hbar$ = Planck's constant divided by 2π, ϵ = elementary electric charge)

$$\phi'_i = \phi_i - i\frac{\hbar c}{\epsilon}\frac{\partial f}{\partial x^i}, \qquad \psi' = \psi e^{if(x)} \tag{1}$$

which is in close analogy with the transformation (477) of the original theory of Weyl, but with an imaginary exponent in ψ instead of the real exponent in g_{ik}. Moreover, the connexion of the conservation law for the electric charge with the new group is the same as with the old one.

Both London,[‡] and Weyl[||] himself, recognized this fact immediately

† O. Chamberlain, E. Segrè, C. Wiegand and Th. Ypsilantis, *Phys. Rev.*, **100** (1955) 947.
‡ F. London, *Z. Phys.*, **42** (1927) 375.
|| H. Weyl, *Gruppentheorie und Quantenmechanik* (Leipzig 1928; 2nd edn., 1931); *Z. Phys.* **56** (1929) 330; Rouse Ball lecture 'Geometry and Physics', *Naturwissenschaften*, **19** (1931), pp. 49–58. Report '50 Jahre Relativitätstheorie', *Naturwissenschaften*, **38** (1951) 73.

224 Supplementary Notes

after the discovery of wave mechanics. Since this time, the name "gauge group" has been the usual one for the group (1) in wave mechanics, indicating in this way Weyl's theory with a non-integrable length as its historical origin.

However, there was now no longer any reason to believe in a non-integrability of length and Weyl himself explicitly stated the failure of his old theory. There seems to be now a general agreement that the g_{ik} themselves, and not only their quotients, are determined and that they should be unchanged when a gradient is added to the electromagnetic potentials.†

Note 23. (pp. 205–206) Other attempts of unified field theories

Before we refer in more detail to some proposals for the "unification" of field theories, some basic remarks are necessary on the range of applicability of classical continuous physics in the explanation of the duality of properties of matter, characterized by the intuitive perception of "waves" and of "particles" and described by the new type of statistical laws established in quantum mechanics (or wave mechanics) since 1927.‡ Most physicists, including the author, agree with the analysis of Bohr and Heisenberg in their judgement of the epistemological situation produced by these developments, and therefore hold a complete solution of the open problems of physics through a return to the classical field concepts to be impossible.

On the other hand Einstein, after he had revolutionized the way of thinking in physics with general methods which are also fundamental for quantum mechanics and its interpretation, maintained until his death the hope that even the quantum-features of atomic phenomena could in principle be explained on the lines of the classical physics of fields. While the concept of physical reality in atomic physics has been generalized by Bohr's concept of complementarity in such a way that the whole experimental arrangement is an essential part of the theoretically described phenomena, Einstein wanted to keep to the ideal of classical

† Independent of the transformation of the electromagnetic potentials one can consider the conformal transformation $g'_{ik} = \lambda g_{ik}$ with an arbitrary function $\lambda(x)$ [see pp. 81 and 199 for the Maxwell equations]. As was shown by R. Bach, *Math. Z.* 9 (1921) 110 (see also C. Lanczos, *Ann. Math., Princeton*, 39 (1938) 842), it is mathematically possible to construct field equations with this invariance property, by using in the action principle a scalar density of the second degree in the components of the conform curvature tensor [for the latter see H. Weyl, *Nachr. Ges. Wiss. Göttingen, math.-naturw. Kl.*, (1921) 99].

At one time, Einstein [*S.B. preuss. Akad. Wiss.* (1921) 261] too, took gravitational equations with conformal invariance into consideration. This point of view, however, was soon abandoned by him and others, as it does not seem to have any physical meaning.

‡ It has to be emphasized that not only the particle concept of classical mechanics but also the wave concept of classical field theory has undergone a fundamental change in quantum mechanics. Indeed, as was shown by Schrödinger, systems of interacting particles can only be described by waves in a multidimensional configuration space and not by waves in ordinary space-time. In cases where particles are generated and annihilated (change of the total number of particles in time) sets of such configuration spaces with different numbers of dimensions are needed. Equivalent to this is the so-called "field-quantization," in which the amplitudes of the wave-fields in ordinary space-time are replaced by suitably chosen operators. See P. Jordan and O. Klein, *Z. Phys.*, 45 (1927) 751; P. Jordan and E. Wigner, *Z. Phys.*, 47 (1928) 631; V. Fock, *Z. Phys.*, 75 (1932) 622.

celestial mechanics that the objective physical state of a system must be entirely independent of the way in which it is observed.

Although Einstein frankly admitted that his hopes for a total solution on these lines had not yet been fulfilled and that the possibility of such a theory had not yet been proved by him, he considered this to be an open question. If he speaks of a "unified field theory", he therefore has in mind this ambitious programme of a theory which solves all problems regarding the elementary particles of matter with the help of classical fields which are everywhere regular (free of singularities).

The physicists who follow the Heisenberg–Bohr interpretation of quantum mechanics take into consideration the unification of classical fields, like the gravitational and the electromagnetic field, only in a restricted way, as long as the sources of the fields, such as masses and electric charges, are not accounted for. For the description of the sources and their properties, matter wave fields and their quantization with a statistical interpretation are assumed.† But even this other kind of programme seems to be still far from its realization.

The reader of the original text of § 67 will see that I was already at that time very doubtful regarding the possibility of explaining the atomism of matter, and particularly of electric charge, with the help of classical concepts of continuous fields alone. In this connexion it should be remembered that the atomicity of electric charge had already found its expression in the specific numerical value of the fine structure constant, a theoretical understanding of which is still missing today. Particularly, I felt rather strongly the fundamental character of the duality (or, as one says since 1927, complementarity) between the measured field and the test body used as measuring instrument. This question has since been discussed by N. Bohr at the eighth Solvay Conference of Physics in 1948 [see report on this Congress (Bruxelles 1950) pp. 376–380].

After these more general introductory remarks we shall discuss in the following two types of attempts at unification of fields, which formally generalize the original relativity theory of Einstein in different directions.

(a) *Theories with unsymmetrical g_{ik} and $\Gamma^l{}_{ik}$.*‡ The theories in question exist in two versions: the earlier one, in which symmetrical or non-symmetrical $\Gamma^l{}_{ik}$ have been considered as the only primary quantities and the later one, in which both unsymmetrical $\Gamma^l{}_{ik}$ and unsymmetrical g_{ik} or g^{ik} are considered as independent variables. In the former theories the metric tensor is assumed to be proportional to the symmetric part R_{ik} of the contracted curvature tensor.

This assumption is only justified if a cosmological term of the field equation is existent. As this is no longer justified, one is left with the second type of theory in which the unsymmetric $\Gamma^l{}_{ik}$ and g_{ik} are considered as

---

† See footnote ‡, p. 224.

‡ Compare A. S. Eddington, *The Mathematical Theory of Relativity* (Cambridge 1924); different papers by A. Einstein in *S.B. preuss. Akad. Wiss.*, (1923–1925); E. Schrödinger, *Space–Time-Structure* (Cambridge 1950), where the author's papers in *Proc. Roy. Irish Acad.* (1943–1948) and the equation of A. Einstein and E. G. Straus, *Ann. Math., Princeton*, (2) **47** (1946) 731, are summarized. See also A. Einstein, *Ann. Math., Princeton*, (2) **46** (1945) 538.

independent variables. Consequently, only this second type of theory has later been considered by Einstein.

All these theories are exposed to the objection that they are in disagreement with the *principle that only irreducible quantities should be used in field theories*. This principle is indeed satisfactory from a formal point of view and has been verified empirically without exception in physics until now. Therefore, I believe† that *cogent mathematical reasons*, (for instance invariance postulates of a wider group of transformations) *have to be given why a decomposition of the reducible quantities used in the theory (for instance R_{ik}, g_{ik} and $\Gamma^l{}_{ik}$) does not occur*. This has not been done at all in the earlier literature.‡

Einstein, however, was well aware of this objection, which he weighed carefully in his later work.‖

In order to explain the point of view and results of Einstein and Kaufman, we notice first that the correct expression for the contracted curvature tensor R_{ik} in terms of the non-symmetrical $\Gamma^l{}_{ik}$ is¶

$$R_{ik} = \Gamma^s{}_{ik,s} - \Gamma^s{}_{is,k} - \Gamma^s{}_{it}\Gamma^t{}_{sk} + \Gamma^s{}_{ik}\Gamma^t{}_{st}, \tag{1}$$

where now the order of the lower indices in the Γ's is important. The authors further point out that this expression is invariant with respect to *λ-transformations*, defined by

$$\Gamma^{l\prime}{}_{ik} = \Gamma^l{}_{ik} + \delta^l{}_i\,\lambda_{,k} \tag{2}$$

where $\lambda(x)$ is an arbitrary function. They now introduce the postulate that all equations should be invariant with respect to these *λ-transformations* (λ-invariance). *Formally this postulate makes the use of symmetrical Γ's impossible*.

As a second postulate Einstein and Kaufman introduce the *transposition invariance*. It states *that all equations stay valid if all quantities A_{ik} are replaced by their transpose $A^T{}_{ik} = A_{ki}$*. The R_{ik} expressed by the $\Gamma^l{}_{ik}$ are not transposition-invariant. The invariance can, however, be obtained if new quantities, defined by

$$\begin{aligned} U^l{}_{ik} &= \Gamma^l{}_{ik} - \Gamma^t{}_{it}\delta^l{}_k, \\ \Gamma^l{}_{ik} &= U^l{}_{ik} - \tfrac{1}{3}U^t{}_{it}\delta^l{}_k, \end{aligned} \tag{3}$$

are introduced. The contracted curvature tensor, expressed in terms of the $U^l{}_{ik}$ is given by

$$R_{ik}(U) = U^s{}_{ik,s} - U^s{}_{it}U^t{}_{sk} + \tfrac{1}{3}U^s{}_{is}U^t{}_{tk} \tag{4}$$

† The same view is held by H. Weyl [*Naturwissenschaften*, **38** (1951) 73, and Proceedings of the Berne Congress, 1955].

‡ Already in a theory on the symmetrical $\Gamma^l{}_{ik}$ as the only independent field variables it is for instance arbitrary to use $\sqrt{(-\det|R_{ik}|)}$ as density in the action integral. The split of R_{ik} into its symmetrical and antisymmetrical part gives many more possibilities (see Note 7).

‖ A. Einstein and B. Kaufman, *Ann. Math.*, Princeton, **62** (1955) 128; see also *The Meaning of Relativity*, (5th edn., Princeton 1955), Appendix II.

¶ In the following, the operation $(\ldots)_{,k}$ always means ordinary differentiation with respect to x^k. The total sign of R_{ik}, chosen by Einstein and Kaufman and reproduced here, is opposite to that used elsewhere in this book.

and is now transposition-invariant. The λ-transformation of the $U^l{}_{ik}$ is given by

$$U^{l'}{}_{ik} = U^l{}_{ik} + (\delta^l{}_j \lambda_{,k} - \delta^l{}_k \lambda_{,i}).$$ (5)

For the transformation law for the $U^l{}_{ik}$ under coordinate transformations we refer to the paper quoted in footnote ||, p. 226. One obtains the field equations by variation of the action integral with *respect to g^{ik} and $U^l{}_{ik}$ as independent variables.*

Instead of the g^{ik} one can also use the tensor density with components $\mathfrak{g}^{ik}$, which in four-dimensional space-time are given by

$$\mathfrak{g}^{ik} = \frac{g^{ik}}{\sqrt{(-\det|g^{ik}|)}}, \qquad g^{ik} = \frac{\mathfrak{g}^{ik}}{\sqrt{(-\det|\mathfrak{g}^{ik}|)}}.$$ (6)

It is in the spirit of ordinary general relativity (see p. 48) that one restricts the scalar density $\mathfrak{L}$ to be used in the variation principle by the postulate that it should contain no derivatives of g^{ik}, only the first derivatives of the $U^l{}_{ik}$, and that it should depend linearly on the latter.† This postulate together with the postulates of λ-invariance and transposition invariance already mentioned, leads to an $\mathfrak{L}$ which is linear in the R_{ik}, expressed in terms of the $U^l{}_{ik}$. If a "cosmological" term independent of R_{ik} is avoided, then a suitable definition of the field g^{ik}, with $\mathfrak{g}^{ik}$ defined by (6), leads to Einstein's choice of the scalar density

$$\mathfrak{L} = \mathfrak{g}^{ik} R_{ik}$$ (7)

as integrand of the action integral which fulfils all the above postulates.

We refer to the quoted literature for the field equations following from it and for the identities between them. The special case, in which the antisymmetric parts of g_{ik} and $\Gamma^l{}_{ik}$ vanish, leads back to the usual field equations of general relativity without matter (see Note 8).

Whether the field equations of this theory, which are based on the formal postulates of λ-invariance and of transposition invariance without any obvious geometrical and physical meaning, can actually be connected with physics at all, is rather doubtful.

A leading physical principle like the principle of equivalence in general relativity, which is based on general empirical evidence, is entirely missing in this "unified field theory". Moreover, in ordinary general relativity it is the line element, and with it the quadratic form $g_{ik} dx^i dx^k$, which has a direct physical meaning, rather than the pseudotensor $\Gamma^l{}_{ik}$ which governs the parallel displacement of vectors.

In the following, we shall consider other kinds of attempts to obtain "unified field theories", in which only irreducible quantities are used.

(b) *Five-dimensional and projective theories.*‡ Kaluza|| found an interesting geometrical representation of the generally covariant form of Maxwell's

---

† The discussion of the possible densities of the purely affine theories in Note 7 was made without use of these restrictive postulates.

‡ The reader will find a summary of the theories discussed here in P. G. Bergmann, *An Introduction to the Theory of Relativity* (New York 1942), Chaps. XVII and XVIII.

|| Th. Kaluza, *S.B. preuss. Akad. Wiss.* (1921) 966.

electrodynamics [§§ 23(a) and 54], which was later improved and
generalized by Klein.†

One considers a five-dimensional space with a "cylindrical" metric
given by

$$ds^2 = \gamma_{\mu\nu} dx^\mu dx^\nu \tag{8}$$

(in the following, Greek indices μ, ν run from 1 to 5, Latin indices i, k, ...
from 1 to 4). The condition of cylindricity can best be described in a
particular coordinate system‡ in which the $\gamma_{\mu\nu}$ are independent of x^5,

$$\frac{\partial \gamma_{\mu\nu}}{\partial x^5} = 0. \tag{9}$$

Moreover, Kaluza and Klein originally assumed

$$\gamma_{55} = 1. \tag{10}$$

The positive sign of γ_{55} implies that the fifth dimension is metrically
space-like. The reason for this choice will become obvious later. Besides
the general coordinate transformations of the four coordinates x^k, as
used in general relativity, the preferred coordinate systems in question
permit the group

$$x'^5 = x^5 + f(x^1, ..., x^4). \tag{11}$$

Writing (8) in the form

$$ds^2 = (dx^5 + \gamma_{i5} dx^i)^2 + g_{ik} dx^i dx^k, \tag{12}$$

one sees that the g_{ik} are invariant under the transformation (11),

$$g'_{ik} = g_{ik}, \tag{13}$$

while

$$\gamma'_{i5} = \gamma_{i5} - \frac{\partial f}{\partial x^i}. \tag{14}$$

The comparison of (8) and (12) gives

$$\gamma_{ik} = g_{ik} + \gamma_{i5} \gamma_{k5}. \tag{15}$$

If g^{ik} is as usual the reciprocal matrix of g_{ik}, $\gamma^{\mu\nu}$ the reciprocal matrix of
$\gamma_{\mu\nu}$, one easily finds

$$\det|\gamma_{\mu\nu}| = \det|g_{ik}|,$$
$$\gamma^{55} = 1 + \gamma^{ik} \gamma_{i5} \gamma_{k5}, \qquad \gamma^{i5} = -g^{ik} \gamma_{k5}, \qquad \gamma^{ik} = g^{ik}. \tag{16}$$

The form of the transformation (14), which is analogous to the gauge

† O. Klein, *Nature, Lond.*, **118** (1926) 516; *Z. Phys.* **37** (1926) 895 [already in these
papers a periodic dependence of the metric on the fifth co-ordinate is taken into considera-
tion]; *Z. Phys.* **46** (1928) 188; *Ark. Mat. Astr. Fys.* **34** (1946) 1; see also Klein's report in
the Proceedings of the Berne Congress, 1955.

‡ For the formulation of the theory in a general coordinate system see Bergmann, footnote
‡, p. 227, *ibid.*

group, suggests the *identification of γ_{i5} with the electromagnetic potential, ϕ_i,* apart from a constant factor of proportionality. The antisymmetric tensor

$$\frac{\partial \gamma_{k5}}{\partial x^i} - \frac{\partial \gamma_{i5}}{\partial x^k} = f_{ik}, \tag{17}$$

which is invariant with respect to the "gauge transformation" (14), is then proportional to the electromagnetic field strength. We shall come back to the definition of the factor of proportionality later.

The *geodesics* of the metric (8) or (12) can also be interpreted physically along these lines. From the independence of the $\gamma_{\mu\nu}$ of x_5 it follows easily that for geodesics the two expressions

$$\frac{dx^5}{ds} + \gamma_{i5} \frac{dx^i}{ds} = \text{const.} = C, \tag{18}$$

$$g_{ik} \frac{dx^i}{ds} \frac{dx^k}{ds} = \text{const.} = -1, \tag{18a}$$

are constant separately for a suitable choice of the parameter s. The constant in (18a) can be normalized to -1. For the equations of the geodetics one obtains

$$\frac{d}{ds}\left(g_{ik}\frac{dx^k}{ds}\right) - \frac{1}{2}\frac{\partial g_{rs}}{\partial x^i}\frac{dx^r}{ds}\frac{dx^s}{ds} = C \cdot f_{ik}\frac{dx^k}{ds}. \tag{19}$$

This is, however, the equation for the orbit of a charged particle in external gravitational and electromagnetic fields. The integration constant C is thereby proportional to the quotient e/m of charge and mass of the particle.

We mention here briefly another equivalent way to formulate the geometrization of the gravitational and the electromagnetic field, namely the *projective* formulation. Many authors have contributed to it, among them Veblen and Hoffmann, Schouten and van Dantzig and myself.† Bergmann† has shown, however, that—in contrast to what I believed myself for a while—this formulation is *not* more general than Kaluza's, and that it is easy to pass from either of these two formulations to the other. Introducing the homogeneous coordinates X^ν by

$$X^\nu = f^\nu(x^i)e^{x^5} \tag{20}$$

(with arbitrary functions f^ν) with the inversion

$$x^i = g^i\left(\frac{X^1}{X^5}, ..., \frac{X^4}{X^5}\right)$$

$$x^5 = \log\left\{X^5 F\left(\frac{X^1}{X^5}, ..., \frac{X^4}{X^5}\right)\right\} = \log H^{(1)}(X^1, ... X^5), \tag{20a}$$

where $H^{(1)}$ is a homogeneous function of the degree 1, one easily sees that the "gauge" transformation (11), combined with the general transformations of the x^k corresponds exactly to the *group of all homogeneous*

† For literature see, besides Bergmann, footnote ‡, p. 227, *ibid.*, for instance C. Ludwig-*Fortschritte der projektiven Relativitätstheorie*, (Braunschweig 1951).

transformations of the first degree of the X^ν. It is the latter which is considered in the projective formulation. Because of its one-to-one correspondence with Kaluza's† formulation we shall not discuss the projective form any further here.

Kaluza's geometric form of the generally covariant laws of the electromagnetic field, as represented here, *is in no way a "unification" of the electromagnetic and the gravitational field*. On the contrary, every theory which is generally covariant and gauge-invariant can also be formulated in Kaluza's form. In the case of the absence of electric charges (currents), the generally covariant form of Maxwell's equations follows by variation of an action integral with the density [see Eq. (231 b), p. 158 and Eqs. (403), (404), p. 162]

$$\mathfrak{L} = \sqrt{(-g)}\left(R + \frac{\kappa}{2}F_{ik}F^{ik}\right), \tag{21}$$

if F_{ik} is the electromagnetic field strength. A more complicated dependence on the field strength of the scalar density in the action integral would, however, equally well be compatible with a cylindrically symmetric five-dimensional metric.

Kaluza and Klein derived, however, a further interesting result. They computed the scalar P of the curvature tensor, which corresponds to the particular five-dimensional metric given by (8) or (12) and found

$$P = R + \tfrac{1}{4}f_{ik}f^{ik} \tag{22}$$

where R is the curvature tensor derived from the four-dimensional metric corresponding to $ds^2 = g_{ik}\,dx^i\,dx^k$, and f_{ik} is defined by (17). This becomes identical with (21) if one puts

$$f_{ik} = \sqrt{(2\kappa)}F_{ik}, \qquad \gamma_{i5} = \sqrt{(2\kappa)}\phi_i. \tag{23}$$

Here, it has to be further noticed that the sign of the second term of the right-hand side of (22) would reverse if we had chosen a time-like character for the fifth dimension ($\gamma_{55} = -1$) instead of a space-like one. The space-like character of the fifth dimension had to be chosen so that in (22) the sign of the right-hand side is the same as in (21). One can also say that, for the choice of P as the invariant in the action integral, the empirical sign of the gravitational constant is represented by the space-like sign of γ_{55}.

There is, however, no justification for the particular choice of the five-dimensional curvature scalar P as integrand of the action integral, from the standpoint of the restricted group of the cylindrical metric. The open problem of finding such a justification seems to point to an amplification of the transformation group. This is connected with the possibilities for generalizing Kaluza's formalism, which we shall now discuss briefly.

One of the generalizations of the Kaluza formalism consists in retaining (9) but in getting rid of the condition (10), $\gamma_{55} = 1$. From the standpoint

† For the metric tensor $\Gamma_{\mu\nu}$ corresponding to the X^ν one has, according to (20),

$$\gamma_{55} = \Gamma_{\mu\nu}\frac{\partial X^\mu}{\partial x^5}\frac{\partial X^\nu}{\partial x^5} = \Gamma_{\mu\nu}\dot{X}^\mu\dot{X}^\nu.$$

f the transformation group of general relativity, γ_{55} is then a new scalar eld which is still assumed to be independent of x^5. Putting

$$\gamma_{55} = J, \qquad \gamma_{i5} = Jf_i, \qquad \gamma_{ik} = g_{ik} + Jf_if_k, \tag{24}$$

ne obtains

$$ds^2 = \gamma_{\mu\nu}\,dx^\mu\,dx^\nu = J(dx^5 + f_i\,dx^i)^2 + g_{ik}\,dx^i\,dx^k, \tag{25}$$

ith the "gauge"-group

$$x'^5 = x^5 + f(x^i), \qquad f'_i = f_i - \frac{\partial f}{\partial x^i}. \tag{26}$$

ordan,† reviving earlier ideas of Dirac,‡ tried in an interesting way to se this new field J in order to obtain a theory in which the gravitational onstant of the usual theory is replaced by a time-dependent field. The nathematical side of this theory has also been investigated independently y Thiry.§ As was shown by M. Fierz,‖ the introduction of matter into the heory adds further assumptions, without which the time dependence of tandard lengths derived from atomic dimensions and of the gravitational ctions between material particles are not yet defined. We shall not enter ere into a discussion of the strength of the empirical evidence for this heory.

Another more fundamental generalization of Kaluza's formalism onsists in abandoning the condition (9) of cylindricity. Already in his arly papers of 1926, Klein had discussed *a periodic dependence of all field ariables on x^5.* Normalizing the period to 2π, *this assumption* I *("All omponents $\gamma_{\mu\nu}$ are periodic functions of x^5 with the period 2π")* can also e expressed by a Fourier decomposition

$$\gamma_{\mu\nu}(x^5, x^i) = \sum_{n=-\infty}^{+\infty} \gamma_{\mu\nu}^{(n)}(x^i)e^{inx^5}, \tag{27}$$

with the usual reality condition

$$\gamma_{\mu\nu}^{(-n)} = \left(\gamma_{\mu\nu}^{(n)}\right)^*. \tag{27a}$$

Geometrically one can then interpret x^5 as an angle variable, so that all values of x^5 differing by an integral multiple of 2π correspond to the same point of the five-dimensional space if the values of the x^i are the same. From this assumption alone the existence of a closed geodesic without discontinuity of its directions does not follow. Einstein and Bergmann¶ particularly investigated the consequences of the additional *assumption* II: through each point of the five-dimensional spaces passes exactly one geodesic curve which returns to the same point with continuous direction.

They showed that, in this case, there always exists a particular

---

† P. Jordan, *Schwerkraft und Weltall* (2nd edn., 1955). Originally he formulated his theory in the projective form.

‡ P. A. M. Dirac, *Nature, Lond.*, **139** (1937) 323; *Proc. Roy. Soc.* A**165** (1938) 199.

§ Y. R. Thiry, *Thèse,* (Paris 1951); cf. also A. Lichnerowicz, *Théories rélativistes de la gravitation et de l'électromagnétisme* (Paris 1955).

‖ *Helv. Phys. Acta*, **29** (1956) 128.

¶ A. Einstein and P. G. Bergmann, *Ann. Math., Princeton*, **39** (1938) 683; also A. Einstein V. Bargmann and P. G. Bergmann, *Th. Kármán Anniversary Volume* (Pasadena 1941), p. 212 and Bergmann, footnote ‡, *ibid.*, p. 227.

coordinate system in which

$$\gamma_{55} = 1, \qquad \frac{\partial \gamma_{5i}}{\partial x^5} = 0. \qquad (28)$$

The transformation group stays the same as in the original formalism of Kaluza [see (15), (16)], but the g_{ik} can now depend periodically on x^5.

The authors then established the most general invariant consistent with the transformation group in question and fulfilling the same general condition with respect to the order of differentiation as in ordinary general relativity theory (namely linearity in the second derivatives of the field without any higher derivatives). The corresponding field equations are, in general, integro-differential equations.

While, with these assumptions, one does not get an interpretation or justification of the particular choice of P for the scalar in the action principle, the situation is essentially changed if one retains assumption I alone and drops assumption II. The transformation group is then

$$x'^5 = x^5 + p^5(x^5, x^k), \qquad x'^i = p^i(x^5, x^k) \qquad (29)$$

where the p^ν are arbitrary periodic functions of x^5 with period 2π. This general group has also been considered by Klein, but its mathematical and physical consequences need further investigation.

It is true that the only scalar constructed from the $\gamma_{\mu\nu}$ by the usual process of differentiation only (with the restrictions concerning the order of differentiations usually assumed in general relativity) is now the curvature scalar P of the five-dimensional metric. But it is still an unsolved problem whether or not there exist other invariants in the large, which could be expressed as integrals over suitably chosen closed curves and could also be used in the action principle.[†]

Besides this mathematical problem there is the other difficult problem of the physical interpretation of general functions periodic in x^5 as given by (27). This leads into wave mechanics, and therefore also to the problem of field quantization.[‡] Tensors like $\gamma_{\mu\nu}{}^{(n)}(x^i)$ correspond to a spin value 2, which, by the way, has never been observed in nature and from which alone the observed spin value $\frac{1}{2}$ can never be obtained by composition.

From our standpoint (see the introduction to this note) it is clear therefore that besides the $\gamma_{\mu\nu}(x^5, x^i)$-field there must be other wave-mechanical fields such as, for instance, spinor fields describing particles of low mass.[‡]

The question of whether Kaluza's formalism has any future in physics is thus leading to the more general unsolved main problem of accomplishing a synthesis between the general theory of relativity and quantum mechanics.

† Dr. P. Bergmann kindly drew my attention to the problem whether, in a five-dimensional manifold with the topology of a cylinder infinitely extended in the space described by $x^1, ..., x^4$ and with a metric fulfilling the assumption I, there always exists a particular coordinate system for which

$$\partial \gamma_{\mu 5}/\partial x^5 = 0 \qquad \text{for} \quad \mu = 1, ..., 5.$$

‡ See footnote ‡, p. 224.
‖ See footnote †, p. 228, O. Klein, *ibid.*

AUTHOR INDEX

SUBJECT INDEX

Aberration of light, 17, 20, 74n., 114.
Action function,
  in five-dimensional theory, 230, 232.
  for gravitational field equations, 162, 182, 215, 227.
  for Maxwell's equations, 199.
  mechanical–electrodynamical, 88–9, 158, 216.
  in Mie's theory, 189–90, 192, 223.
  in relativistic mechanics, 119–20.
  thermodynamical, 135–6.
  in Weyl's theory, 200–1.
Action principle, see Action function.
Aether, 4–7.
Affine
  connection, 197, 209, 212.
  geometry, 209.
  space, 209–12.
  tensors, 60–2.
  transformations, 23–7, 36, 81.
    tensor calculus for, 24–7.
Age of universe, 222.
Angle measurement,
  in Euclidean geometry, 28.
  in Riemannian geometry, 36.
Antiproton, 223.
Antisymmetric tensors, see Tensors.

Bianchi identities, 212.
Black-body radiation, 138–9.
Bolyai–Lobachevski geometry, 74, 97.

Canonical equations, 121.
Cayley system of measurement, 74n.
Centrifugal force, see Force.
Charge
  conservation, see Conservation laws.
  distribution, in electron, 185–7, 192.
  total, of particle, 77, 191.
Charged particles, equations of motion, 82–4, 191; see also Particles and Point-charge.
Christoffel symbols, see Geodesic components.
Clock
  paradox, 13, 72–3, 152.
  synchronization, 9.
Closed system, 5.
  electromagnetic, 117.
  gravitational, energy–momentum law for, 176–7.
Complementarity, 224.
Conduction current,
  electron theory, 106.
  Minkowski's electrodynamics, 101–3.
Conductors, moving, 113.
Conservation laws,
  angular momentum, 125.
  charge, 76–8, 200–1.
  energy, 85–6, 107, 162, 189, 200–1.
  energy–momentum, see Energy-momentum law
  momentum, 85–6, 162.
Contraction factor, 2, 3, 9–10.

Contravariant components,
  of tensors, 25.
  of vectors, 27–8.
Convection current, 101–2, 104, 111.
Co-ordinate system,
  Gaussian, 149.
  geodesic, 38, 44.
  Riemannian, 44–8, 50–1.
Coriolis force, see Force.
Cosmological
  models, static (Einstein, de Sitter), 179–83.
    non-static (Lemaître, Friedmann), 219–22.
  term, 181, 204, 219–20.
Covariance,
  of Maxwell's equations, 78–81, 156–7.
  of physical laws, 3, 5, 144, 149–50, 205.
Covariant
  components, of tensors, 25.
    of vectors, 29.
  differentiation, 57, 59.
    in affine space, 209–10.
Curl, 55–6.
Current vector, 76–7.
  in electron theory, 104.
  in Minkowski's electrodynamics, 100–1.
  in Weyl's theory, 201.
Curvature,
  Gaussian, 45–6.
  of Riemannian space, 41–4, 46, 69.
  sectional, 46.
  of space (Weyl's theory), 201.
  space of constant, 49–52.
Curvature invariant, 44, 48.
  in Einstein's theory, 204.
  in five-dimensional theory, 230, 232.
  in space of constant curvature, 49.
  weight of, 197–8.
Curvature tensor, 42–3.
  in affine space, 210–12.
  Bianchi identities, 212.
  contracted, 44, 165.
  geometrical interpretation, 47.
  in non-static universe, 221.
  non-symmetrical, 211.
  in space of constant curvature, 49, 220.
  in terms of non-symmetrical Γ's, 226.
  weight of, 197.
  covariant, 43, 213.
  in affine space, 211.
  in Euclidean space, 48.
  in space of constant curvature, 49.
  weight of, 197.
  weight of, 197.

D'Alembertian, 56.
De Broglie wave-length, 220.
Determinant of orthogonal linear transformations, 23.
Dielectric, moving, 111–13.
Differential operations, invariant, 56–60.
Dipole, moving, 97–8.
Distance between world points, 21, 22; see also Line element.
Divergence (Div and $\mathfrak{D}\mathfrak{i}\mathfrak{v}$), 55–9, 210.

A CATALOG OF SELECTED
DOVER BOOKS
IN SCIENCE AND MATHEMATICS

A CATALOG OF SELECTED
DOVER BOOKS
IN SCIENCE AND MATHEMATICS

QUALITATIVE THEORY OF DIFFERENTIAL EQUATIONS, V.V. Nemytskii and V.V. Stepanov. Classic graduate-level text by two prominent Soviet mathematicians covers classical differential equations as well as topological dynamics and ergodic theory. Bibliographies. 523pp. 5⅜ × 8½. 65954-2 Pa. $10.95

MATRICES AND LINEAR ALGEBRA, Hans Schneider and George Phillip Barker. Basic textbook covers theory of matrices and its applications to systems of linear equations and related topics such as determinants, eigenvalues and differential equations. Numerous exercises. 432pp. 5⅜ × 8½. 66014-1 Pa. $10.95

QUANTUM THEORY, David Bohm. This advanced undergraduate-level text presents the quantum theory in terms of qualitative and imaginative concepts, followed by specific applications worked out in mathematical detail. Preface. Index. 655pp. 5⅜ × 8½. 65969-0 Pa. $14.95

ATOMIC PHYSICS (8th edition), Max Born. Nobel laureate's lucid treatment of kinetic theory of gases, elementary particles, nuclear atom, wave-corpuscles, atomic structure and spectral lines, much more. Over 40 appendices, bibliography. 495pp. 5⅜ × 8½. 65984-4 Pa. $12.95

ELECTRONIC STRUCTURE AND THE PROPERTIES OF SOLIDS: The Physics of the Chemical Bond, Walter A. Harrison. Innovative text offers basic understanding of the electronic structure of covalent and ionic solids, simple metals, transition metals and their compounds. Problems. 1980 edition. 582pp. 6⅛ × 9¼. 66021-4 Pa. $15.95

BOUNDARY VALUE PROBLEMS OF HEAT CONDUCTION, M. Necati Özisik. Systematic, comprehensive treatment of modern mathematical methods of solving problems in heat conduction and diffusion. Numerous examples and problems. Selected references. Appendices. 505pp. 5⅜ × 8½. 65990-9 Pa. $12.95

A SHORT HISTORY OF CHEMISTRY (3rd edition), J.R. Partington. Classic exposition explores origins of chemistry, alchemy, early medical chemistry, nature of atmosphere, theory of valency, laws and structure of atomic theory, much more. 428pp. 5⅜ × 8½. (Available in U.S. only) 65977-1 Pa. $11.95

A HISTORY OF ASTRONOMY, A. Pannekoek. Well-balanced, carefully reasoned study covers such topics as Ptolemaic theory, work of Copernicus, Kepler, Newton, Eddington's work on stars, much more. Illustrated. References. 521pp. 5⅜ × 8½. 65994-1 Pa. $12.95

PRINCIPLES OF METEOROLOGICAL ANALYSIS, Walter J. Saucier. Highly respected, abundantly illustrated classic reviews atmospheric variables, hydrostatics, static stability, various analyses (scalar, cross-section, isobaric, isentropic, more). For intermediate meteorology students. 454pp. 6⅛ × 9¼. 65979-8 Pa. $14.95

RELATIVITY, THERMODYNAMICS AND COSMOLOGY, Richard C. Tolman. Landmark study extends thermodynamics to special, general relativity; also applications of relativistic mechanics, thermodynamics to cosmological models. 501pp. 5⅜ × 8½. 65383-8 Pa. $13.95

APPLIED ANALYSIS, Cornelius Lanczos. Classic work on analysis and design of finite processes for approximating solution of analytical problems. Algebraic equations, matrices, harmonic analysis, quadrature methods, much more. 559pp. 5⅜ × 8½. 65656-X Pa. $13.95

AN INTRODUCTION TO THE PHILOSOPHY OF SCIENCE, Rudolf Carnap. Stimulating, thought-provoking text clearly and discerningly makes accessible such topics as probability, structure of space, causality and determinism, theoretical concepts and much more. 320pp. 5⅜ × 8½. 28318-6 Pa. $8.95

INTRODUCTION TO ANALYSIS, Maxwell Rosenlicht. Unusually clear, accessible coverage of set theory, real number system, metric spaces, continuous functions, Riemann integration, multiple integrals, more. Wide range of problems. Undergraduate level. Bibliography. 254pp. 5⅜ × 8½. 65038-3 Pa. $7.95

INTRODUCTION TO QUANTUM MECHANICS With Applications to Chemistry, Linus Pauling & E. Bright Wilson, Jr. Classic undergraduate text by Nobel Prize winner applies quantum mechanics to chemical and physical problems. Numerous tables and figures enhance the text. Chapter bibliographies. Appendices. Index. 468pp. 5⅜ × 8½. 64871-0 Pa. $11.95

ASYMPTOTIC EXPANSIONS OF INTEGRALS, Norman Bleistein & Richard A. Handelsman. Best introduction to important field with applications in a variety of scientific disciplines. New preface. Problems. Diagrams. Tables. Bibliography. Index. 448pp. 5⅜ × 8½. 65082-0 Pa. $12.95

MATHEMATICS APPLIED TO CONTINUUM MECHANICS, Lee A. Segel. Analyzes models of fluid flow and solid deformation. For upper-level math, science and engineering students. 608pp. 5⅜ × 8½. 65369-2 Pa. $14.95

ELEMENTS OF REAL ANALYSIS, David A. Sprecher. Classic text covers fundamental concepts, real number system, point sets, functions of a real variable, Fourier series, much more. Over 500 exercises. 352pp. 5⅜ × 8½. 65385-4 Pa. $10.95

PHYSICAL PRINCIPLES OF THE QUANTUM THEORY, Werner Heisenberg. Nobel Laureate discusses quantum theory, uncertainty, wave mechanics, work of Dirac, Schroedinger, Compton, Wilson, Einstein, etc. 184pp. 5⅜ × 8½.
60113-7 Pa. $5.95

INTRODUCTORY REAL ANALYSIS, A.N. Kolmogorov, S.V. Fomin. Translated by Richard A. Silverman. Self-contained, evenly paced introduction to real and functional analysis. Some 350 problems. 403pp. 5⅜ × 8½. 61226-0 Pa. $10.95

PROBLEMS AND SOLUTIONS IN QUANTUM CHEMISTRY AND PHYSICS, Charles S. Johnson, Jr. and Lee G. Pedersen. Unusually varied problems, detailed solutions in coverage of quantum mechanics, wave mechanics, angular momentum, molecular spectroscopy, scattering theory, more. 280 problems plus 139 supplementary exercises. 430pp. 6½ × 9¼. 65236-X Pa. $13.95

ASYMPTOTIC METHODS IN ANALYSIS, N.G. de Bruijn. An inexpensive, comprehensive guide to asymptotic methods—the pioneering work that teaches by explaining worked examples in detail. Index. 224pp. 5⅜ × 8½.   64221-6 Pa. $6.95

OPTICAL RESONANCE AND TWO-LEVEL ATOMS, L. Allen and J.H. Eberly. Clear, comprehensive introduction to basic principles behind all quantum optical resonance phenomena. 53 illustrations. Preface. Index. 256pp. 5⅜ × 8½.
65533-4 Pa. $8.95

COMPLEX VARIABLES, Francis J. Flanigan. Unusual approach, delaying complex algebra till harmonic functions have been analyzed from real variable viewpoint. Includes problems with answers. 364pp. 5⅜ × 8½.   61388-7 Pa. $8.95

ATOMIC SPECTRA AND ATOMIC STRUCTURE, Gerhard Herzberg. One of best introductions; especially for specialist in other fields. Treatment is physical rather than mathematical. 80 illustrations. 257pp. 5⅜ × 8½.   60115-3 Pa. $6.95

APPLIED COMPLEX VARIABLES, John W. Dettman. Step-by-step coverage of fundamentals of analytic function theory—plus lucid exposition of five important applications: Potential Theory; Ordinary Differential Equations; Fourier Transforms; Laplace Transforms; Asymptotic Expansions. 66 figures. Exercises at chapter ends. 512pp. 5⅜ × 8½.   64670-X Pa. $12.95

ULTRASONIC ABSORPTION: An Introduction to the Theory of Sound Absorption and Dispersion in Gases, Liquids and Solids, A.B. Bhatia. Standard reference in the field provides a clear, systematically organized introductory review of fundamental concepts for advanced graduate students, research workers. Numerous diagrams. Bibliography. 440pp. 5⅜ × 8½.   64917-2 Pa. $11.95

UNBOUNDED LINEAR OPERATORS: Theory and Applications, Seymour Goldberg. Classic presents systematic treatment of the theory of unbounded linear operators in normed linear spaces with applications to differential equations. Bibliography. 199pp. 5⅜ × 8½.   64830-3 Pa. $7.95

LIGHT SCATTERING BY SMALL PARTICLES, H.C. van de Hulst. Comprehensive treatment including full range of useful approximation methods for researchers in chemistry, meteorology and astronomy. 44 illustrations. 470pp. 5⅜ × 8½.   64228-3 Pa. $11.95

CONFORMAL MAPPING ON RIEMANN SURFACES, Harvey Cohn. Lucid, insightful book presents ideal coverage of subject. 334 exercises make book perfect for self-study. 55 figures. 352pp. 5⅜ × 8¼.   64025-6 Pa. $9.95

OPTICKS, Sir Isaac Newton. Newton's own experiments with spectroscopy, colors, lenses, reflection, refraction, etc., in language the layman can follow. Foreword by Albert Einstein. 532pp. 5⅜ × 8½.   60205-2 Pa. $11.95

GENERALIZED INTEGRAL TRANSFORMATIONS, A.H. Zemanian. Graduate-level study of recent generalizations of the Laplace, Mellin, Hankel, K. Weierstrass, convolution and other simple transformations. Bibliography. 320pp. 5⅜ × 8½.   65375-7 Pa. $8.95

CATALOG OF DOVER BOOKS

THE ELECTROMAGNETIC FIELD, Albert Shadowitz. Comprehensive undergraduate text covers basics of electric and magnetic fields, builds up to electromagnetic theory. Also related topics, including relativity. Over 900 problems. 768pp. 5⅜ × 8¼. 65660-8 Pa. $18.95

FOURIER SERIES, Georgi P. Tolstov. Translated by Richard A. Silverman. A valuable addition to the literature on the subject, moving clearly from subject to subject and theorem to theorem. 107 problems, answers. 336pp. 5⅜ × 8½. 63317-9 Pa. $8.95

THEORY OF ELECTROMAGNETIC WAVE PROPAGATION, Charles Herach Papas. Graduate-level study discusses the Maxwell field equations, radiation from wire antennas, the Doppler effect and more. xiii + 244pp. 5⅜ × 8½. 65678-0 Pa. $6.95

DISTRIBUTION THEORY AND TRANSFORM ANALYSIS: An Introduction to Generalized Functions, with Applications, A.H. Zemanian. Provides basics of distribution theory, describes generalized Fourier and Laplace transformations. Numerous problems. 384pp. 5⅜ × 8½. 65479-6 Pa. $11.95

THE PHYSICS OF WAVES, William C. Elmore and Mark A. Heald. Unique overview of classical wave theory. Acoustics, optics, electromagnetic radiation, more. Ideal as classroom text or for self-study. Problems. 477pp. 5⅜ × 8½. 64926-1 Pa. $12.95

CALCULUS OF VARIATIONS WITH APPLICATIONS, George M. Ewing. Applications-oriented introduction to variational theory develops insight and promotes understanding of specialized books, research papers. Suitable for advanced undergraduate/graduate students as primary, supplementary text. 352pp. 5⅜ × 8½. 64856-7 Pa. $8.95

A TREATISE ON ELECTRICITY AND MAGNETISM, James Clerk Maxwell. Important foundation work of modern physics. Brings to final form Maxwell's theory of electromagnetism and rigorously derives his general equations of field theory. 1,084pp. 5⅜ × 8½. Two-vol. set. Vol. I: 60636-8 Pa. $11.95 Vol. II: 60637-6 Pa. $11.95

AN INTRODUCTION TO THE CALCULUS OF VARIATIONS, Charles Fox. Graduate-level text covers variations of an integral, isoperimetrical problems, least action, special relativity, approximations, more. References. 279pp. 5⅜ × 8½. 65499-0 Pa. $7.95

HYDRODYNAMIC AND HYDROMAGNETIC STABILITY, S. Chandrasekhar. Lucid examination of the Rayleigh-Benard problem; clear coverage of the theory of instabilities causing convection. 704pp. 5⅜ × 8¼. 64071-X Pa. $14.95

CALCULUS OF VARIATIONS, Robert Weinstock. Basic introduction covering isoperimetric problems, theory of elasticity, quantum mechanics, electrostatics, etc. Exercises throughout. 326pp. 5⅜ × 8½. 63069-2 Pa. $8.95

DYNAMICS OF FLUIDS IN POROUS MEDIA, Jacob Bear. For advanced students of ground water hydrology, soil mechanics and physics, drainage and irrigation engineering and more. 335 illustrations. Exercises, with answers. 784pp. 6⅛ × 9¼. 65675-6 Pa. $19.95

CATALOG OF DOVER BOOKS

NUMERICAL METHODS FOR SCIENTISTS AND ENGINEERS, Richard Hamming. Classic text stresses frequency approach in coverage of algorithms, polynomial approximation, Fourier approximation, exponential approximation, other topics. Revised and enlarged 2nd edition. 721pp. 5⅜ × 8½.
65241-6 Pa. $15.95

THEORETICAL SOLID STATE PHYSICS, Vol. I: Perfect Lattices in Equilibrium; Vol. II: Non-Equilibrium and Disorder, William Jones and Norman H. March. Monumental reference work covers fundamental theory of equilibrium properties of perfect crystalline solids, non-equilibrium properties, defects and disordered systems. Appendices. Problems. Preface. Diagrams. Index. Bibliography. Total of 1,301pp. 5⅜ × 8½. Two-vol. set.   Vol. I: 65015-4 Pa. $14.95
Vol. II: 65016-2 Pa. $14.95

OPTIMIZATION THEORY WITH APPLICATIONS, Donald A. Pierre. Broad-spectrum approach to important topic. Classical theory of minima and maxima, calculus of variations, simplex technique and linear programming, more. Many problems, examples. 640pp. 5⅜ × 8½.   65205-X Pa. $14.95

THE CONTINUUM: A Critical Examination of the Foundation of Analysis, Hermann Weyl. Classic of 20th-century foundational research deals with the conceptual problem posed by the continuum. 156pp. 5⅜ × 8½.   67982-9 Pa. $5.95

ESSAYS ON THE THEORY OF NUMBERS, Richard Dedekind. Two classic essays by great German mathematician: on the theory of irrational numbers; and on transfinite numbers and properties of natural numbers. 115pp. 5⅜ × 8½.
21010-3 Pa. $5.95

THE FUNCTIONS OF MATHEMATICAL PHYSICS, Harry Hochstadt. Comprehensive treatment of orthogonal polynomials, hypergeometric functions, Hill's equation, much more. Bibliography. Index. 322pp. 5⅜ × 8½.   65214-9 Pa. $9.95

NUMBER THEORY AND ITS HISTORY, Oystein Ore. Unusually clear, accessible introduction covers counting, properties of numbers, prime numbers, much more. Bibliography. 380pp. 5⅜ × 8½.   65620-9 Pa. $9.95

THE VARIATIONAL PRINCIPLES OF MECHANICS, Cornelius Lanczos. Graduate level coverage of calculus of variations, equations of motion, relativistic mechanics, more. First inexpensive paperbound edition of classic treatise. Index. Bibliography. 418pp. 5⅜ × 8½.   65067-7 Pa. $11.95

MATHEMATICAL TABLES AND FORMULAS, Robert D. Carmichael and Edwin R. Smith. Logarithms, sines, tangents, trig functions, powers, roots, reciprocals, exponential and hyperbolic functions, formulas and theorems. 269pp. 5⅜ × 8½.   60111-0 Pa. $6.95

THEORETICAL PHYSICS, Georg Joos, with Ira M. Freeman. Classic overview covers essential math, mechanics, electromagnetic theory, thermodynamics, quantum mechanics, nuclear physics, other topics. First paperback edition. xxiii + 885pp. 5⅜ × 8½.   65227-0 Pa. $19.95

HANDBOOK OF MATHEMATICAL FUNCTIONS WITH FORMULAS, GRAPHS, AND MATHEMATICAL TABLES, edited by Milton Abramowitz and Irene A. Stegun. Vast compendium: 29 sets of tables, some to as high as 20 places. 1,046pp. 8 × 10½. 61272-4 Pa. $24.95

MATHEMATICAL METHODS IN PHYSICS AND ENGINEERING, John W. Dettman. Algebraically based approach to vectors, mapping, diffraction, other topics in applied math. Also generalized functions, analytic function theory, more. Exercises. 448pp. 5⅜ × 8¼. 65649-7 Pa. $10.95

A SURVEY OF NUMERICAL MATHEMATICS, David M. Young and Robert Todd Gregory. Broad self-contained coverage of computer-oriented numerical algorithms for solving various types of mathematical problems in linear algebra, ordinary and partial, differential equations, much more. Exercises. Total of 1,248pp. 5⅜ × 8½. Two-vol. set.　　　　　Vol. I: 65691-8 Pa. $14.95
Vol. II: 65692-6 Pa. $14.95

TENSOR ANALYSIS FOR PHYSICISTS, J.A. Schouten. Concise exposition of the mathematical basis of tensor analysis, integrated with well-chosen physical examples of the theory. Exercises. Index. Bibliography. 289pp. 5⅜ × 8½.
65582-2 Pa. $8.95

INTRODUCTION TO NUMERICAL ANALYSIS (2nd Edition), F.B. Hildebrand. Classic, fundamental treatment covers computation, approximation, interpolation, numerical differentiation and integration, other topics. 150 new problems. 669pp. 5⅜ × 8½. 65363-3 Pa. $15.95

INVESTIGATIONS ON THE THEORY OF THE BROWNIAN MOVEMENT, Albert Einstein. Five papers (1905–8) investigating dynamics of Brownian motion and evolving elementary theory. Notes by R. Fürth. 122pp. 5⅜ × 8½.
60304-0 Pa. $4.95

CATASTROPHE THEORY FOR SCIENTISTS AND ENGINEERS, Robert Gilmore. Advanced-level treatment describes mathematics of theory grounded in the work of Poincaré, R. Thom, other mathematicians. Also important applications to problems in mathematics, physics, chemistry and engineering. 1981 edition. References. 28 tables. 397 black-and-white illustrations. xvii + 666pp. 6⅛ × 9¼.
67539-4 Pa. $16.95

AN INTRODUCTION TO STATISTICAL THERMODYNAMICS, Terrell L. Hill. Excellent basic text offers wide-ranging coverage of quantum statistical mechanics, systems of interacting molecules, quantum statistics, more. 523pp. 5⅜ × 8½. 65242-4 Pa. $12.95

ELEMENTARY DIFFERENTIAL EQUATIONS, William Ted Martin and Eric Reissner. Exceptionally clear, comprehensive introduction at undergraduate level. Nature and origin of differential equations, differential equations of first, second and higher orders. Picard's Theorem, much more. Problems with solutions. 331pp. 5⅜ × 8½. 65024-3 Pa. $8.95

STATISTICAL PHYSICS, Gregory H. Wannier. Classic text combines thermodynamics, statistical mechanics and kinetic theory in one unified presentation of thermal physics. Problems with solutions. Bibliography. 532pp. 5⅜ × 8½.
65401-X Pa. $12.95

ORDINARY DIFFERENTIAL EQUATIONS, Morris Tenenbaum and Harry Pollard. Exhaustive survey of ordinary differential equations for undergraduates in mathematics, engineering, science. Thorough analysis of theorems. Diagrams. Bibliography. Index. 818pp. 5⅜ × 8½. 64940-7 Pa. $18.95

STATISTICAL MECHANICS: Principles and Applications, Terrell L. Hill. Standard text covers fundamentals of statistical mechanics, applications to fluctuation theory, imperfect gases, distribution functions, more. 448pp. 5⅜ × 8½. 65390-0 Pa. $11.95

ORDINARY DIFFERENTIAL EQUATIONS AND STABILITY THEORY: An Introduction, David A. Sánchez. Brief, modern treatment. Linear equation, stability theory for autonomous and nonautonomous systems, etc. 164pp. 5⅜ × 8¼. 63828-6 Pa. $5.95

THIRTY YEARS THAT SHOOK PHYSICS: The Story of Quantum Theory, George Gamow. Lucid, accessible introduction to influential theory of energy and matter. Careful explanations of Dirac's anti-particles, Bohr's model of the atom, much more. 12 plates. Numerous drawings. 240pp. 5⅜ × 8½. 24895-X Pa. $6.95

THEORY OF MATRICES, Sam Perlis. Outstanding text covering rank, non-singularity and inverses in connection with the development of canonical matrices under the relation of equivalence, and without the intervention of determinants. Includes exercises. 237pp. 5⅜ × 8½. 66810-X Pa. $7.95

GREAT EXPERIMENTS IN PHYSICS: Firsthand Accounts from Galileo to Einstein, edited by Morris H. Shamos. 25 crucial discoveries: Newton's laws of motion, Chadwick's study of the neutron, Hertz on electromagnetic waves, more. Original accounts clearly annotated. 370pp. 5⅜ × 8½. 25346-5 Pa. $10.95

INTRODUCTION TO PARTIAL DIFFERENTIAL EQUATIONS WITH AP-PLICATIONS, E.C. Zachmanoglou and Dale W. Thoe. Essentials of partial differential equations applied to common problems in engineering and the physical sciences. Problems and answers. 416pp. 5⅜ × 8½. 65251-3 Pa. $10.95

BURNHAM'S CELESTIAL HANDBOOK, Robert Burnham, Jr. Thorough guide to the stars beyond our solar system. Exhaustive treatment. Alphabetical by constellation: Andromeda to Cetus in Vol. 1; Chamaeleon to Orion in Vol. 2; and Pavo to Vulpecula in Vol. 3. Hundreds of illustrations. Index in Vol. 3. 2,000pp. 6⅛ × 9¼. Three-vol. set.
Vol. I: 23567-X Pa. $13.95
Vol. II: 23568-8 Pa. $13.95
Vol. III: 23673-0 Pa. $13.95

CHEMICAL MAGIC, Leonard A. Ford. Second Edition, Revised by E. Winston Grundmeier. Over 100 unusual stunts demonstrating cold fire, dust explosions, much more. Text explains scientific principles and stresses safety precautions. 128pp. 5⅜ × 8½. 67628-5 Pa. $5.95

AMATEUR ASTRONOMER'S HANDBOOK, J.B. Sidgwick. Timeless, comprehensive coverage of telescopes, mirrors, lenses, mountings, telescope drives, micrometers, spectroscopes, more. 189 illustrations. 576pp. 5⅜ × 8¼. (Available in U.S. only) 24034-7 Pa. $11.95

SPECIAL FUNCTIONS, N.N. Lebedev. Translated by Richard Silverman. Famous Russian work treating more important special functions, with applications to specific problems of physics and engineering. 38 figures. 308pp. 5⅜ × 8½.
60624-4 Pa. $8.95

OBSERVATIONAL ASTRONOMY FOR AMATEURS, J.B. Sidgwick. Mine of useful data for observation of sun, moon, planets, asteroids, aurorae, meteors, comets, variables, binaries, etc. 39 illustrations. 384pp. 5⅜ × 8¼. (Available in U.S. only)
24033-9 Pa. $8.95

INTEGRAL EQUATIONS, F.G. Tricomi. Authoritative, well-written treatment of extremely useful mathematical tool with wide applications. Volterra Equations, Fredholm Equations, much more. Advanced undergraduate to graduate level. Exercises. Bibliography. 238pp. 5⅜ × 8½.
64828-1 Pa. $7.95

POPULAR LECTURES ON MATHEMATICAL LOGIC, Hao Wang. Noted logician's lucid treatment of historical developments, set theory, model theory, recursion theory and constructivism, proof theory, more. 3 appendixes. Bibliography. 1981 edition. ix + 283pp. 5⅜ × 8½.
67632-3 Pa. $8.95

MODERN NONLINEAR EQUATIONS, Thomas L. Saaty. Emphasizes practical solution of problems; covers seven types of equations. ". . . a welcome contribution to the existing literature. . . ."—*Math Reviews.* 490pp. 5⅜ × 8½. 64232-1 Pa. $11.95

FUNDAMENTALS OF ASTRODYNAMICS, Roger Bate et al. Modern approach developed by U.S. Air Force Academy. Designed as a first course. Problems, exercises. Numerous illustrations. 455pp. 5⅜ × 8½.
60061-0 Pa. $9.95

INTRODUCTION TO LINEAR ALGEBRA AND DIFFERENTIAL EQUATIONS, John W. Dettman. Excellent text covers complex numbers, determinants, orthonormal bases, Laplace transforms, much more. Exercises with solutions. Undergraduate level. 416pp. 5⅜ × 8½.
65191-6 Pa. $10.95

INCOMPRESSIBLE AERODYNAMICS, edited by Bryan Thwaites. Covers theoretical and experimental treatment of the uniform flow of air and viscous fluids past two-dimensional aerofoils and three-dimensional wings; many other topics. 654pp. 5⅜ × 8½.
65465-6 Pa. $16.95

INTRODUCTION TO DIFFERENCE EQUATIONS, Samuel Goldberg. Exceptionally clear exposition of important discipline with applications to sociology, psychology, economics. Many illustrative examples; over 250 problems. 260pp. 5⅜ × 8½.
65084-7 Pa. $8.95

LAMINAR BOUNDARY LAYERS, edited by L. Rosenhead. Engineering classic covers steady boundary layers in two- and three-dimensional flow, unsteady boundary layers, stability, observational techniques, much more. 708pp. 5⅜ × 8½.
65646-2 Pa. $18.95

LECTURES ON CLASSICAL DIFFERENTIAL GEOMETRY, Second Edition, Dirk J. Struik. Excellent brief introduction covers curves, theory of surfaces, fundamental equations, geometry on a surface, conformal mapping, other topics. Problems. 240pp. 5⅜ × 8½.
65609-8 Pa. $8.95

ROTARY-WING AERODYNAMICS, W.Z. Stepniewski. Clear, concise text covers aerodynamic phenomena of the rotor and offers guidelines for helicopter performance evaluation. Originally prepared for NASA. 537 figures. 640pp. 6⅛ × 9¼.
64647-5 Pa. $15.95

DIFFERENTIAL GEOMETRY, Heinrich W. Guggenheimer. Local differential geometry as an application of advanced calculus and linear algebra. Curvature, transformation groups, surfaces, more. Exercises. 62 figures. 378pp. 5⅜ × 8½.
63433-7 Pa. $8.95

INTRODUCTION TO SPACE DYNAMICS, William Tyrrell Thomson. Comprehensive, classic introduction to space-flight engineering for advanced undergraduate and graduate students. Includes vector algebra, kinematics, transformation of coordinates. Bibliography. Index. 352pp. 5⅜ × 8½. 65113-4 Pa. $8.95

A SURVEY OF MINIMAL SURFACES, Robert Osserman. Up-to-date, in-depth discussion of the field for advanced students. Corrected and enlarged edition covers new developments. Includes numerous problems. 192pp. 5⅜ × 8½.
64998-9 Pa. $8.95

ANALYTICAL MECHANICS OF GEARS, Earle Buckingham. Indispensable reference for modern gear manufacture covers conjugate gear-tooth action, gear-tooth profiles of various gears, many other topics. 263 figures. 102 tables. 546pp. 5⅜ × 8½. 65712-4 Pa. $14.95

SET THEORY AND LOGIC, Robert R. Stoll. Lucid introduction to unified theory of mathematical concepts. Set theory and logic seen as tools for conceptual understanding of real number system. 496pp. 5⅜ × 8¼. 63829-4 Pa. $12.95

A HISTORY OF MECHANICS, René Dugas. Monumental study of mechanical principles from antiquity to quantum mechanics. Contributions of ancient Greeks, Galileo, Leonardo, Kepler, Lagrange, many others. 671pp. 5⅜ × 8½.
65632-2 Pa. $14.95

FAMOUS PROBLEMS OF GEOMETRY AND HOW TO SOLVE THEM, Benjamin Bold. Squaring the circle, trisecting the angle, duplicating the cube: learn their history, why they are impossible to solve, then solve them yourself. 128pp. 5⅜ × 8½. 24297-8 Pa. $4.95

MECHANICAL VIBRATIONS, J.P. Den Hartog. Classic textbook offers lucid explanations and illustrative models, applying theories of vibrations to a variety of practical industrial engineering problems. Numerous figures. 233 problems, solutions. Appendix. Index. Preface. 436pp. 5⅜ × 8½. 64785-4 Pa. $10.95

CURVATURE AND HOMOLOGY, Samuel I. Goldberg. Thorough treatment of specialized branch of differential geometry. Covers Riemannian manifolds, topology of differentiable manifolds, compact Lie groups, other topics. Exercises. 315pp. 5⅜ × 8½. 64314-X Pa. $9.95

HISTORY OF STRENGTH OF MATERIALS, Stephen P. Timoshenko. Excellent historical survey of the strength of materials with many references to the theories of elasticity and structure. 245 figures. 452pp. 5⅜ × 8½. 61187-6 Pa. $11.95

CATALOG OF DOVER BOOKS

GEOMETRY OF COMPLEX NUMBERS, Hans Schwerdtfeger. Illuminating, widely praised book on analytic geometry of circles, the Moebius transformation, and two-dimensional non-Euclidean geometries. 200pp. 5⅜ × 8¼.
63830-8 Pa. $8.95

MECHANICS, J.P. Den Hartog. A classic introductory text or refresher. Hundreds of applications and design problems illuminate fundamentals of trusses, loaded beams and cables, etc. 334 answered problems. 462pp. 5⅜ × 8½. 60754-2 Pa. $9.95

TOPOLOGY, John G. Hocking and Gail S. Young. Superb one-year course in classical topology. Topological spaces and functions, point-set topology, much more. Examples and problems. Bibliography. Index. 384pp. 5⅜ × 8¼.
65676-4 Pa. $9.95

STRENGTH OF MATERIALS, J.P. Den Hartog. Full, clear treatment of basic material (tension, torsion, bending, etc.) plus advanced material on engineering methods, applications. 350 answered problems. 323pp. 5⅜ × 8½. 60755-0 Pa. $8.95

ELEMENTARY CONCEPTS OF TOPOLOGY, Paul Alexandroff. Elegant, intuitive approach to topology from set-theoretic topology to Betti groups; how concepts of topology are useful in math and physics. 25 figures. 57pp. 5⅜ × 8½.
60747-X Pa. $3.50

ADVANCED STRENGTH OF MATERIALS, J.P. Den Hartog. Superbly written advanced text covers torsion, rotating disks, membrane stresses in shells, much more. Many problems and answers. 388pp. 5⅜ × 8½. 65407-9 Pa. $9.95

COMPUTABILITY AND UNSOLVABILITY, Martin Davis. Classic graduate-level introduction to theory of computability, usually referred to as theory of recurrent functions. New preface and appendix. 288pp. 5⅜ × 8½. 61471-9 Pa. $7.95

GENERAL CHEMISTRY, Linus Pauling. Revised 3rd edition of classic first-year text by Nobel laureate. Atomic and molecular structure, quantum mechanics, statistical mechanics, thermodynamics correlated with descriptive chemistry. Problems. 992pp. 5⅜ × 8½. 65622-5 Pa. $19.95

AN INTRODUCTION TO MATRICES, SETS AND GROUPS FOR SCIENCE STUDENTS, G. Stephenson. Concise, readable text introduces sets, groups, and most importantly, matrices to undergraduate students of physics, chemistry, and engineering. Problems. 164pp. 5⅜ × 8½. 65077-4 Pa. $6.95

THE HISTORICAL BACKGROUND OF CHEMISTRY, Henry M. Leicester. Evolution of ideas, not individual biography. Concentrates on formulation of a coherent set of chemical laws. 260pp. 5⅜ × 8½. 61053-5 Pa. $7.95

THE PHILOSOPHY OF MATHEMATICS: An Introductory Essay, Stephan Körner. Surveys the views of Plato, Aristotle, Leibniz & Kant concerning propositions and theories of applied and pure mathematics. Introduction. Two appendices. Index. 198pp. 5⅜ × 8½. 25048-2 Pa. $7.95

THE DEVELOPMENT OF MODERN CHEMISTRY, Aaron J. Ihde. Authoritative history of chemistry from ancient Greek theory to 20th-century innovation. Covers major chemists and their discoveries. 209 illustrations. 14 tables. Bibliographies. Indices. Appendices. 851pp. 5⅜ × 8½. 64235-6 Pa. $18.95

CATALOG OF DOVER BOOKS

DE RE METALLICA, Georgius Agricola. The famous Hoover translation of greatest treatise on technological chemistry, engineering, geology, mining of early modern times (1556). All 289 original woodcuts. 638pp. 6¾ × 11.
60006-8 Clothbd. $18.95

SOME THEORY OF SAMPLING, William Edwards Deming. Analysis of the problems, theory and design of sampling techniques for social scientists, industrial managers and others who find statistics increasingly important in their work. 61 tables. 90 figures. xvii + 602pp. 5⅜ × 8½.
64684-X Pa. $15.95

THE VARIOUS AND INGENIOUS MACHINES OF AGOSTINO RAMELLI: A Classic Sixteenth-Century Illustrated Treatise on Technology, Agostino Ramelli. One of the most widely known and copied works on machinery in the 16th century. 194 detailed plates of water pumps, grain mills, cranes, more. 608pp. 9 × 12.
28180-9 Pa. $24.95

LINEAR PROGRAMMING AND ECONOMIC ANALYSIS, Robert Dorfman, Paul A. Samuelson and Robert M. Solow. First comprehensive treatment of linear programming in standard economic analysis. Game theory, modern welfare economics, Leontief input-output, more. 525pp. 5⅜ × 8½.
65491-5 Pa. $14.95

ELEMENTARY DECISION THEORY, Herman Chernoff and Lincoln E. Moses. Clear introduction to statistics and statistical theory covers data processing, probability and random variables, testing hypotheses, much more. Exercises. 364pp. 5⅜ × 8½.
65218-1 Pa. $9.95

THE COMPLEAT STRATEGYST: Being a Primer on the Theory of Games of Strategy, J.D. Williams. Highly entertaining classic describes, with many illustrated examples, how to select best strategies in conflict situations. Prefaces. Appendices. 268pp. 5⅜ × 8½.
25101-2 Pa. $7.95

MATHEMATICAL METHODS OF OPERATIONS RESEARCH, Thomas L. Saaty. Classic graduate-level text covers historical background, classical methods of forming models, optimization, game theory, probability, queueing theory, much more. Exercises. Bibliography. 448pp. 5⅜ × 8¼.
65703-5 Pa. $12.95

CONSTRUCTIONS AND COMBINATORIAL PROBLEMS IN DESIGN OF EXPERIMENTS, Damaraju Raghavarao. In-depth reference work examines orthogonal Latin squares, incomplete block designs, tactical configuration, partial geometry, much more. Abundant explanations, examples. 416pp. 5⅜ × 8¼.
65685-3 Pa. $10.95

THE ABSOLUTE DIFFERENTIAL CALCULUS (CALCULUS OF TENSORS), Tullio Levi-Civita. Great 20th-century mathematician's classic work on material necessary for mathematical grasp of theory of relativity. 452pp. 5⅜ × 8½.
63401-9 Pa. $11.95

VECTOR AND TENSOR ANALYSIS WITH APPLICATIONS, A.I. Borisenko and I.E. Tarapov. Concise introduction. Worked-out problems, solutions, exercises. 257pp. 5⅜ × 8¼.
63833-2 Pa. $8.95

CATALOG OF DOVER BOOKS

THE FOUR-COLOR PROBLEM: Assaults and Conquest, Thomas L. Saaty and Paul G. Kainen. Engrossing, comprehensive account of the century-old combinatorial topological problem, its history and solution. Bibliographies. Index. 110 figures. 228pp. 5⅜ × 8½.   65092-8 Pa. $6.95

CATALYSIS IN CHEMISTRY AND ENZYMOLOGY, William P. Jencks. Exceptionally clear coverage of mechanisms for catalysis, forces in aqueous solution, carbonyl- and acyl-group reactions, practical kinetics, more. 864pp. 5⅜ × 8½.   65460-5 Pa. $19.95

PROBABILITY: An Introduction, Samuel Goldberg. Excellent basic text covers set theory, probability theory for finite sample spaces, binomial theorem, much more. 360 problems. Bibliographies. 322pp. 5⅜ × 8½.   65252-1 Pa. $9.95

LIGHTNING, Martin A. Uman. Revised, updated edition of classic work on the physics of lightning. Phenomena, terminology, measurement, photography, spectroscopy, thunder, more. Reviews recent research. Bibliography. Indices. 320pp. 5⅜ × 8¼.   64575-4 Pa. $8.95

PROBABILITY THEORY: A Concise Course, Y.A. Rozanov. Highly readable, self-contained introduction covers combination of events, dependent events, Bernoulli trials, etc. Translation by Richard Silverman. 148pp. 5⅜ × 8¼.
    63544-9 Pa. $6.95

AN INTRODUCTION TO HAMILTONIAN OPTICS, H. A. Buchdahl. Detailed account of the Hamiltonian treatment of aberration theory in geometrical optics. Many classes of optical systems defined in terms of the symmetries they possess. Problems with detailed solutions. 1970 edition. xv + 360pp. 5⅜ × 8½.
    67597-1 Pa. $10.95

STATISTICS MANUAL, Edwin L. Crow, et al. Comprehensive, practical collection of classical and modern methods prepared by U.S. Naval Ordnance Test Station. Stress on use. Basics of statistics assumed. 288pp. 5⅜ × 8½.
    60599-X Pa. $7.95

DICTIONARY/OUTLINE OF BASIC STATISTICS, John E. Freund and Frank J. Williams. A clear concise dictionary of over 1,000 statistical terms and an outline of statistical formulas covering probability, nonparametric tests, much more. 208pp. 5⅜ × 8½.   66796-0 Pa. $6.95

STATISTICAL METHOD FROM THE VIEWPOINT OF QUALITY CONTROL, Walter A. Shewhart. Important text explains regulation of variables, uses of statistical control to achieve quality control in industry, agriculture, other areas. 192pp. 5⅜ × 8½.   65232-7 Pa. $7.95

THE INTERPRETATION OF GEOLOGICAL PHASE DIAGRAMS, Ernest G. Ehlers. Clear, concise text emphasizes diagrams of systems under fluid or containing pressure; also coverage of complex binary systems, hydrothermal melting, more. 288pp. 6½ × 9¼.   65389-7 Pa. $10.95

STATISTICAL ADJUSTMENT OF DATA, W. Edwards Deming. Introduction to basic concepts of statistics, curve fitting, least squares solution, conditions without parameter, conditions containing parameters. 26 exercises worked out. 271pp. 5⅜ × 8½.   64685-8 Pa. $8.95

CATALOG OF DOVER BOOKS

TENSOR CALCULUS, J.L. Synge and A. Schild. Widely used introductory text covers spaces and tensors, basic operations in Riemannian space, non-Riemannian spaces, etc. 324pp. 5⅜ × 8¼. 63612-7 Pa. $8.95

A CONCISE HISTORY OF MATHEMATICS, Dirk J. Struik. The best brief history of mathematics. Stresses origins and covers every major figure from ancient Near East to 19th century. 41 illustrations. 195pp. 5⅜ × 8½. 60255-9 Pa. $7.95

A SHORT ACCOUNT OF THE HISTORY OF MATHEMATICS, W.W. Rouse Ball. One of clearest, most authoritative surveys from the Egyptians and Phoenicians through 19th-century figures such as Grassman, Galois, Riemann. Fourth edition. 522pp. 5⅜ × 8½. 20630-0 Pa. $11.95

HISTORY OF MATHEMATICS, David E. Smith. Nontechnical survey from ancient Greece and Orient to late 19th century; evolution of arithmetic, geometry, trigonometry, calculating devices, algebra, the calculus. 362 illustrations. 1,355pp. 5⅜ × 8½. Two-vol. set. Vol. I: 20429-4 Pa. $12.95
Vol. II: 20430-8 Pa. $11.95

THE GEOMETRY OF RENÉ DESCARTES, René Descartes. The great work founded analytical geometry. Original French text, Descartes' own diagrams, together with definitive Smith-Latham translation. 244pp. 5⅜ × 8½. 60068-8 Pa. $7.95

THE ORIGINS OF THE INFINITESIMAL CALCULUS, Margaret E. Baron. Only fully detailed and documented account of crucial discipline: origins; development by Galileo, Kepler, Cavalieri; contributions of Newton, Leibniz, more. 304pp. 5⅜ × 8½. (Available in U.S. and Canada only) 65371-4 Pa. $9.95

THE HISTORY OF THE CALCULUS AND ITS CONCEPTUAL DEVELOPMENT, Carl B. Boyer. Origins in antiquity, medieval contributions, work of Newton, Leibniz, rigorous formulation. Treatment is verbal. 346pp. 5⅜ × 8½. 60509-4 Pa. $9.95

THE THIRTEEN BOOKS OF EUCLID'S ELEMENTS, translated with introduction and commentary by Sir Thomas L. Heath. Definitive edition. Textual and linguistic notes, mathematical analysis. 2,500 years of critical commentary. Not abridged. 1,414pp. 5⅜ × 8½. Three-vol. set. Vol. I: 60088-2 Pa. $9.95
Vol. II: 60089-0 Pa. $9.95
Vol. III: 60090-4 Pa. $9.95

GAMES AND DECISIONS: Introduction and Critical Survey, R. Duncan Luce and Howard Raiffa. Superb nontechnical introduction to game theory, primarily applied to social sciences. Utility theory, zero-sum games, n-person games, decision-making, much more. Bibliography. 509pp. 5⅜ × 8½. 65943-7 Pa. $12.95

THE HISTORICAL ROOTS OF ELEMENTARY MATHEMATICS, Lucas N.H. Bunt, Phillip S. Jones, and Jack D. Bedient. Fundamental underpinnings of modern arithmetic, algebra, geometry and number systems derived from ancient civilizations. 320pp. 5⅜ × 8½. 25563-8 Pa. $8.95

CALCULUS REFRESHER FOR TECHNICAL PEOPLE, A. Albert Klaf. Covers important aspects of integral and differential calculus via 756 questions. 566 problems, most answered. 431pp. 5⅜ × 8½. 20370-0 Pa. $8.95

CATALOG OF DOVER BOOKS

CHALLENGING MATHEMATICAL PROBLEMS WITH ELEMENTARY SOLUTIONS, A.M. Yaglom and I.M. Yaglom. Over 170 challenging problems on probability theory, combinatorial analysis, points and lines, topology, convex polygons, many other topics. Solutions. Total of 445pp. 5⅜ × 8½. Two-vol. set.
Vol. I 65536-9 Pa. $7.95
Vol. II 65537-7 Pa. $6.95

FIFTY CHALLENGING PROBLEMS IN PROBABILITY WITH SOLUTIONS, Frederick Mosteller. Remarkable puzzlers, graded in difficulty, illustrate elementary and advanced aspects of probability. Detailed solutions. 88pp. 5⅜ × 8½.
65355-2 Pa. $4.95

EXPERIMENTS IN TOPOLOGY, Stephen Barr. Classic, lively explanation of one of the byways of mathematics. Klein bottles, Moebius strips, projective planes, map coloring, problem of the Koenigsberg bridges, much more, described with clarity and wit. 43 figures. 210pp. 5⅜ × 8½.
25933-1 Pa. $6.95

RELATIVITY IN ILLUSTRATIONS, Jacob T. Schwartz. Clear nontechnical treatment makes relativity more accessible than ever before. Over 60 drawings illustrate concepts more clearly than text alone. Only high school geometry needed. Bibliography. 128pp. 6⅛ × 9¼.
25965-X Pa. $7.95

AN INTRODUCTION TO ORDINARY DIFFERENTIAL EQUATIONS, Earl A. Coddington. A thorough and systematic first course in elementary differential equations for undergraduates in mathematics and science, with many exercises and problems (with answers). Index. 304pp. 5⅜ × 8½.
65942-9 Pa. $8.95

FOURIER SERIES AND ORTHOGONAL FUNCTIONS, Harry F. Davis. An incisive text combining theory and practical example to introduce Fourier series, orthogonal functions and applications of the Fourier method to boundary-value problems. 570 exercises. Answers and notes. 416pp. 5⅜ × 8½.
65973-9 Pa. $11.95

THE THEORY OF BRANCHING PROCESSES, Theodore E. Harris. First systematic, comprehensive treatment of branching (i.e. multiplicative) processes and their applications. Galton-Watson model, Markov branching processes, electron-photon cascade, many other topics. Rigorous proofs. Bibliography. 240pp. 5⅜ × 8½.
65952-6 Pa. $6.95

AN INTRODUCTION TO ALGEBRAIC STRUCTURES, Joseph Landin. Superb self-contained text covers "abstract algebra": sets and numbers, theory of groups, theory of rings, much more. Numerous well-chosen examples, exercises. 247pp. 5⅜ × 8½.
65940-2 Pa. $7.95
